Lecture Notes in Computer Science 12557

More information about this subseries at http://www.springer.com/series/7407

Min Han · Sitian Qin · Nian Zhang (Eds.)

Advances in Neural Networks – ISNN 2020

17th International Symposium on Neural Networks, ISNN 2020
Cairo, Egypt, December 4–6, 2020
Proceedings

 Springer

Editors
Min Han ⓘD
Faculty of Electronic Information
and Electrical Engineering
Dalian University of Technology
Dalian, China

Sitian Qin ⓘD
School of Science
Harbin Institute of Technology
Weihai, China

Nian Zhang ⓘD
School of Engineering and Applied Sciences
University of the District of Columbia
Washington, DC, USA

ISSN 0302-9743   ISSN 1611-3349 (electronic)
Lecture Notes in Computer Science
ISBN 978-3-030-64220-4   ISBN 978-3-030-64221-1 (eBook)
https://doi.org/10.1007/978-3-030-64221-1

LNCS Sublibrary: SL1 – Theoretical Computer Science and General Issues

This Springer imprint is published by the registered company Springer Nature Switzerland AG
The registered company address is: Gewerbestrasse 11, 6330 Cham, Switzerland

Preface

This volume of *Lecture Notes in Computer Science* (LNCS) constitutes the proceedings of the 17th International Symposium on Neural Networks (ISNN 2020), held during December 4–6, 2020, in Cairo, Egypt, and over the Internet. Thanks to the success of the previous events, ISNN has become a well-established series of popular and high-quality conferences on the theory and methodology of neural networks and their applications. This year's symposium was postponed for more than two months due to the COVID-19 pandemic. But it still achieved great success. ISNN aims at providing a high-level international forum for scientists, engineers, educators, and students to gather, present, and discuss the latest progress in neural network research and applications in diverse areas. This symposium encouraged open discussion and exchange of ideas. We believed that it would extensively promote research in the fields of neural networks and applications.

This year, the conference received 39 submissions, much less submissions than previous years, due to an obvious reason. Each submission was reviewed by at least three, and on average, four Program Committee members. After the rigorous peer reviews, the committee decided to accept 26 papers for publication in the LNCS proceedings with an acceptance rate of two thirds. These papers cover many topics of neural network-related research, including computational intelligence, neurodynamics, stability analysis, deep learning, pattern recognition, image processing, and so on. In addition to the contributed papers, the ISNN 2020 technical program included two plenary speeches by world renowned scholars: Prof. Tingwen Huang (IEEE Fellow) from the Texas A&M University, Qatar, and Prof. Dongbin Zhao (IEEE Fellow) from the Institute of Automation, Chinese Academy of Sciences, China.

Many organizations and volunteers made great contributions toward the success of this symposium. We would like to express our sincere gratitude to the British University in Egypt and The City University of Hong Kong for their sponsorship, as well as the International Neural Network Society and the Asian Pacific Neural Network Society for their technical co-sponsorship. We would also like to sincerely thank all the committee members for their great efforts in organizing the symposium. Special thanks to the Program Committee members and reviewers whose insightful reviews and timely feedback ensured the high quality of the accepted papers and the smooth flow of the symposium. We would also like to thank Springer for their cooperation in publishing the proceedings in the prestigious LNCS series. Finally, we would like to thank all the speakers, authors, and participants for their support.

October 2020

Min Han
Sitian Qin
Nian Zhang

Organization

General Chairs

Samir Abou El-Seoud   The British University in Egypt, Egypt
Omar H. Karam   The British University in Egypt, Egypt

Advisory Chairs

Ahmad Hamad   The British University in Egypt, Egypt
Derong Liu   University of Illinois at Chicago, USA

Steering Chairs

Haibo He   University of Rhode Island, USA
Jun Wang   The City University of Hong Kong, Hong Kong

Organizing Chairs

Yehia Bahei-El-Din   The British University in Egypt, Egypt
Jun Wang   The City University of Hong Kong, Hong Kong

Program Chairs

Min Han   Dalian University of Technology, China
Sitian Qin   Harbin Institute of Technology, China
Nian Zhang   University of the District of Columbia, USA

Special Sessions Chairs

Tasos Dagiuklasis   London South Bank University, UK
Andreas Pester   Carinthia University of Applied Sciences, Austria
Zhihui Zhan   South China University of Technology, China
Bo Zhao   Beijing Normal University, China

Publicity Chairs

Tingwen Huang   Texas A&M University at Qatar, Qatar
Wing W. Y. Ng   South China University of Technology, China
Zhigang Zeng   Huazhong University of Science and Technology, China

Publications Chairs

Hangjun Che   Southwest University, China
Xinyi Le   Shanghai Jiao Tong University, China
Man-Fai Leung   The Open University of Hong Kong, Hong Kong

Registration Chairs

Shenshen Gu   Shanghai University, China
Qingshan Liu   Southeast University, China
Ka Chun Wong   The City University of Hong Kong, Hong Kong

Local Arrangements Chair

Samy Ghoniemy   The British University in Egypt, Egypt

Webmaster

Jiasen Wang   The City University of Hong Kong, Hong Kong

Program Committee

Sabri Arik   Istanbul University, Turkey
Wei Bian   Harbin Institute of Technology, China
Zhaohui Cen   Qatar Environment and Energy Research Institute,
                               Qatar
Hangjun Che   Southwest University, China
Jie Chen   Sichuan University, China
Long Cheng   Institute of Automation, Chinese Academy of Sciences,
                               China
Fengyu Cong   Dalian University of Technology, China
Ruxandra Liana Costea   Polytechnic University of Bucharest, Romania
Xianguang Dai   Chongqing Three Gorges University, China
Mingcong Deng   Tokyo University of Agriculture and Technology,
                               Japan
Jianchao Fan   National Marine Environmental Monitoring Center,
                               China
Wai-Keung Fung   Robert Gordon University, UK
Shenshen Gu   Shanghai University, China
Zhenyuan Guo   Hunan University, China
Ping Guo   Beijing Normal University, China
Min Han   Dalian University of Technology, China
Wangli He   East China University of Science and Technology,
                               China
Xing He   Southwest University, China
Zhenan He   Sichuan University, China

Xiaolin Hu	Tsinghua University, China
Jinglu Hu	Waseda University, Japan
He Huang	Soochow University, China
Min Jiang	Xiamen University, China
Danchi Jiang	University of Tasmania, Australia
Long Jin	Lanzhou University, China
Feng Jiqiang	Shenzhen University, China
Sungshin Kim	Pusan National University, South Korea
Chiman Kwan	Signal Processing, Inc., USA
Rushi Lan	Guilin University of Electronic Technology, China
Xinyi Le	Shanghai Jiao Tong University, China
Man Fai Leung	The Open University of Hong Kong, Hong Kong
Jie Lian	Dalian University of Technology, China
Cheng Lian	Wuhan University of Technology, China
Hualou Liang	Drexel University, USA
Qingshan Liu	Southeast University, China
Ju Liu	Shandong University, China
Zhi-Wei Liu	Wuhan University, China
Wenlian Lu	Fudan University, China
Bao-Liang Lu	Shanghai Jiao Tong University, China
Jinwen Ma	Peking University, China
Nankun Mu	Southwest University, China
Xi Peng	Institute for Infocomm Research, Singapore
Sitian Qin	Harbin Institute of Technology, China
Hong Qu	University of Electronic Science and Technology of China, China
Qiankun Song	Chongqing Jiaotong University, China
Yang Tang	East China University of Science and Technology, China
Alexander Tuzikov	United Institute of Informatics Problems, National Academy of Sciences of Belarus, Belarus
Feng Wan	University of Macau, Macau
Jiasen Wang	The City University of Hong Kong, Hong Kong
Jian Wang	China University of Petroleum, China
Dianhui Wang	La Trobe University, Australia
Xiaoping Wang	Huazhong University of Science and Technology, China
Guanghui Wen	Southeast University, China
Nian Zhang	University of the District of Columbia, USA
Xiaojun Zhou	Central South University, China
Bo Zhou	Southwest University, China

Additional Reviewers

Gang Bao
Jiayi Cai
Yiyuan Chai
Xujia Chang
Xiufang Chen
Zhiyi Chen
Sanbo Ding
Yangming Dou
Xiaohao Du
Yuming Feng
Yuan Gao
Ming-Feng Ge
Chuanyu Geng
Jingyi He
Li He
Shuang He
Huifen Hong
Dengzhou Hu
Hongxiang Hu
Haoen Huang
Yaoting Huang
Chi Li
Jiachang Li
Min Li
Suibing Li
Tingting Liang
Jinliang Liu

Lei Liu
Na Liu
Xiangyuan Lu
Dexiu Ma
Yimeng Qi
Jiahu Qin
Zhongbo Sun
Chufeng Tang
Zhengwen Tu
Limin Wang
Peijun Wang
Xinzhe Wang
Lin Wei
Shiping Wen
Yanming Wu
Zhengtai Xie
Jing-Zhe Xu
Jingkun Yan
Qiang Yang
Xujun Yang
Liufu Ying
Jiahui Yu
Wei-Jie Yu
Zengyun Wang
Gang Zhang
Jiazheng Zhang
Xiaoyan Zhang

Contents

Optimization Algorithms

Parameters Identification of Solar Cells Based on Classification Particle Swarm Optimization Algorithm

Haijie Bao[1], Chuyi Song[1], Liyan Xu[1], and Jingqing Jiang[1,2(✉)]

[1] College of Mathematics and Physics, Inner Mongolia University for Nationalities,
Tongliao 028000, China
`3089484347@qq.com`, `songchuyi@sina.com`, `389736834@qq.com`,
`jiangjingqing@aliyun.com`
[2] College of Computer Science and Technology, Inner Mongolia University for Nationalities,
Tongliao 028000, China

Abstract. As a new type of clean energy, solar energy has been widely used in many fields. The core component of photovoltaic power generation system, photovoltaic module, is composed of solar cells in series connection and parallel connection. Photovoltaic module directly converts the light energy of the sun into electric energy. The construction of solar cell model needs precise parameters to support. This paper proposes using classification particle swarm optimization algorithm (CPSO) to identify the parameters of the solar cell for single diode equivalent circuit model and the double diode equivalent circuit model. The simulation results show that the performance of CPSO is better than or similar to other algorithms.

Keywords: Solar cell model · Parameter identification · Classification particle swarm optimization algorithm

1 Introduction

With the rapid deterioration of the global environment and the depletion of non-renewable energy, the global energy structure has been changed and transformed. The demand for renewable energy is continuing increase. Solar energy is an important renewable energy. Photovoltaic power generation system is the main part of solar energy, and solar cell plays an important role in photovoltaic power generation system. Scholars have proposed many equivalent circuit models to simulate the I-V curve of solar cell in different ways [1]. Two models are practically used. The first one is the single diode circuit model which contains five parameters $(I_{ph}, I_{SD}, n, R_s, R_{sh})$. The second model is a double diode circuit model which contains seven parameters $(I_{ph}, I_{SD1}, I_{SD2}, n_1, n_2, R_s, R_{sh})$. These parameters can be used to understand the intrinsic characteristics of the cell model.

Nowadays, there are many ways to optimize the parameters of the equivalent circuit model of solar cells. The methods are classified into traditional algorithms and intelligent

© Springer Nature Switzerland AG 2020
M. Han et al. (Eds.): ISNN 2020, LNCS 12557, pp. 3–12, 2020.
https://doi.org/10.1007/978-3-030-64221-1_1

algorithms. The traditional algorithm includes analytic method [2], numerical solution method [3]. The intelligent algorithms includes artificial bee swarm optimization algorithm [4] (ABSO), simulated annealing algorithm [5] (SA), genetic algorithm [6] (GA) and particle swarm optimization algorithm [7] (PSO). The intelligent algorithms are meta-heuristic algorithms. They are suitable for solar cell parameter extraction because the derivative is not required. The particle swarm optimization algorithm is a swarm intelligence algorithm developed by Kennedy and Eberhart in 1995 [8]. Lately, many different versions of particle swarm algorithms have been proposed. In this paper, the classification particle swarm algorithm is applied to identify the commercial silicon cell parameters with a diameter of 57 mm.

2 Solar Cell Models

There are two practical equivalent circuit models: single and double diode models.

2.1 Single Diode Model

The solar single diode equivalent circuit consists of a current source, a diode, a shunt resistor and a series resistor, as shown in Fig. 1:

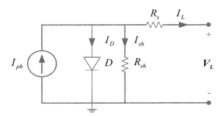

Fig. 1. Equivalent circuit model for single diode solar cell

According to Fig. 1, the I-V characteristic equation of the single diode model of solar cells can be expressed as follows:

$$I_L = I_{ph} - I_{SD}\left\{exp\left[\frac{q(V_L + I_LR_s)}{nKT}\right] - 1\right\} - \frac{V_L + I_LR_s}{R_{sh}} \tag{1}$$

In this formula, I_L is the output current, V_L is the output voltage, I_{ph} is the Cell-Generated photocurrent, I_{SD} is the reverse saturation current of diode, n is the diode ideality factor, R_s is the series resistors, R_{sh} is the shunt resistors, K is the Boltzmann's constant ($1.381 \times 10^{-23} J/K$), q is the Electronic charge ($1.602 \times 10^{-19}C$), T is the cell temperature.

2.2 Double Diode Model

The solar double diode equivalent circuit consists of a current source, two parallel diodes, a shunt resistor and a series resistor, as shown in Fig. 2.

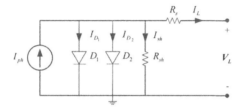

Fig. 2. Equivalent circuit model for double diode solar cell

According to Fig. 2, the I-V characteristic equation of the double diode model of solar cells can be expressed as follows:

$$I_L = I_{ph} - I_{SD1}\left\{exp\left[\frac{q(V_L + I_L R_s)}{n_1 KT}\right] - 1\right\} - I_{SD2}\left\{exp\left[\frac{q(V_L + I_L R_s)}{n_2 KT}\right] - 1\right\} - \frac{V_L + I_L R_s}{R_{sh}} \qquad (2)$$

In this formula, I_L is the output current, V_L is the output voltage, I_{ph} is the Cell-Generated photocurrent, I_{SD1} and I_{SD2} are the diode reverse saturation current of diode, n_1 and n_2 are the diode ideality factor, R_s is the series resistors, R_{sh} is the shunt resistors, K, q and T have the same meaning as the single diode model.

2.3 Solar Cell Parameter Identification Problem

The five parameters involved in the single diode equivalent circuit of solar cells and the seven parameters involved in the double diode equivalent circuit of solar cells are particularly important. Through these parameters, we can understand the intrinsic characteristics of the cell model and make the use of solar cells more efficient. Identifying these parameters is the problem that should be solved.

3   Classification Particle Swarm Optimization Algorithm (CPSO)

The particle swarm optimization algorithm converges slowly. An improved algorithm based on classification was proposed by Tong [9]. This algorithm is used to identify the parameters of solar cell in this paper.

3.1   Particle Swarm Optimization Algorithm (PSO)

PSO algorithm initializes the position and velocity of particles, and then each particle changes its velocity V and position X according to its current best position (Pbest) and the current global best position (Gbest) of the population. The optimal solution (the global best position) is found by iteration. In each iteration, each particle updates its position X and velocity V by formula (3) and formula (4).

$$V^{i+1} = \omega \times V^i + c_1 \times rand \times \left(pbest - X^i\right) + c_2 \times rand \times \left(gbest - X^i\right) \qquad (3)$$

$$X^{i+1} = X^i + V^{i+1} \qquad (4)$$

Where ω is the inertia coefficient. c_1 and c_2 are the learning factor. Pbest is the individual optimal position and Gbest is the global optimal position. i is the number of iteration.

3.2 Steps for Classification Particle Swarm Optimization Algorithm

The PSO algorithm has some disadvantages such as slow convergence speed and falling into local optimum easily. So, the PSO algorithm can't achieve better performance in the parameters identification of solar cell model. An improved particle swarm algorithm based on classification idea (Classification Particle Swarm Optimization, CPSO) is proposed [9].

The searching region is divided into three domains: the rejection domain, close proximity domain and reasonable domain. The division criteria is the comparison of the difference between the fitness function values f_i of the particle i and the mean fitness function value f_V of all particles with the standard deviation O_i of all particles' fitness function values. If $f_i - f_V > O_i$, then $\omega = 0.9, c_1 = 3, c_2 = 1$. The particle locates on the rejection domain. If $f_i - f_V < -O_i$, then $\omega = 0.4, c1 = 1, c2 = 3$. The particle locates on close proximity domain. If $|f_i - f_V| < |O_i|$, then $\omega = (\omega_{max} - \omega_{min} - d_1) \times exp\left(\frac{1}{1+d_2 \times \frac{t}{ger}}\right), c1 = 3\cos\left(\frac{\pi t}{2ger}\right), c2 = 3\sin\left(\frac{\pi t}{2ger}\right)$. The particle locates on the reasonable domain. Where $\omega_{max} = 0.9, \omega_{min} = 0.4, d_1 = 0.2, d_2 = 0.7$, t is the number of current iteration number, ger is the maximum iteration number [10]. It is found that the updating formula of inertia weight adopts the nonlinear decreasing strategy in a reasonable region could improve the performance of PSO.

The steps of the classification particle swarm optimization are as follows:

Step 1: Initialize the parameters of particle. The position range refers to [11–13] and the velocity range is $V_{max} = 0.15 \times (X_{max} - X_{min})$, $V_{min} = -V_{max}$.

Step 2: Initialize the position and velocity of each particle randomly according to the position range and velocity range. The initialization formula of position and velocity are $X = X_{min} + (X_{max} - X_{min}) \times rand(N, 1)$ and $V = V_{min} + (V_{max} - V_{min}) \times rand(N, 1)$. The fitness function value of each particle is calculated using the initial position.

Step 3: Initializes the individual optimal position pbest, the individual optimal fitness value, and the global optimal position gbest and the optimal fitness value according to the fitness function.

Step 4: Update inertia coefficient ω and learning factor c_1, c_2 according to the location of the particle.

Step 5: Update the velocity and position according to the formula (3) and (4). Calculate the fitness function value again. Update the individual optimal position, the individual optimal fitness value, the global optimal position and the optimal fitness value.

Step 6: If the maximum number of iterations is reached, the global optimal fitness value and the global optimal position are output, and the algorithm ends. Otherwise goto step 4.

3.3 Objective Function

The purpose of solar parameter identification is to obtain more accurate current value. So the measured current value is compared with the calculated current value. In order to minimize the error, a fitness function is established. In this paper, the RMSE formula is used as the fitness function. The fitness function is shown as follows:

$$fitness(\theta) = RMSE(\theta) = \sqrt{\frac{1}{N}\sum_{i=1}^{N}f(V_L, I_L, \theta)^2} \qquad (5)$$

Where θ is the unknown parameter vector, N is the number of measured data, the error function f is shown in formula (6) and (7). (6) is the error function of the single diode model, and (7) is the error function of the double diode.

$$f_{SDM}(V_L, I_L, \theta) = I_{ph} - I_{SD}\left\{exp\left[\frac{q(V_L + I_L R_s)}{KnT}\right] - 1\right\} - \frac{V_L + I_L R_s}{R_{sh}} - I_L \qquad (6)$$

$$f_{DDM}(V_L, I_L, \theta) = I_{ph} - I_{SD1}\left\{exp\left[\frac{q(V_L + I_L R_s)}{n_1 KT}\right] - 1\right\} - I_{SD2}\left\{exp\left[\frac{q(V_L + I_L R_s)}{n_2 KT}\right] - 1\right\} - \frac{V_L + I_L R_s}{R_{sh}} - I_L \qquad (7)$$

4   Simulation Results and Analysis

Some simulation experiments have done to verify the performance of using the CPSO to identify the parameters of the solar cell model.

4.1   Parameters for CPSO

Before using particle swarm optimization algorithm to identify the parameters of solar cells, some parameters should be determined. These parameters include the population size, space dimension, maximum number of iterations, position and velocity range, inertia coefficient ω and learning factor c_1, c_2 and initial position and velocity, the historical optimal position of each individual pbest and the historical optimal position of the population gbest, the historical optimal fitness of each individual and the optimal fitness of the population. In this paper, when identifying the parameters of solar cell single diode model, the population size of particle swarm optimization algorithm is 100, the space dimension is 5. The position of particle represents the parameters of solar cells, which are I_{ph}, I_{SD}, n, R_s, R_{sh}. The maximum number of iterations is 500, the position range is defined by reference [11–13], and the velocity range is set according to the position range. The corresponding formula is $V_{max} = 0.15 \times (X_{max} - X_{min})$, $V_{min} = -V_{max}$. The inertia coefficient and the learning factor are updated according to the description in 3.2. There are also three constants in fitness function, which are Boltzmann's constant $K = 1.381 \times 10^{-23}$ J/K, electronic charge $q = 1.602 \times 10^{-19}$ C, solar cell kelvin temperature $T = 306$ K. In the identification of solar cell double diode model parameters, the space dimension is 7. The position of particle represents the parameters of solar cells, which are I_{ph}, I_{SD1}, I_{SD2}, n_1, n_2, R_s, R_{sh}. The others are the same as the single diode model.

4.2   Parameter Range for Solar Cell Model

The ranges of the parameters in this paper are set according to references [11–13] shown in Table 1 and Table 2.

Table 1. Parameters range of single diode model

	R_s/Ω	R_{sh}/Ω	I_{ph}/A	I_{SD}/A	n
Minimum value	0	0	0	0	1
Maximum value	0.5	100	1	1×10^{-6}	2

Table 2. Parameters range of double diode model

	R_s/Ω	R_{sh}/Ω	I_{ph}/A	I_{SD1}/A	I_{SD2}/A	n_1	n_2
Minimum value	0	0	0	0	0	1	1
Maximum value	0.5	100	1	1×10^{-6}	1×10^{-6}	2	2

Table 3. Experimental measurements I_L-V_L data

Number	V_L/V	I_L/A	Number	V_L/V	I_L/A
1	−0.2057	0.7640	14	0.4137	0.7280
2	−0.1291	0.7620	15	0.4373	0.7065
3	−0.0588	0.7605	16	0.4590	0.6755
4	0.0057	0.7605	17	0.4784	0.6320
5	0.0646	0.7600	18	0.4960	0.5730
6	0.1185	0.7590	19	0.5119	0.4990
7	0.1678	0.7570	20	0.5265	0.4130
8	0.2132	0.7570	21	0.5398	0.3165
9	0.2545	0.7555	22	0.5521	0.2120
10	0.2924	0.7540	23	0.5633	0.1035
11	0.3269	0.7505	24	0.5736	−0.0100
12	0.3585	0.7465	25	0.5833	−0.1230
13	0.3873	0.7385	26	0.5900	−0.21

4.3 Experimental Date

A commercial silicon cell with a diameter of 57 mm is used in this paper. The measured I_L and V_L refer to references [14]. They are shown in Table 3.

4.4 Experimental Result

We compared the parameters and the root mean square error (RMSE) obtained by classification particle swarm optimization algorithm (CPSO) with the simulation results of other algorithms in [12] which are shown in Table 4 and Table 5:

Table 4. Parameter identification results of single diode model

	R_s/Ω	R_{sh}/Ω	I_{ph}/A	I_{SD}/A	n	$RMSE$
CPSO	0.0363771	53.7185	0.760776	3.23021×10^{-7}	1.48137	9.8602×10^{-4}
FPSO	0.03637	53.71852	0.76077552	3.2302×10^{-7}	1.48110817	9.8602×10^{-4}
PSO	0.0354	59.18	0.7607	4.0×10^{-7}	1.5033	1.3×10^{-3}
ABSO	0.03659	52.2903	0.76080	3.0623×10^{-7}	1.47583	9.9124×10^{-4}
SA	0.0345	43.1034	0.7620	4.798×10^{-7}	1.5712	1.9×10^{-2}

Table 5. Parameter identification results of double diode model

	R_s/Ω	R_{sh}/Ω	I_{ph}/A	I_{SD1}/A	I_{SD2}/A	n_1	n_2	$RMSE$
CPSO	0.0367403	55.4845	0.760781	2.25971×10^{-7}	7.48632×10^{-7}	1.4512	2.0000	9.8249×10^{-4}
FPSO	0.036737	55.3923	0.76078	2.2731×10^{-7}	7.2786×10^{-7}	1.45160	1.99969	9.8253×10^{-4}
PSO	0.0325	43.1034	0.7623	4.767×10^{-7}	0.1×10^{-7}	1.5172	2.0000	1.660×10^{-2}
ABSO	0.03657	54.6219	0.76078	2.6713×10^{-7}	3.8191×10^{-7}	1.46512	1.98152	9.8344×10^{-4}
SA	0.0345	43.1034	0.7623	4.767×10^{-7}	0.1×10^{-7}	1.5172	2.0000	1.664×10^{-2}

From Table 4, we can see that RMSE obtained by the classification particle swarm optimization algorithm is the same as that obtained by flexible particle swarm algorithm in reference [12], and it is obviously superior to other intelligent algorithms. From Table 5, we can see that the RMSE obtained by the classification particle swarm optimization algorithm is smaller than that obtained by the flexible particle swarm algorithm in reference [12], and it is better than that obtained by other intelligent algorithms. Compared with the same type of intelligent algorithm can highlight the advantages of the algorithm used in this paper. And literature [12] was published in 2019, which has a high comparative value.

Therefore, the classification particle swarm optimization algorithm has a better performance on identification parameters of double diode model.

The left and right sides of Fig. 3 are the convergence diagram of the parameter identification process use CPSO for the single diode model and the double diode model, respectively. It can be seen that both the single diode model and the double diode model convergent rapidly.

Tables 6 and Table 7 are the current values of the single diode model and the double diode model calculated using the explicit calculation current equation [15] obtained from the Taylor expansion. Compared with the measured experimental values in Table 3, it

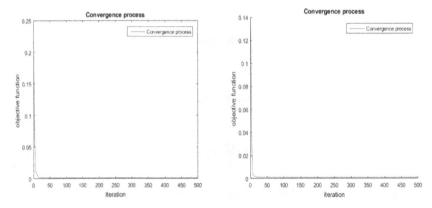

Fig. 3. Convergence curve for Single Diode (left) and Double Diode (right)

can be found that the current values calculated using the simulation parameters are close to the measured values.

Table 6. Calculation of current by single diode model

Number	V_L/V	I_{Lc}/A	Number	V_L/V	I_{Lc}/A
1	−0.2057	0.764088	14	0.4137	0.727523
2	−0.1291	0.762663	15	0.4373	0.707155
3	−0.0588	0.761355	16	0.4590	0.675577
4	0.0057	0.760155	17	0.4784	0.631222
5	0.0646	0.759056	18	0.4960	0.572419
6	0.1185	0.758044	19	0.5119	0.499764
7	0.1678	0.757092	20	0.5265	0.413664
8	0.2132	0.756143	21	0.5398	0.317293
9	0.2545	0.75509	22	0.5521	0.212118
10	0.2924	0.753672	23	0.5633	0.102718
11	0.3269	0.751404	24	0.5736	−0.00925
12	0.3585	0.747384	25	0.5833	−0.12438
13	0.3873	0.740167	26	0.5900	−0.20916

The left and right sides of Fig. 4 are the I-V fitting curves of the single diode model and the double diode model. From the image, the calculated I-V curve is almost exactly consistent with the measured I-V curve.

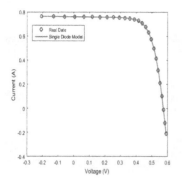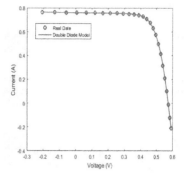

Fig. 4. I-V Fitting curve for single diode model (left) and double diode model (right)

Table 7. Calculation of current by double diode model

Number	V_L/V	I_{Lc}/A	Number	V_L/V	I_{Lc}/A
1	−0.2057	0.763983	14	0.4137	0.727395
2	−0.1291	0.762604	15	0.4373	0.707045
3	−0.0588	0.761337	16	0.4590	0.675525
4	0.0057	0.760174	17	0.4784	0.631242
5	0.0646	0.759108	18	0.4960	0.572496
6	0.1185	0.758122	19	0.5119	0.49986
7	0.1678	0.757189	20	0.5265	0.413738
8	0.2132	0.756245	21	0.5398	0.317321
9	0.2545	0.75518	22	0.5521	0.212097
10	0.2924	0.75373	23	0.5633	0.102667
11	0.3269	0.751412	24	0.5736	−0.0093
12	0.3585	0.747332	25	0.5833	−0.12439
13	0.3873	0.740063	26	0.5900	−0.20912

5   Conclusion

It is particularly important to develop the renewable energy for protecting environment. Many intelligent algorithms have been used to identifying the parameters of solar cell. The improvement of particle swarm optimization algorithm, classification particle swarm optimization algorithm, achieves better performance on these parameters simulation. It converges rapidly and the calculated current values using parameters obtained by CPSO are almost exactly consistent with the I-V curve of the measured experimental values.

Acknowledgement. This work was supported by The National Natural Science Foundation of China (Project No. 61662057, 61672301).

References

1. Cheng, Z., Dong, M.N., Yang, T.K., Han, L.J.: Extraction of solar cell model parameters based on self-adaptive chaos particle swarm optimization algorithm. J. Electr. Technol. **29**(09), 245–252 (2014)
2. Guo, X.K.: Parameter Extraction of Solar Cell Equivalent Circuit Model. Henan Agricultural University, China (2015)
3. Li, M., et al.: Numerical extraction method for solar cell parameters. J. Beijing Norm Univ. (Nat. Sci.) **02**, 215–219 (1999)
4. Alireza, A., Alireza, R.: Artificial bee swarm optimization for parameter identification of solar cells models. Appl. Energy **102**, 943–949 (2013)
5. El-Naggar, K.M., AlRashidi, M.R., AlHajri, M.F., Al-Othman, A.K.: Simulated annealing algorithm for photovoltaic parameters identification. Sol. Energy **86**, 266–274 (2012)
6. AIRashidi, M.R., AIHajri, M.F., EI-Naggar, K.M., AI-Othman, A.K.: A new estimation approach for determining the I-V characteristics of solar cells. Sol Energy **85**, 1543–1550(2011)
7. Huang, W., Jiang, C., Xue, L.Y.: Extracting solar cell model parameters based on chaos particle swarm algorithm. In: International Conference on Electric Information and Control Engineering (ICEICE), pp. 398–402 (2011)
8. Kennedy, J., Eberhart, R.: Particle swarm optimization. Process. IEEE Int. Conf. Neural Networks **4**, 1942–1948 (1995)
9. Tong, Q.J., Li, M., Zhao, Q.: An improved particle swarm optimization algorithm based on classification idea. Mod. Electron. Technol. **42**(19), 11–14 (2019)
10. Li, H.R., Gao, Y.L., Li, J.M.: A particle swarm optimization algorithm for nonlinear decreasing inertial weight strategy. J. Shangluo Univ. **21**(4), 16–20 (2007)
11. Jian, X.Z., Wei, K., Guo, Q.: Application of artificial bee swarm algorithm in solar cell parameters identification. Meas. Control Technol. **34**(07), 132–135+139 (2015)
12. Ebrahimi, S. M., Salahshour, E., Malekzadeh, M., Gordillo, F.: Parameters identification of PV solar cells and modules using flexible particle swarm optimization algorithm. Energy **179**, 358–372 (2019)
13. Mughal, M.A., Ma, Q.S., Xiao, C.X.: Photovoltaic cell parameter estimation using hybrid particle swarm optimization and simulated annealing. Energies **10**, 1213 (2017)
14. Hamid, N.F.A., Rahim, N.A., Selvaraj, J.: Solar cell parameters extraction using particle swarm optimization algorithm. In: 2013 IEEE Conference on Clean Energy and Technology (CEAT), pp. 461–465(2013)
15. Du, C.J.: The Research of I-V Characteristic Explicit Expression for the Photovoltaic Cells. Bohai Univ, Grad School (2014)

A Quantum-Inspired Genetic K-Means Algorithm for Gene Clustering

Chun Hua[✉] and Narengerile Liu

College of Computer Sciences and Technology, Inner Mongolia University for Nationalities, Tongliao 028043, People's Republic of China
chunhua99018074@163.com

Abstract. K-means is a widely used classical clustering algorithm in pattern recognition, image segmentation, bioinformatics and document clustering. But, it is easy to fall into local optimum and is sensitive to the initial choice of cluster centers. As a remedy, a popular trend is to integrate the swarm intelligence algorithm with K-means to obtain hybrid swarm intelligence K-means algorithms. Such as, K-means is combined with particle swarm optimization algorithm to obtain particle swarm optimization K-means, and is also combined with genetic mechanism to obtain Genetic K-means algorithms. The classical K-means clustering algorithm requires the number of cluster centers in advance. In this paper, an automatic quantum genetic K-means algorithm for unknown K (AQGUK) is proposed to accelerate the convergence speed and improve the global convergence of AGUK. In AQGUK, a Q-bit based representation is employed for exploration and exploitation in discrete 0–1 hyperspace using rotation operation of quantum gate as well as the genetic operations (selection, crossover and mutation) of Q-bits. The length of Q-bit individuals is variable during the evolution, which is Different from the typical genetic algorithms. Without knowing the exact number of clusters beforehand, AQGUK can obtain the optimal number of clusters and provide the optimal cluster centroids. Five gene datasets are used to validate AQGUK, AGUK and K-means. The experimental results show that AQGUK is effective and promising.

Keywords: Quantum computing · Genetic algorithm · Gene clustering · K-means

1 Introduction

Clustering analysis plays an important role in data mining areas, such as, pattern recognition, image segmentation, bioinformatics and document clustering. Clustering is the process of grouping a set of n objects into K groups according to some standard. Therefore, similar objects will be classified into the same cluster and dissimilar objects willt be in different clusters. Clustering is an unsupervised machine learning algorithm without a training process. Clustering analysis does not need the cluster labels of any samples in advance and utilizes an algorithm to divide a group of samples with unknown categories

© Springer Nature Switzerland AG 2020
M. Han et al. (Eds.): ISNN 2020, LNCS 12557, pp. 13–24, 2020.
https://doi.org/10.1007/978-3-030-64221-1_2

into several categories. Clustering analysis clusters similar things together and does not care too much what is it.

K-means is proposed by Steinhaus [1] and Macqueen [2] from different fields. Then, MacQueen summarized the previous researcher's achievements and gave the detailed steps of the K-means algorithm. K-means is widely used in various research fields because of it's simplicity and fast convergence. But, the disadvantages of K-means is easy fall into local optimum resulting in poor clustering results and is sensitive to initial cluster centers. In order to improve the shortcomings of K-means, researchers usually combined the K-means with swarm intelligence algorithm to obtain intelligence K-means algorithms. The hybrid intelligence K-means algorithms can help to avoid K-means trapping into local optimum. To accelerate the convergence speed and improve the global convergence of intelligence K-means algorithms, researchers usually combined intelligence K-means with quantum computing. For example, CS capabilities [3] is extended by using nonhomogeneous update inspired by the quantum theory in order to deal with the cuckoo search clustering problem in terms of global search ability. A quantum inspired genetic algorithm for k-means clustering (KMQGA) is proposed in [4]. KMQGA employed a Q-bit based representation for exploration and exploitation in discrete 0–1 hyperspace using rotation operation of quantum gate as well as the typical genetic algorithm operations of Q-bits.

In this paper, we attempt to combine the quantum computing with genetic K-means algorithm to remedy the drawbacks of the k-means mentioned above. An improved automatic genetic K-means clustering algorithm based on quantum-inspired algorithm (AQGUK) is proposed. AQGUK employed variable length of Q-bit individuals in the evolution and used rotation operation of quantum gate as well as the genetic operations (selection, crossover and mutation) of Q-bits. AQGUK accelerates the convergence speed and improves the global convergence of AGUK by introducing quantum computing, and attempts to obtain the optimal number of clusters and provides the optimal cluster centroids.

The rest part of this paper is structured as follows. Section 2 first briefly introduced K-means clustering technique and the background of quantum computing. Section 3 described the proposed hybrid algorithm (AQGUK) in detail. In Sect. 4, Experimental results and analysis are shown on K-means, AGUK and AQGUK. In Sect. 5, some short conclusions are drawn.

2   Related Works

In this Section, first, we mainly introduced the genetic algorithm with clonal reproduction employed in this paper, K-means clustering method and quantum computing theory. Then, the common clustering evaluation criteria are described.

2.1   K-Means Clustering Technique

An important branch of unsupervised machine learning is clustering analysis. In which, the samples without category mark is divided into several subsets according to some criteria, so the clustering analysis groups the similar samples together as much as possible, groups dissimilar samples into different clusters. K-means is a well-known partitioning method on account of its simplicity and fast convergence speed. Which minimize (or maximize) the certain criterion function's values in the iteration process.

In the classical K-means, K points are randomly selected as the initial clustering center. Then the distance between each data and each clustering center are calculated. Finally, the samples are grouped into the cluster that is closest to it. Therefore, the new clustering centers are calculated, and the samples are reassigned into the new clusters. The iteration process is repeated until termination condition is met. The main steps of the K-means are described as follows (Table 1):

Table 1. Main steps of K-means

Input parameters:
K: the number of cluster, T: the maximum number of iterations
Output parameters: the partition of samples' set

For $i = 1, 2, \cdots, T$
For every x_i(for all samples)
calculate the distance (x_i, c_k);
group the x_i into the cluster that is nearest to it;
end For
update all cluster centers;
if$(i \leq T)$
continue;
else
break;
end For

The classical K-means requires users input number of cluster K in advance, and some hybrid K-means clustering algorithm also require the number of cluster in advance. Such as, GKA [5] and PK-means [6]. In this article, we proposed a quantum automatic genetic algorithm based K-means for unknown K called (AQGUK).

2.2 Quantum Computing

Before describing AQGUK, we briefly introduce the basic concept of quantum computing. In which, the smallest information representation is called quantum bit (Q-bit) [7]. The quantum bit will have three state, "0" state, "1" state or any superposition of the two respectively. Therefore, the state of Q-bit can be represented as follows:

$$|\psi> = \alpha|0> + \beta|1>, \tag{1}$$

where α and β are complex numbers that specify the probability amplitudes of the corresponding states. Thus, $|\alpha|^2$ and $|\beta|^2$ denote probabilities that the Q-bit will be found in the "0" state and the "1" state, respectively. Normalization of the state is as follows:

$$|\alpha|^2 + |\beta|^2 = 1 \tag{2}$$

Therefore, a Q-bit can represent the linear superposition of the two binary genes (0 and 1). The following is a representation of m Q-bits individual.

$$\begin{bmatrix} \alpha_1 & \alpha_2 & \cdots & \alpha_m \\ \beta_1 & \beta_2 & \cdots & \beta_m \end{bmatrix} \tag{3}$$

Thus, string of m Q-bits can represent 2^m states. Quantum gate operation can change the state of Q-bit, such as, the NOT gate, the rotation gate, etc. However, the superposition of "0" state and "1" state must be collapse to a single state in the action of quantum state, that is, a quantum state have to be the "0" state or "1" state. In the evolutionary computing, the Q-bit representation can enhance the diversity of population than other representations because of it's linear superposition of states probabilistically. More examples of Q-bit representation can be found in [8]. AQGUK is designed with this novel Q-bit representation.

3   The Proposed Algorithm (AQGUK)

In this Section, we described main process of our proposed algorithm (AQGUK). The specific steps of AQGUK are as follows:

3.1   Main Operation of AQGUK

Genetic algorithm is an intelligent heuristic search algorithm based on Darwin's evolutionary theory of survival of the fittest. Genetic algorithm includes three main operations: selection, crossover and mutation. The detailed description of main operators employed in this paper is as follows:

Selection. Selection operator can select good individuals from the current population according to selection probability. Therefore, selection operator gives opportunities for good individuals to be parents then breed next generation population. The criterion for judging the individual's good or bad is their fitness. The higher fitness of individuals, the greater the chance be selected. The frequently used selection method includes roulette wheel selection, stochastic universal selection, local selection, truncation selection and tournament selection. In this paper, we employed the roulette wheel selection operations. The roulette wheel selection method can also be called proportional selection, in which the probability of an individual being selected is related to its fitness. Specific steps of roulette wheel selection are as follows:

(1) Calculate the fitness value of each individual.
(2) Calculate the probability of each individual being selected according to the fitness value.

$$P(I_i) = f(I_i) / \sum_{j=1}^{N} f(I_j) \tag{4}$$

where, $f(I_i)$ is fitness value of individual I_i.

(3) Calculate the cumulative probability of each individual according to the following formula.

$$q(I_i) = \sum_{j=1}^{i} P(I_j) \tag{5}$$

(4) Generate a random number r in the interval $[0,1]$.
(5) If $r < q[1]$, individual 1 is chosen, otherwise, individual k is chosen according to $q[k-1] < q[k]$.
(6) Repeat (4) and (5) N times.

Clonal Reproduction. This is main process of clonal selection algorithm. The main purpose of introducing the clonal reproduction operation in our proposed algorithm (AQGUK) is to further enhance diversity of population. The central idea of clonal reproduction is proportional replication of individuals based on it's fitness values. Namely, individuals with large fitness values are copied more than individuals with small fitness values.

Crossover Operation. In this stage, first, the individuals are randomly paired in the pairing pool, and then set a crossover point for paired individuals. Finally, the paired individuals exchange genes with each other. Crossover methods commonly used includes single-point crossover, multiple-point crossover, uniform crossover, shuffle crossover and crossover with reduced surrogate. The crossover operation employed in this paper is simple single-point crossover operation.

Mutation. Individuals in current population change the value of one or more genes with mutation probability. In order to explore different good solutions, the proposed algorithm randomly changes some chromosomes of current population according to mutation probability.

The basic idea of the mutation operation employed in this paper is described as follows. Chromosomes with a low fitness have a high probability of getting a random change, while, chromosomes with a high fitness have a low probability. The mutation probability of i-th chromosome is calculated according to following formula.

$$
M_{I_i} = \begin{cases} k_1 * \dfrac{fmax - f_i}{fmax - \bar{f}}, f_i > \bar{f} \\ k_2, f_i \leq \bar{f} \end{cases} \tag{6}
$$

where, k_1 and k_2 are equal to 0.5, $fmax$ is the maximum fitness of a chromosome in current population, $\bar{f}$ is average fitness of the chromosomes of current population, and f_i is the fitness of the i-th chromosome. Each gene of chromosome be selected for mutation is changed and modified the attribute value randomly.

Rotation Operation. The main purpose of rotation operation is to adjust the probability amplitudes of each Q-bit. A rotation gate $U(\theta)$ is employed to update a Q-bit individual as follows:

$$
\begin{bmatrix} \alpha_i' \\ \beta_i' \end{bmatrix} = U(\theta) \times \begin{bmatrix} \alpha_i \\ \beta_i \end{bmatrix} = \begin{bmatrix} \cos\theta_i & -\sin\theta_i \\ \sin\theta_i & \cos\theta \end{bmatrix} \times \begin{bmatrix} \alpha_i \\ \beta_i \end{bmatrix}, \tag{7}
$$

where $\begin{bmatrix} \alpha_i \\ \beta_i \end{bmatrix}$ is the i-th Q-bit and θ_i is the rotation angle of each Q-bit toward either 0 or 1 state depending on its sign. Quantum gate $U(\theta)$ is a function of $\theta_i = s(\alpha_i, \beta_i) \times \Delta\theta_i$, where $s(\alpha_i, \beta_i)$ is the sign of θ_i, which determines the direction and $\Delta\theta_i$ is the magnitude of rotation angle [8].

Catastrophe Operation. When the best individual does not change in a certain number of consecutive generations, the Catastrophe operation is performed. AQGUK records the fitness of best individual for each iteration and compares the best fitness in current iteration with the best fitness.

3.2 Evaluation Strategies

In our numerical experiments, we will use the sum squared error (SSE), the Xie–Beni index (XB) [9], the Davies–Bouldin index (DB) [10, 17], and the separation index (S) [11]. Choosing the optimal centers c_k's and the optimal label matrix W are the aim of the clustering algorithms discussed in this paper.

SSE is defined by

$$
SSE = \sum_{k=1}^{K} \sum_{i=1}^{n} w_{ik} ||x_i - c_k||^2 \tag{8}
$$

Generally speaking, the lower the SSE value, the better the clustering result. The XB index [9] is defined as follows:

$$XB = \frac{SSE}{n * d_{min}},$$ (9)

where, d_{min} is the shortest distance between cluster centers. Higher d_{min} implies better clustering result and SSE is the lower the better, as we mentioned above. Therefore, the lower XB value, the better clustering results.

To define the DB index [10], we first defined the within-cluster separation S_k and R_k as follows:

$$S_k = (\frac{1}{|c_k|} \sum_{x_i \in c_k} ||x_i - c_k||^2)^{\frac{1}{2}},$$ (10)

where c_k (resp. $|c_k|$) denotes the set (resp. the number) of the samples belonging to the cluster k.

$$R_k = \max_{j,j \neq k} \frac{S_k + S_j}{||c_k - c_j||}.$$ (11)

Then, the DB index is defined as

$$DB = \frac{1}{K} \sum_{k=1}^{K} R_k.$$ (12)

Generally speaking, lower DB implies better clustering results.
The separation index S [11] is defined as follows:

$$S = \frac{1}{\sum_{k,j=1;k \neq j}^{K} |c_k||c_j|} \sum_{k,j=1;k \neq j}^{K} |c_k||c_j|||c_k - c_j||$$ (13)

Generally speaking, the higher S values, the better clustering results .

3.3 Main Steps of Proposed Algorithm

See Table 2.

Table 2. Main steps of proposed algorithm

Input: the maximum number of iterations T
Output: the number of cluster center and the result of cluster

1. Initialization: Set the population size N, the maximum number of iterations T, the mutation probability Pm and the error tolerance E_{tol}. Let t = 0, and choose the initial population P(0). In addition, choose the best individual from P(0) and denote it as super individual $L^*(0)$.
2. Selection: Select a new population from P(t) according to formula (1), and denote it by $P_1(t)$.
3. Clonal reproduction: Perform a clone operation based on the fitness value of the individuals and get a new population denoted by $P_2(t)$.
4. Crossover: Paired individuals exchange genes with each other, and get a new population denoted by $P_3(t)$.
5. Mutation: Mutate each individual in $P_2(t)$ according to formula (3), and get a new population denoted by P(t + 1).
6. Update the super individual with the quantum rotation operation: choose the best individual from P(t + 1) and compare it with $L^*(t)$ to get $L^*(t + 1)$. If the best individual does not change in a certain number of consecutive generations, perform catastrophe operation.
7. Stop if either t = T, otherwise go to 2 with t ← t + 1.

4 Experimental Evaluation and Results

4.1 Data Sets and Parameters

Four gene expression data sets shown in Table 3 are used for evaluating our algorithms. The first three data sets are Sporulation [12], Yeast Cell Cycle [13], Lymphoma [14], and the other two are Yeast and Ecoli.

As shown in Table 3, there are some sample vectors with missing component values in Yeast Cell Cycle, Sporulation and Lymphoma data sets. To rectify these defective data, the strategy adopted in this paper is as follows: First, the sample vectors with more than 20% missing components are removed from the data sets. Then, for the sample vectors with less than 20% missing components, the missing component values are estimated by the KNN algorithm with the parameter k = 15 as in [16], where k is the number of the neighboring vectors used to estimate the missing component value (see [15, 16, 17] for details).

The values of the parameters used in the computation are set as follows:
Population size N = 30, Crossover probability Pc = 0.3,
Mutation probability Pm = 0.1, T = 150.

4.2 Experimental Results and Discussion

This section is divided into two parts. The first part is the performances of the algorithms in terms of SSE, S, DB and XB. The second part demonstrates the clustering accuracy of three algorithms for five data sets.

Each of the three algorithms conducted fifty trials on the five data sets. The averages over the fifty trials for the four evaluation criteria (SSE, S, DB and XB) are listed in Tables 4. We shall pay our attention mainly on the comparison of AGUK and AQGUK,

Table 3. Data sets used in experiments

Data sets	No. of vectors n	No. of vectors with missing components <20%	No. of vectors with missing components ≥20%	No. of attributes D	No. of classes K
Sporulation	6023	413	198	7	16
Yeast cell cycle	6078	5498	680	77	256
Lymphoma	4022	3166	3	96	150
Yeast	1484	0	0	8	10
Ecoli	336	0	0	8	7

so as to show the benefit of the introduction of Q-bit representation of individuals and quantum gate operation. The clustering accuracy of three algorithms for five data sets is shown in Fig. 5.

Table 4. Average SSE, S, DB and XB on the five gene data sets

Data sets	Algorithm	SSE (lower the better)	S (higher the better)	DB (lower the better)	XB (lower the better)
Sporulation	K-means	5.413×10^3	3.162	2.678	5.019×10^{-4}
	AGUK	5.392×10^3	3.254	2.653	4.901×10^{-4}
	AQGUK	5.218×10^3	3.381	2.638	4.725×10^{-4}
Yeast cell cycle	K-means	1.413×10^4	2.732	1.569	3.085×10^{-4}
	AGUK	1.392×10^4	2.785	1.542	2.765×10^{-4}
	AQGUK	1.381×10^4	2.801	1.467	2.542×10^{-4}
Lymphoma	K-means	1.948×10^4	7.192	2.846	7.462×10^{-4}
	AGUK	1.865×10^4	7.345	2.436	6.766×10^{-4}
	AQGUK	1.541×10^4	7.422	2.137	4.346×10^{-4}
Yeast	K-means	238.923	0.321	1.692	9.642×10^{-4}
	AGUK	229.452	0.332	1.564	8.564×10^{-4}
	AQGUK	221.343	0.343	1.465	7.432×10^{-4}
Ecoli	K-means	97.234	3.312	0.936	7.800×10^{-3}
	AGUK	96.321	3.436	0.921	7.543×10^{-3}
	AQGUK	94.387	3.509	0.876	6.781×10^{-3}

From Tables 4, we can see that the proposed algorithm AQGUK achieves the lowest SSE, DB, XB and highest S, for all the five gene data sets. Therefore, AQGUK performs better than the two four algorithms.

Figures 1, 2, 3, 4 show clearly the overall performance for SSE, DB, XB and S evaluations respectively. Four figures clearly show that the proposed AQGUK outperforms the other two algorithms, in the sense of average performance.

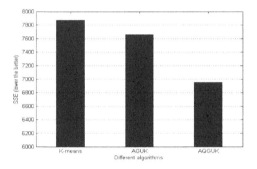

Fig. 1. Average SSE on five data sets

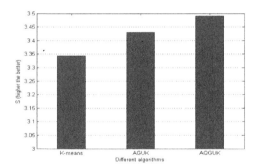

Fig. 2. Average S on five data sets

From Fig. 5, we can see clearly the clustering accuracy based on SSE for five data sets respectively, and it shows that AQGUK outperforms AGUK and K-means, in the sense of clustering accuracy.

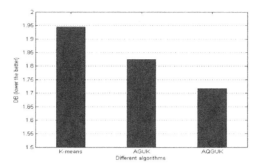

Fig. 3. Average DB on five data sets

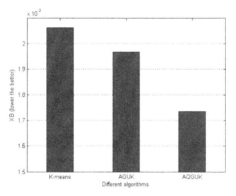

Fig. 4. Average XB on five data sets

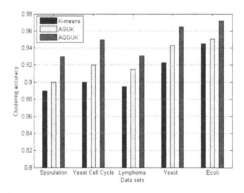

Fig. 5. Average clustering accuracy on five data sets

5   Conclusions

Numerical experiments are carried out on the comparison of K-means, AGUK and AQGUK algorithm. The evaluation tools include the sum squared error (SSE), the Davies–Bouldin index (DB), Xie–Beni index (XB) and the separation index (S).

The conclusions we draw from the simulation results are as follows: the overall performances of AGUK in terms of the four indexes outperform those of K-means, and the overall performances of AQGUK outperform those of AGUK and K-means. This shows the effectiveness of the utilization of the quantum Q-bit representation for individual, clonal reproduction and quantum gate operation.

References

1. Steinhaus, H.: Sur la division des corp materiels en parties. Bull Acad Polon Sci. **3**, 801–804 (1956)
2. Macqueen, J.: Some methods for classification and analysis of multivariate observations. In Proceedings of the 5th Berkeley Symposium on Mathematical Statistics and Probability, pp. 281–297 (1967)
3. Ishak Boushaki, S., Kamel, N.: A new quantum chaotic cuckoo search algorithm for data clustering. Expert Syst. Appl. **96**, 358–372 (2018)
4. Xiao, J., Yan, Y.P.: A quantum-inspired genetic algorithm for k-means clustering. Expert Syst. Appl. **37**, 4966–4973 (2010)
5. Krishna, K., Murty, M.N.: Genetic k-means algorithm. IEEE Trans. Syst. Man Cybern. **29**, 433–439 (1999)
6. Du, Z., Wang, Y.: PK-means: a new algorithm for gene clustering. Comput. Biol. Chem. **32**, 243–247 (2008)
7. Hey, T.: Quantum computing: an introduction. Comput. Control Eng. J. **10**(3), 105–112 (1999)
8. Han, K.H., Kim, J.H.: Quantum-inspired evolutionary algorithm for a class of combinatorial optimization. IEEE Trans. Evol. Comput. **6**(6), 580–593 (2002)
9. Xie, X.L., Beni, G.: A validity measure for fuzzy clustering. IEEE Trans. Pattern Anal. Mach. Intell. **13**, 841–847 (1991)
10. Liu, Y.G.: Automatic clustering using genetic algorithms. Appl. Math. Comput. **218**, 1267–1279 (2011)
11. Chu, S., DeRisi, J.: The transcriptional program of sporulation in budding yeast. Science **282**, 699–705 (1998)
12. Spellman, P.T.: Comprehensive identification of cell cycle-regulated genes of the yeast saccharomyces cerevisiae by microarray hybridization. Mol. Biol. **9**, 3273–3297 (1998)
13. Alizadeh, A.A., Eisen, M.B.: Distinct types of diffuse large b-cell lymphoma identified by gene expression profiling. Nature **403**, 503–511 (2000)
14. Yoon, D., Lee, E.K.: Robust imputation method for missing values in microarray data. BMC Bioinform. **8**, 6–12 (2007)
15. Troyanskaya, O., Cantor, M.: Missing value estimation methods for DNA microarrays. Bioinformatics **17**, 520–525 (2001)
16. Chun, H., Feng, L.: A genetic xk-means algorithm with empty cluster reassignment. Symmetry **11**(744), 49–65 (2019)

Spark Parallel Acceleration-Based Optimal Scheduling for Air Compressor Group

Long Chen[1(✉)], Xiaojuan Zhang[2], Jun Zhao[1], Long Chen[1], and Wei Wang[1]

[1] Faculty of Electronic Information and Electrical Engineering, Dalian University of Technology, Dalian 116023, China
chenlong@dlut.edu.cn
[2] Faculty of Mechanical Engineering,
Dalian Institute of Science and Technology, Dalian 116052, China

Abstract. For the air compressor system in iron and steel enterprises, an optimal scheduling method for it was proposed, where the predicted value of the production load and the equipment capacity are treated as the model constraints, aiming to reduce the optimal economic cost and improve the energy conversion efficiency. In addition, an optimization method, combining the hierarchical search and the adaptive particle swarm optimization algorithm, was proposed to fully consider the performance of the air compressors, resulting in the great improvement of the search efficiency. In order to further accelerate the computation process of the model, a parallel acceleration algorithm based on Spark framework was also designed. The experimental results show that the proposed method exhibits good performance in the optimization of the air compressor group scheduling problem. In addition, under the premise of algorithm stability, good acceleration effect could be obtained by the Spark parallel algorithm.

Keywords: Air compressor · Optimal scheduling · Hierarchical search · Spark parallel acceleration

1 Introduction

The steel industry has always been a high-energy-consuming industry in China, whose energy consumption accounts for more than 10% of China's total energy consumption [1]. Compressed air, as one of the important power sources of iron and steel enterprises, plays an important role in iron and steel enterprises. It is estimated that the energy consumption of the compressed air system of China's iron and steel enterprises accounts for about 10% to 15% of the total industrial energy consumption [2]. Therefore, the optimal scheduling of compressed air energy systems in iron and steel enterprises urgently needs to be resolved.

In recent years, researchers have paid more and more attention to the energy efficiency evaluation of com-pressed air systems and the optimal scheduling of equipment and systems. In [3], an optimal control method was proposed for oxygen generators based on dynamic programming algorithm, which effectively reduced the operating cost, while

© Springer Nature Switzerland AG 2020
M. Han et al. (Eds.): ISNN 2020, LNCS 12557, pp. 25–36, 2020.
https://doi.org/10.1007/978-3-030-64221-1_3

this method suffered from the low calculation efficiency with large-scale air compressor group. Reference [4] reported a mixed integer nonlinear programming for the compressed gas storage operation in traditional gas power generation companies. In [5], an improved particle swarm optimization algorithm with nonlinear dynamic adjustment of inertia weights was proposed, which achieved the goals of energy saving, emission reduction and balanced scheduling of the air compressor control system. In addition, the methods in the above literature did not consider the difference on the performance of the air compressors. The air compressors participating in the scheduling are all manually confirmed, resulting in the fact that the optimality of the solution cannot be ensured.

In practice, enterprises have put forward higher requirements for the computation efficiency. In recent years, the parallel acceleration algorithms have gradually shifted from the Hadoop framework [6] to the Spark [7] framework based on in-memory computing. Since the latter one can effectively avoid frequent I/O access, the algorithm has a good performance on parallel acceleration. In [8] and [9], the K-nearest neighbor method and Bayesian network based on the Spark computing framework were studied respectively, which were used to solve the big data classification problem. In [10] and [11], the iterative calculation acceleration of genetic algorithm and ant colony algorithm were studied respectively in the Spark framework. At present, the Spark is mostly applied in data mining for users by Internet companies, but there are few applications in the traditional industrial enterprises.

By determining the constraint boundary with the predicted value of total compressed air demand and the production capacity of the equipment, we established the scheduling model, whose objective function is the economic cost or energy conversion efficiency. Considering that the traditional method does not take the performance of each air compressors into account, a hierarchical search algorithm is proposed to determine the optimal group combination of the air compressor by production plan, air compressor maintenance plan and compressed air demand as constraints. On this basis, in order to improve the accuracy and convergence speed of the solution, we employed an improved adaptive particle swarm algorithm (APSO) to solve the optimization problem under different open combinations. Besides, an APSO parallel acceleration algorithm based on the Spark framework is also designed. Through experiments on the scheduling problem of the air compressor system of a domestic iron and steel enterprise, the effectiveness of the proposed method is verified. On the premise of meeting the needs of the enterprise, the operating cost of the enterprise's air compressor group is reduced.

2   Problem Description and Optimization Model

2.1   Problem Description

The compressed air system of iron and steel enterprises is mainly composed of three parts, including generating system, storage system and consuming users. Generally, the air compressor unit can be divided into multiple air compressor stations according to the area and their functions. A schematic diagram of the structure of a compressed air system of a steel enterprise is shown in Fig. 1. Compressed air is generated via an air compressor station and transported along the pipe network to various parts of the steel plant. It contains four air compressor stations. Each station comprises of 6 sets of air compressor equipment and one set of auxiliary equipment. Each set of air compressor equipment includes an air compressor, a gas storage tank, a dryer and a filter. Compressed air consumption equipment mainly includes a sintering workshop, a rotary hearth furnace and a blast furnace.

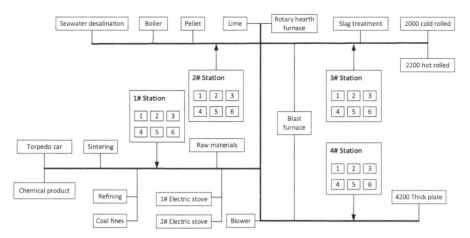

Fig. 1. Schematic diagram of the compressed air system of a steel company

At present, in most of the iron and steel enterprises in China, the air compressor systems are basically dispatched manually, according to the actual production conditions of the enterprise. The status of the air compressor in the process is completely determined manually, based on the equipment maintenance plan, however, the performance of the equipment has not been involved as an evaluation condition. It can be seen that the manual scheduling method has the disadvantages of large error and low degree of automation. In addition, as for the scheduling methods in the existing literature, the performance difference of air compressor equipment is not considered.

2.2 Optimal Scheduling Model

In view of the many shortcomings of the manual scheduling method, in this paper, the safe operation of air compressor equipment is treated as a constraint to establish an optimal scheduling model of the economically optimal air compressor cluster. The objective function is generally composed of the sum of the operating cost and the start-stop cost of the air compressor in the period $t_1 \sim t_2$. The objective function is given by:

$$
\begin{cases}
\min_x J_{\cos t} = \varsigma \left(J_{Air} + J_{Drying} + J_{others} \right) \\
J_{Air} = \int_{t1}^{t2} \sum_{i=1}^{m} \sum_{j=1}^{S_i} W_{ij}^{run} dt + \\
\quad \sum_{i=1}^{m} \sum_{j=1}^{S_i} \left(W_{ij}^{start} + W_{ij}^{stop} \right) \\
J_{Drying} = \int_{t1}^{t2} \sum_{i=1}^{m} \tilde{W}_{ij}^{run} (Q_{S''}) dt \\
W_{ij}^{run} = \alpha_{ij} * x_{ij} + \beta_{ij}, \\
\quad i = 1, 2, \ldots, m; \ j = 1, 2, \ldots, |S_j|
\end{cases}
\tag{1}
$$

where J_{Air}, J_{Drying} and J_{others} denote the energy consumption of the air compressor, the combined drying unit and other auxiliary facilities during the $t_1 \sim t_2$ period respectively, J_{others} is generally set to a fixed value; ς is the electricity price; m is the number of air compressor stations; S_i and S_i'' are the number of open compressors and combined drying units of the i-th air compressor station; W_{ij}^{start} and W_{ij}^{stop} denote the energy consumption of starting and stopping the j-th air compressor in the i-th station respectively, both of which are a fixed value; $\tilde{W}_{ij}^{run}(Q_{S''})$ is the relationship between the outlet flow $Q_{S''}$ of the air compressor station and the energy consumption of the combined dryer unit when the i-th air compressor station combined drying unit working at opening strategy S''. In actual working conditions, in order to facilitate calculation, it is generally set to a fixed value; α_{ij} and β_{ij} are the parameters of the curve between opening degree and energy consumption of the j-th air compressor in the ith station; x_{ij} is the opening value, and it is also the only independent variable of the optimization problem.

The constraints of the objective function are described as follows:

$$
s.t \begin{cases}
\underline{R}_{ij} \leq x_{ij} \leq \overline{R}_{ij} & \textcircled{1} \\
Q_{\min} \leq Q_{ij} \leq Q_{\max} & \textcircled{2} \\
\sum_{i=1}^{m} \sum_{j=1}^{S_i} Q_{ij} \geq Q_{need} & \textcircled{3}
\end{cases}
\tag{2}
$$

The meaning of each constraint is listed as below.

① Air compressor intake valve opening constraints: $\underline{R}_{ij}$ and $\overline{R}_{ij}$ are the maximum and minimum constraints of air compressor intake valve flow opening.
② Air compressor capacity constraints: Q_{max} and Q_{min} represent the maximum and minimum constraints of air compressor capacity, respectively.
③ Gas production and gas consumption matching constraints: $\sum_{i=1}^{m} \sum_{j=1}^{S_i} Q_{ij}$ is the total gas production of m air compressors, and Q_{need} is the user's air demand.

3 Hierarchical Search and APSO Algorithm Based on Spark Framework

Without taking the performance difference of individual air compressor into account, the solution of the model in Sect. 2.2 cannot be guaranteed to be the optimal solution.

The performance of an air compressor can be described by a curve of the opening value and energy conversion efficiency which is fitted by the curve of opening value and flow value and the curve of opening value and energy consumption. It can be seen from Fig. 2 that the performance of each air compressor is different under different opening values, thus it is necessary to select an optimal configuration of the air compressor group.

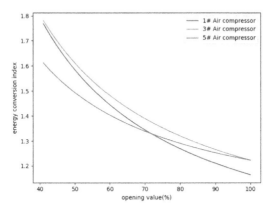

Fig. 2. Performance curves of 1#, 3# and 5# air compressors of 1# station

Considering the difference on the performance of air compressors, we propose to employ the hierarchical search to determine the combination set of air compressors. Then, the APSO algorithm based on the Spark framework is designed to solve the optimal problem with less computational burden. Finally, the optimal solution is obtained by evaluating the optimal value under different combinations. The general flow of the optimized dispatch of compressed air system is shown in Fig. 3.

3.1 Hierarchical Search

The hierarchical search mainly determines the combination set of the air compressor through the information of the air compressor equipment and the demand for compressed air, which is divided into three layers, i.e., a human-computer interaction layer, a strategy search layer, and a device selection layer.

(1) Human-computer interaction layer: Determine the status information of the air compressor equipment based on the database and manual correction.
(2) Strategy search layer: According to the state of the air compressor determined by the human-machine interaction layer, determine the opening strategy $\{S_1, S_2, \ldots, S_m\}$ of the air compressor unit, where S_i is the number of air compressors opened at the i-th air compressor station. The opening strategy must satisfy the condition that the maximum gas production is not less than the predicted value of the compressed air demand, otherwise continue to search for a new opening strategy until the conditions are met.

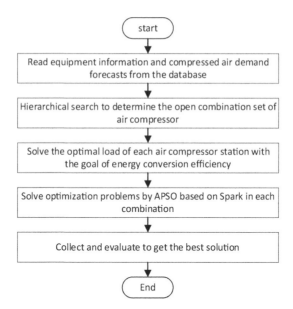

Fig. 3. Schematic diagram of the manual scheduling method of the air compressor system

(3) Equipment selection layer: According to the search results of the air compressor station opening strategy, all possible combinations are arranged and combined. For example, if the number of air compressors in the air compressor station is $\{Q_1, Q_2, \ldots, Q_m\}$, the number of open compressor combinations for each air compressor station is $\left\{C_{Q_1}^{S_1}, C_{Q_2}^{S_2}, \ldots, C_{Q_m}^{S_m}\right\}$, the total is n, and the i-th combination is $\pi_i = \{\pi_1^i, \pi_2^i, \ldots, \pi_m^i\}$. Among them, the opening combination must meet the following load balance constraints:

$$\left| \sum_{j=1}^{\left\lfloor \frac{Q_i}{2} \right\rfloor} \sigma_{ij} - \sum_{j=\left\lfloor \frac{Q_i}{2} \right\rfloor}^{Q_i} \sigma_{ij} \right| \leq 1, \; i = 1, 2, \ldots, m \tag{3}$$

$$\sigma_{ij} = \begin{cases} 1, & compressor(i, j) \; participate \; in \\ 0, & otherwise \end{cases} \tag{4}$$

where compressor (i, j) is the j-th air compressor in the i-th station.

After the combination is determined, in order to ensure the load balance between the air compressor stations, the load between the stations must also be distributed. The upper limit of the output of each air compressor station is also determined, and the load distribution between air compressor stations can be described as the following optimization problem:

$$
\max_{Q(\pi_i,t)} \frac{\sum_{i=1}^{m} \lambda_1\lambda_2 \int_{t_1}^{t_2} Q(\pi_i,t)+h(\pi_i,t)dt}{\sum_{i=1}^{m} \int_{t_1}^{t_2} x(\pi_i,t)dt}
$$

$$
s.t. \begin{cases} \sum_{i=1}^{m} Q(\pi_i,t) \geq Q \\ 0 \leq Q(\pi_i,t) \leq Q_{\pi_i}^{max}, i = 1, 2, \ldots, m \end{cases}
$$

(5)

where λ_1 and λ_2 are the standard coal conversion coefficient and the equivalent electric conversion coefficient respectively; $Q(\pi_i, t)$, $h(\pi_i, t)$ and $x(\pi_i, t)$ are the output, residual energy recovery and input quantity (electric energy) of the i-th air compressor station when taking the combination π_i at time t; $x(\pi_i, t) = g(Q(\pi_i, t))$, The specific expression of $g(\bullet)$ is determined according to the air compressor equipment parameters under π_i; Q is the measured value of the total compressed air demand; $Q_{\pi_i}^{max}$ is the upper limit of the output of the ith air compressor station under π_i. To simplify the calculation in actual operation, the value of $t_2 - t_1$ can be fixed to unit time.

The difference between the load distribution of the air compressor station and the actual output of each station is added as a penalty term to the original objective function. Therefore, the formula (1) is converted as:

$$
\min_{x} J_{\cos t} = \varsigma_1 \left(J_{Air} + J_{Drying} + J_{others} \right)
$$

$$
+ \sum_{i=1}^{m} \left| Q_{\pi_i} - \sum_{j=1}^{S_i} Q_{ij} \right|_2^2
$$

(6)

The proposed optimization model is solved by using the APSO algorithm for determining the i-th combination π_i in N combinations. The objective function is the suboptimization problem of formula (6), after aggregating N results, the air compressor's opening value x_{ij} under π_i with formula (1) as the objective function are selected.

3.2 Adaptive Particle Swarm Optimization

Since the PSO algorithm has the advantages of fast search speed and easy implementation, this algorithm based on adaptive inertia weight is used to solve the optimization problem, and the Gaussian mutation mechanism [12] is introduced to improve the global search capability for the premature convergence phenomenon. The position of the particle is composed of the opening value a of the air compressor x_{ij}, which is a one-dimensional vector. The update formula of the particle is given by:

$$
x_i(t + 1) = x_i(t) + v_i(t + 1)
$$

(7)

$$
v_i(t + 1) = \omega_t v_i(t) + r_1 c_1 (p_{id} - x_i(t)) + r_2 c_2 (p_{gd} - x_i(t))
$$

(8)

where ω_t is the inertia weight of the t-th iteration; r_1 and r_2 are a random number within $0 \sim 1$; c_1 and c_2 are the acceleration constant, generally 0.5; p_{gd} is the global optimal particle; p_{id} is the historical optimal particle.

We set the inertia weight as a function of the fitness of the current optimal value in t iterations, as follows:

$$\omega_t = 0.5 \times \left[1 + \tanh\left(\frac{F(P_{gd}^t)}{F(P_{gd}^1)}\right) \right] \tag{9}$$

where $\tanh(\cdot)$ is the hyperbolic tangent function; $F(\cdot)$ is the fitness function. From formula (9), one can see that $0 \le \omega_t \le 1$. If the adaptability of the current optimal solution is not significantly improved, the inertia weight decreases slowly since it still requires global search capabilities, otherwise it is conducive to better local search.

For enhancing the global search ability, the Gaussian random disturbance is added to the particles in the mutation mechanism:

$$x_{ij} = x_{ij} + M \times \beta_{ij} \tag{10}$$

where x_{ij} is the j-th component of the i-th particle; $\beta_{ij} \sim N(0, 1)$; M is a variable step.

From formula (10), M determines the strength of the mutation operator's global search ability and local search ability. The variable asynchronous length in this paper can be adaptively changed according to the fitness value of particles. Therefore, M is based on the fitness of the current optimal solution. During the t-th iteration it is defined by:

$$M_t = x_{max} \times \tanh\left(\frac{F(P_{gd}^t)}{F(P_{gd}^1)}\right) \tag{11}$$

where x_{max} is the maximum value of the particle component. As the iteration process progresses, M_t will gradually decrease as the global optimal particle adaptation value decreases.

3.3 Optimized Scheduling Algorithm Based on Spark Parallel Acceleration

Hierarchical search will generate a large number of combinations, and the APSO algorithm is executed for each combination needs, which suffers from the extremely high calculation costs. The core process of the APSO algorithm is particle speed and position update. The update path of each particle is completely independent, and the update method is iterative calculation, which is suitable for the Spark framework. Therefore, we use the Spark calculation framework to accelerate APSO calculations. We mainly use map () and mapPartition () in Transformation functions and other methods in Action functions.

In this study, the particles are encapsulated into an RDD, and the update process of the particles is executed in parallel. Using the map function provided by Spark, a series of conversion operations are per-formed on the particles in the RDD according to the algorithm, and the optimal solution is finally obtained. The procedure is shown in Algorithm 1, in which the steps 4, 5, 7 and 8 are executed in the map function.

Algorithm 1: APSO algorithm based on Spark

Input: Compressed air demand Q ; Air compressor station load $\{\pi_1, \pi_2, ...\pi_m\}$ under the combination $\{Q_{\pi_1}, Q_{\pi_2}, ...Q_{\pi_m}\}$

Output: Air compressor value $x_{ij}, i = 1, 2, ..., m; j = 1, 2, ..., |Q_i|$

1   Initialization parameters: learning rate c_1, c_2 ; The maximum number of iterations T ; Particle size n

2   Initialize particle swarm pop , call SparkContext.Parallelize (pop) to convert particle swarm to RDD
     dataset

3   **while** $t < T$ **do**

4   update the fitness value according to formula (6)

5   update P_{id}^i , P_{gd} and P_{gd}

6   update ω_t and M_t according to formula (9) and (11)

7   update particle speed and position by formula (8) and (7)

8   perform mutation operations on each particle

9   Get updated global variables from Redis database

10   **if** not met stop criteria **then**

11   jump to 4

12   **else**

13   terminate the iteration and confirm x_{ij} according to P_{gd}

14   **end**

15   **end**

16   return $x_{ij}, i = 1, 2, ..., m; j = 1, 2, ..., |Q_i|$

Under the traditional programming method, for the global variables required in the update process of the APSO algorithm, such as the mutation operator step, inertia factor, and global optimal particle, the Driver passes to each Executor is only their copy. And the modification made by each Executor is only valid for this Executor, and cannot be shared with other. To address it, this study uses the Spark broadcast variables, Broadcast and Redis database, to store and share global variables in the algorithm. One can define the Broadcast variable on the Driver side as the address and account password of the Redis database. The Redis database stores the global variables of the algorithm in the form of key-value pairs, and then connect to the database to update when it needs to be updated.

4   Experiments and Analysis

To verify the advantage of the hierarchical search method and the acceleration effect of the Spark-based APSO algorithm. This study uses Virtual Box to build 6 virtual machines on a physical machine to form a Spark cluster, and compares the acceleration effect through experiments on a single virtual machine and a Spark cluster.

4.1   Experimental Environment and Calculation Examples

The performance parameters of the air compressor used in the experiment in this paper are all from the actual measured data of an iron and steel enterprise in China. The structure of the compressed air system is shown in Fig. 1.

In order to facilitate the calculation, the energy consumption of some equipment is set to a fixed value. The energy consumption and power of some equipment in the experiment are shown in Table 1. In the experiment, the total compressed air load of the steel company in April 2019 was used as the predicted value of compressed air demand. The maintenance schedule of the air compressor during this period was treated as a manual intervention item.

Table 1. Energy consumption and power of some equipment

Param	Description	Value
ς	Electricity price	0.56 yuan/kWh
$\tilde{W}_{ij}^{run}(Q_{S''})$	Combined drying unit power	450 kW
W_{ij}^{start}	Start-up energy consumption of air compressor	10 kWh
W_{ij}^{stop}	Stopping energy consumption of air compressor	10 kWh
J_{others}	Power of auxiliary equipment of air compressor station	300 kW

4.2 Optimal Scheduling of Air Compressor Group

The manual scheduling method always uses the current combination. If the current combination is not optimal, the optimal solution will never be obtained, and when the current air compressor is maintained, the selection of the replacement air compressor is usually based on human experience. We verify that the solution of the APSO algorithm based on the hierarchical search method is superior to the APSO algorithm.

The operating cost of each scheme is shown in Table 2. It can be seen that the APSO algorithm reduces the electricity cost of the air compression system, however, since its opening combination is the opening combination of the manual scheme, the cost reduction is not large. The APSO algorithm of the hierarchical search method can search the optimal scheduling scheme of the entire air compressor group.

Table 2. Operating costs of various schemes

	Manual	APSO	Ours
yuan/hour	17 421	17 118	**16 121**

4.3 Spark Parallel Accelerated Experiment

We first analyze the effect of the Spark frame-work on the stability of the solution. For the same optimal scheduling problem, serial experiments and Spark parallel experiments were conducted. It can be concluded from the results of the Spark accuracy verification experiment that the opening combination of the parallel experiment and the serial experiment is consistent, and the maximum deviation of the opening value of the air compressor is 1.11%. Considering the random search characteristics of swarm intelligence algorithms, the Spark framework exhibits an acceptable performance on the stability of the solution.

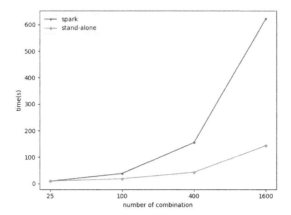

Fig. 4. Time-consuming under different combinations

In order to compare the efficiency of single machine solution and Spark parallel acceleration, we designed experiments based on the same parameters and number of iterations when the total number of open combinations is 25, 100, 400 and 1600.

As can be seen from Fig. 4, when the number of open combinations is small, because of the communication consumption caused by the Spark Driver sending tasks and data to the Executer and the data shuffle after the task is executed, the parallel mode is slower than the stand-alone mode. When the number of open combinations gradually increases, the running time of Spark is significantly reduced compared to that of stand-alone, and the acceleration ratio gradually increases. When the total number of combinations is 1600, the speedup of the parallel algorithm is close to 4.5 times.

5 Conclusion

From the perspective of the actual operating conditions of iron and steel enterprises, considering the difference of the performance of the individual air compressors, a hierarchical search method is designed to generate combinations that satisfy the production plan and production constraints, and the APSO is employed to solve the optimization problem. For speeding up the calculation, we designed a parallel acceleration algorithm based on the Spark framework considering the characteristics of the iterative update of the APSO algorithm. Experimental results show that the proposed hierarchical search method can find a better air compressor group scheduling scheme than that of the traditional methods. On the premise of satisfying the stability of the solution, the use of Spark has contributed to a significant acceleration.

Acknowledgements. This work was supported by the National Key R&D Program of China (2017YFA0700300), the National Natural Sciences Foundation of China (61833003, 61533005), the Fundamental Research Funds for the Central Universities (DUT18TD07, DUT20RC(3)013), and the Outstanding Youth Sci-Tech Talent Program of Dalian (2018RJ01).

References

1. Dong, H.Z., Xue, H.F., Song, H.L., Zhang, Q.: Analysis on the main factors changing iron & steel industry energy consumption intensity. Sci. Res. Manag. **30**(3), 132–138 (2009)
2. Li, H.M.: Energy consumption analysis and energy saving measures for compressed air system in iron and steel enterprises. Metal Mater. Metall. Eng. 57–61 (2016)
3. Hao, Y.S., Peng, X., Li, B., et al.: Key problem research of energy dispatching and optimization based on EMS in iron and steel enterprises. Metall. Ind. Autom. **37**(3), 7–12 (2013)
4. Meng, X.Y., Hao, Y.S., Peng, X., et al.: Control and optimization of air compressor of iron and steel plant. Metall. Power 9–11 (2013)
5. Bing, X., Ping, Q.: Notice of retraction multiobjective evolutionary algorithms applied to compressor stations network optimization scheduling control system. In: Second International Conference on Mechanic Automation & Control Engineering (2011)
6. Abbaspour, M., Satkin, M., Mohammadi-Ivatloo, B., et al.: Optimal operation scheduling of wind power integrated with compressed air energy storage (CAES). Renew. Energy **51**, 53–59 (2013)
7. Ji, L.: Application of improved particle swarm optimization algorithm in air compressor associated controlling system. Light Ind. Mach. **32**(4), 57–60+64 (2014)
8. Shvachko, K., Kuang, H., Radia, S., et al.: The Hadoop distributed file system. In: 2010 IEEE 26th Symposium on Mass Storage Systems and Technologies (MSST), pp. 1–10. IEEE (2010)
9. Yang, Z.W.: The research of recommendation system based on Spark platform. University of Science and Technology of China, Hefei (2015)
10. Maillo, J., Ramírez, S., Triguero, I., et al.: kNN-IS: an iterative Spark-based design of the k-nearest neighbors classifier for big data. Knowl.-Based Syst. **117**, 3–15 (2017)
11. Arias, J., Gamez, J.A., Puerta, J.M.: Learning distributed discrete Bayesian network classifiers under MapReduce with Apache Spark. Knowl.-Based Syst. **117**, 16–26 (2017)
12. Liu, P., Ye, S., Ment, L., et al.: A Spark based parallel genetic algorithm solving multimodal function extremums. Comput. Eng. Sci. **40**(2), 210–217 (2018)
13. Wang, Z.Y., Wang, H.J., Xing, H.L., et al.: Ant colony optimization algorithm based on Spark. J. Comput. Appl. **35**(10), 2777–2780+2797 (2015)
14. Andrews, P.S.: An investigation into mutation operators for particle swarm optimization. In: 2006 IEEE International Conference on Evolutionary Computation, pp. 1044–1051. IEEE (2006)
15. Karau, H., Konwinski, A., Wendell, P., et al.: Learning Spark: Lightning-Fast Big Data Analysis. O'Reilly Media Inc., Sebastopol (2015)

A PSO Based Technique for Optimal Integration of DG into the Power Distribution System

K. Moloi$^{(\boxtimes)}$, J. A. Jordaan$^{(\boxtimes)}$, and Y. Hamam$^{(\boxtimes)}$

Tshwane University of Technology, eMalahleni Campus, Emalahleni, South Africa
moloikt023@gmail.com, {jordaanJA,hamama}@tut.ac.za

Abstract. Integrating renewable energy distributed generation (REDG) into the existing power distribution grid has become a significantly exercise to carry out. This is because of several technical, economic and environmental benefits accruing from it. However, optimal location and sizing of REDG especially photovoltaic (PV) and wind turbine (WT), is still a difficult task due to the natural dependency of these renewable source on meteorological conditions. In this paper, we proposed a technique based on particle swarm optimization (PSO) to solve the location and sizing of REDG problem. The proposed PSO algorithm is used to minimize the power losses and maximize the voltage stability of the power grid distribution system integrated with REDG. The paper also presents a comparison of the proposed method and other related techniques for REDG sizing and location. The proposed method is validated using the IEEE 33 bus systems.

Keywords: Distribution power grid system · Particle swarm optimization · Power loss minimization

1 Introduction

The global electricity and load demand have increased rapidly over the years and this has extensity the pressure on the power generation utilities to increase the supply capacity to meet the demand. The existing distribution system infra-structure is not capable to support a huge demand of electricity [1]. Moreover; the traditional method of electricity generation is depended on fossil fuels such as coal which has environmental impacts [2]. This has detrimental effects on the environment and has contributed negatively to the global warming [3]. Furthermore, large power plants used for electricity generation are usually located far from the load centers, which result in about 15% losses of active power long the transmission and distribution lines. Therefore, there has been a significant drive to find alternative solutions for electricity generation to meet the required demand with minimum impacts on various aspects. This need has led to the exploitation of renewable energy sources (RES) as alternative source of electricity to meet the projected increase in load demand and to solve ecological environmental issues [4].

Distributed generations (DGs) by definition are generation units located at distribution power grid systems close to the load centers. These units are primarily used to abruptly meet the load demand, reduce operating costs during peak time, reduce power

© Springer Nature Switzerland AG 2020
M. Han et al. (Eds.): ISNN 2020, LNCS 12557, pp. 37–46, 2020.
https://doi.org/10.1007/978-3-030-64221-1_4

losses, reduce distribution loading, improve system reliability and increase power quality [5, 6]. Optimal planning of DGs location and sizing is a critical to achieving the maximum technical and economic benefits. The non-optimal location and sizing of DGs may result in increase in power loss and affect the system voltage stability [7]. There has been a great significant research work proposed to finding an optimal solution to REDGs sizing and location in distribution systems. In [8], an analytical method was proposed to determine the optimal position and size of DG to minimize active power losses. An intelligent search technique based on the backtracking search optimization algorithm (BSOA) was proposed by [9]. The BSOA was used to find the optimal location and size of the DG to be connected into the power grid to minimize power losses and improve the voltage profile. A technique based on artificial bee colony (ABC) was proposed in [10], for optimal DG's size and location for minimization of active power losses. In [11], a technique-based firefly algorithm (FA) was proposed to find the optimal location and size of DG to reduce the active power losses. The lengthy discussion of determining the optimal location and size of the REDGs in distribution networks indicates the importance of finding an optimal solution to satisfy technical, environmental and economic benefits. In this paper a technique base on particle swarm optimization (PSO) is proposed to determine the best location and size of REDGs to be connected into the power distribution grid. The paper is organized in the following structure. Section 2, discuss the mathematical modelling of REDGs. Section 3 discuss the problem formulation; this includes technical constrains and objective functions. The results of power loss minimization and voltage improvement are discussed in Sect. 4 and lastly the conclusion and recommendations are presented in Sect. 5.

2   Mathematical Modelling of REDGs

In this section, the mathematical modelling of PV and WT is discussed. Generally, the PV and WT energy sources mainly depend on weather conditions (i.e., temperature, wind speed and heat emission). These conditions bring uncertainties and must be considered when developing the planning and optimization problem of REDGs integration into the distribution power grid.

2.1   Photovoltaic Systems

The generation of electricity using PV has proven to be an effective alternative method for energy sustainability with a benefit of reduction in fossil fuel [12]. The power generated by the PV plant is represented as:

$$P_{PV}(G) = \begin{cases} (P_{PVr} \times G^2)/(S_{STC} \times R) \ for \ G < R_C \\ (P_{PVr} \times G)/S_{STC} \qquad for \ G > R_C \end{cases} \tag{1}$$

where, P_{PVr} is the rated output power of the PV unit, G is the solar irradiance probability, S_{STC} is the solar irradiance at standard test conditions, and R_C is a certain irradiance point.

2.2 Wind Turbine Energy Systems

Generating electricity from (WT) has become one of the most popular and efficient technology widely used [13, 14]. The electrical power generated from the WT is mathematically defined as:

$$P_{WT}(V_W) = \begin{cases} 0 & for\ v_w \leq v_{ci} \\ [(v_w - v_{ci})/(v_r - v_{co})] \times P_{wtr} & for\ v_i < v_w \leq v_n \\ P_{wtr} & for\ v_w \leq v_{ci} \\ 0 & for\ v_w \leq v_{co} \end{cases} \qquad (2)$$

where, P_{WT} is the output power generated by the WT, v_w is the wind speed at the hub height of the WT, P_{wtr} is the rated power of the WT, v_{ci} is the cut-in wind speed of the WT, v_{co} is the cut-out wind speed of the WT and v_n is the nominal speed of the WT.

3 Problem Formulation

In this section, the proposed methodology for power loss minimization and voltage improvement is discussed. The voltage sensitivity index (VSI) technique proposed in [15] is used to determine the weakest voltage at a specific bus bar of a power distribution grid. The VSI technique can be mathematically defined as:

$$VSI = |V_s|^4 - 4 \times \left\{ P_r X_{ij} - Q_r X_{ij} \right\}^2 - 4 \times \left\{ P_r r_{ij} + Q_r X_{ij} \right\}^2 \times \lfloor V_S \rfloor^2 \geq 0 \qquad (3)$$

where, V_S is the sending bus bar voltage, P_r and Q_r represents the active and reactive power at the receiving end respectively, r_{ij} and X_{ij} are the resistance and reactance parameters of the $i-j$ line. If $VSI > 0$ at all busses of the power system grid the system would be termed to be stable. The highest VSI value indicates an optimal location for connecting the REDG (PV and WT).

3.1 Objective Functions

In this study the main objective is to find an optimal location and size of the REDGs to be integrated into the power distribution grid. In order to achieve this objective, the following objective fitness functions OF_1, and OF_2 represents the, power loss, bus voltage sensitivity and line voltage stability respectively, and must be considered:

- Minimisation of active power losses: this is mathematically represented as:

$$OF_1 = Min \left\{ P_a = \sum_{a=1}^{nbr} |I_a|^2 R_a \right\} \qquad (4)$$

where, I_a is the total current flowing in the network, R_a is the total resistance of the network and nbr is the number of branches in the network.

- Improve bus voltage: this is achieved by minimising the voltage sensitive index:

$$OF_2 = min\{VSI\} \qquad (5)$$

The normalized overall fitness function (*OFF*) is the summation of all individual objective fitness functions:

$$OFF = \min(power\ loss) + \min(VSI) \tag{6}$$

3.2 Technical Constrains

In order to optimally locate the correct REDG size certain constrains must be satisfied. Constrains are critical in engineering design as they permit the operational applicability of a system. In this work power flow, voltage limits and capacity constrain are considered.

- The power balance in an electrical power system is mathematically defined as:

$$P_{slak} = \sum_{i=1}^{nb} P_{REDG,i} + \sum_{a=1}^{nbr} P_{LOSS,a} + \sum_{m=1}^{nredg} P_{REDG,m} \tag{7}$$

where, P_{slak} is the active power injected by the slack bus, $P_{REDG,i}$ is the active power injected by the *ith* REDG, $P_{LOSS,a}$ is the active power losses of branch *ath*, $P_{REDG,m}$ is the active power injected by the *mth* REDG, *nb* is the set of the bus system, *nbr* is the set of branches and *nredg* is the set of system buses.

- The variation of voltage may affect the performance of the system. It is therefore imperative that the voltage is maintained within permissible limits.

$$V_{min} \leq V_i \leq V_{max} \quad i\,\forall nb \tag{8}$$

3.3 Particle Swarm Optimization

Particle swarm optimization (PSO) was initially developed in 1995 by Eberhart and Kennedy [16]. The PSO technique is based on the social behavior demonstrated by various species to fill their needs in the search space. The PSO algorithm is formulated by the personal experience (*pbest*) and overall experience (*gbest*). The particle X_i ($i = 1, 2, \ldots, N$) is given by $X_i = [X_{i,1}, X_{i,2}, \ldots X_{i,d}]$, d is the dimension of the search space. Furthermore, the initial velocity of the particle is given by $V_i = [V_{i,1}, V_{i,2}, \ldots V_{i,d}]$. The position and velocity of the particles are updated during each iteration process using Eqs. (9) and (10) respectively.

$$X_{id}^{(t+1)} = X_{id}^{(t)} + V_{id}^{(t+1)} \tag{9}$$

$$V_{id}^{(t+1)} = W.V_{id}^{(t)} + C_1 r_1.(pbest_{id} - X_{id}) + C_2 r_2.(gbest_{id} - X_{id}) \tag{10}$$

where, t is the number of iterations, C_1 and C_2 are constants r_1 and r_2 are random numbers, W is the inertia weight given by:

$$W = W_{min} + \frac{W_{max} - W_{min}}{iter_{max}} \times t \tag{11}$$

where, $W_{min} = 0.3$, $W_{max} = 0.7$, $C_1 = C_2 = 0.5$. In the present work PSO algorithm is utilised to determine the fitness function. This problem is considered to be a multi-objective problem and is mathematically represented as:

$$\min(f) = w_1 \times F_1 + w_2 \times F_2 \tag{12}$$

where, f is the fitness function, and the weight factors are represented by w_1 and w_2 and both equal to 0.5.

4 Results and Discussions

In order to validate the efficiency of the proposed PSO algorithm, the IEEE 33 bus system is utilized. The IEEE network parameters and the initial power flow results are presented in Tables 1 and 2 respectively.

Table 1. Different power systems specifications

Parameters	69 IEEE bus system
nr	69
nbr	68
$V_{sys}(kV)$	22
$S_{Base}(MVA)$	100
$Z_{Base}(\Omega)$	$3.12 + j2.8$

Table 2. Initial power flow analysis results

Parameters	69 IEEE bus system
$P_{Loss}(kW)$	245.6
$Q_{Loss}(kW)$	110.7
$V_{max}(p.u)$	1.0191
$V_{min}(p.u)$	0.9205
Voltage drop $(p.u)$	1.8872

The proposed PSO technique is applied to resolve the problem of optimal location and sizing of REDGs integration into the power grid considering the technical constrains. To evaluate the reliable performance of the proposed method, different penetration scenarios REDGs levels are considered. These scenarios consist of the single and multiple REDGs (WT & PV) to be integrated into the power distribution grid. Due to non-rotational effects from the PV energy only the active power is injected on the system, however the WT injects both the active and reactive power. To achieve power balancing at all load terminal of the power grid, it is assumed that only one REDG is connected into the bus bar. The main objective is to determine the position and size of the REDGs units to minimize power losses and improve the voltage profile ad stability.

To evaluate the performance of the proposed PSO technique, a standard IEEE 33 bus system is used to study different cases (i.e., integration of single and multiple PV and WT generating units). In the first scenario, single REDG units are integrated into the 33 bus IEEE-system. The optimal bus number 6 is selected as an optimal location for connecting the PV and WT. the size of the PV and WT units are 2.56 MW and 3.72 MVA with 0.92 lagging power factor. The results of the optimal location and sizing of REDGs using the proposed PSO technique are presented in Table 3. The proposed PSO technique significant decreases the active power losses when integrating the PV and WT single REDGs units. The losses are decreased from 230.145 kW of the system without REDGs to 100.57 kW and 75.145 KW when the PV and WT are connected respectively. Furthermore, the line and voltage stability index are improved significantly. This improvement enhances the technical performance of the overall power distribution grid. The integration of multiple REDGs (i.e. two PV and two WT units) is considered, and its impacts are examined. In a case of two PV units the optimal bus location of the REDGs is 12 and 32, with the capacity of 0.960 MW and 1.142 MW respectively. From the results obtained, the active power losses have decreased significantly from 230.145 kW to 75.145 kW. The 33 IEEE bus system is presented in Fig. 1.

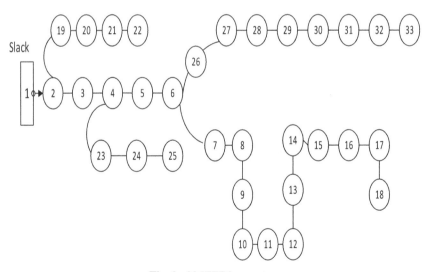

Fig. 1. 33 IEEE bus system

Moreover, the bus voltage and line voltage stability indexes are improved by optimally locating REDGs using the proposed PSO technique. Subsequently, the optimal bus location of the two WT units is 12 and 32, with the capacity of 1.0551 MVA and 1.7030 MVA, with power factor of 0.9055 and 0.825 lagging respectably. From the results obtained the active power losses is drastically reduced from 230.145 kW to

36.15 kW. The WT units inject reactive power on the system which results in a significant improvement on the line and bus voltage compared to the PV units in both single and multiple cases. The general results presented in Table 3, shows that increasing the PV and WT units improve the technical performance of the system. However, it is worth mentioning that the improvements are as a result of optimally selecting the best location and size of REDGs to be integrated into the power grid. This means that an increase of un-optimum external energy sources will not improve the technical performance of the system. Hence, PSO is utilized to determine the optimal solution for best technical performance. The voltage profile, voltage stability and active power losses of a 33 bus IEEE system are presented in Figs. 2, 3 and 4 respectively.

Table 3. Optimal location and sizing of REDGs using PSO (33 IEEE bus system)

Description	Without any REDGs	Single REDG		Multiple REDGs	
		PV	WT	2 PV	2 WT
$P_{loss}(kW)$	230.145	100.57	75.145	89.145	36.15
$V_{min}bus(p.u)$	0.9321	0.9522	0.9715	0.9621	0.9875
$V_{max}bus(p.u)$	1.02	0.9921	1.0253	0.9983	1.0452
REDG size (MW, MVA)	NA	2.560	3.720	0.960	1.0551
				1.142	1.703
REDG location	NA	6	6	12	12
				32	32
$LSVI$	19.221	26.223	31.042	28.112	34.257
VSI	15.332	18.201	21.225	22.052	24.063
Voltage drop $(p.u)$	0.821	0.327	0.441	0.528	0.152

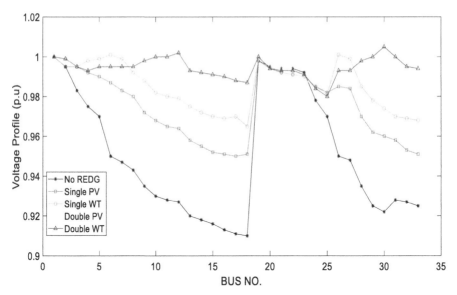

Fig. 2. Voltage profile of a 33 IEEE bus system

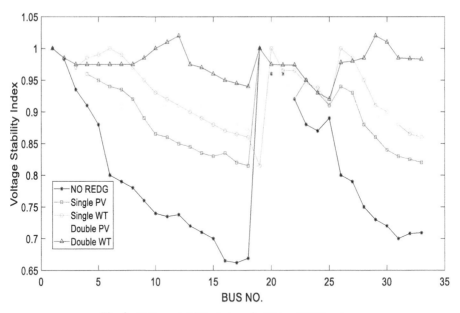

Fig. 3. Voltage stability index of a 33 bus IEEE system

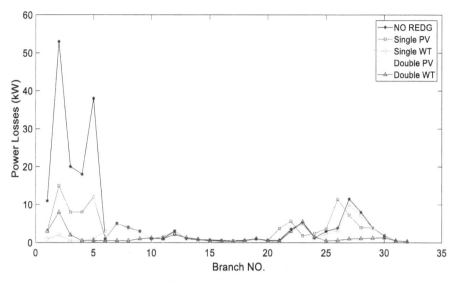

Fig. 4. Active power losses

5   Conclusion

In this present work, an optimal planning technique for REDGs integration into the power grid is proposed. The proposed technique uses the PSO algorithm to evaluate the impact of REDGs and subsequently determine the appropriate size and location which must be connected into the grid. The proposed PSO technique is validated using a 33 IEEE bus system. Various scenarios of REDGs are analyzed and the results are presented. From the presented results the following conclusion may be highlighted:

- Optimising the location and size of the REDG is practically essential for the efficient technical and financially viable performance.
- The increase in the penetration levels of REDGs increases the complexity of the network analysis, it is therefore imperative that the uncertainty of both the PV and WT sources are modelled as accurately as possible to improve the overall model efficiency.
- There is a significant decrease of the active power losses recorded when using PSO compared to the other methods presented.

Future research work will entail the optimization of economic aspects of REDGs integration as-well as the impact of introducing the storage systems.

References

1. Aman, M., Jasmon, G., Bakar, A., Mokhlis, H.: A new approach for optimum DG placement and sizing based on voltage stability maximization and minimization of power losses. Energy Convers. Manag. **70**, 202–210 (2013)

2. Elkadeem, M.R., Wang, S., Sharshir, S.W., Atia, E.G.: Techno-economic design and assessment of grid-isolated hybrid renewable energy system for agriculture sector. In: 14th IEEE Conference on Industrial Electronics and Applications (ICIEA), Xi'an, China (2019)

3. Quek, T.Y.A., Ee, W.L.A., Chen, W., Ng, T.S.A.: Enviromental impacts of transitioning to renewable electricity for Singapore and surrounding region: a life cycle assessment. J. Clean Prod. **214**, 1–11 (2017)

4. Santos, S.F., Fitiwi, D.Z., Shafie-Khan, M., Bizuayehu, A., Catalao, J., Gabbar, H.: Optimal sizing and placement of smart grid enabling technologies for maximising renewable integration. J. Renew. Sustain. Energy Resour. **15**(2), 47–81 (2017)

5. Jordehi, A.R.: Allocation of distributed generation units in electric power system: a review. J. Renew. Sustain. Energy Resour. **85**, 893–905 (2016)

6. Ha, M.P., Huy, P.D., Ramachandaramurthy, V.K.: A review of the optimal allocation of distributed generation: objectives, constrains, methods, and algorithms. J. Renew. Sustain. Energy Resour. **75**, 293–312 (2017)

7. Alrashidi, M.R., Alhajri, M.F.: Optimal planning of multiple distributed generation sources in distribution networks: a new approach. Energy Convet. Manag. **52**, 3301–3308 (2011)

8. Wang, C., Nehrir, M.H.: Analytical approaches for optimal placement of distributed generation sources in power systems. IEEE Trans. Power Syst. **19**(4), 2068–2076 (2004)

9. EI-Fergany, A.: Optimal allocation of multi-type distributed generators using backtracking search optimisation algorithm. Int. J. Electr. Power Energy Syst. **64**, 1197–1205 (2015)

10. Abu-Mouti, F.S., EI-Hawary, M.E.: Optimal distributed generation allocation and sizing in distribution systems via artificial bee colony algorithm. IEEE Trans. Power Deliv. **26**, 2090–2101 (2011)

11. Nadhir, K., Chabane, D., Tarek, B.: Distributed generation location and size determination to reduce power losses of a distribution feeder by Firefly Algorithm. Int. J. Adv. Sci. Technol. **56**(3), 61–72 (2013)

12. Jafarzadeh, M., Sipaut, C.S., Dayou, J., Mansa, R.F.: Recent progresses in solar cells: inside into hollow micro/nano-structures. Renew. Sustain. Energy Rev. **64**(2), 543–568 (2016)

13. Liserre, M., Sauter, T., Hung, J.Y.: Future energy systems: integrating renewable energy sources into the smart power grids through industrial electronics. IEEE Ind. Electron. **4**(1), 18–37 (2010)

14. Mariam, L., Basu, M., Conlon, M.F.: Policy and future trends. Renew. Sustain. Energy Rev. **64**, 477–489 (2016)

15. Chakravorty, M., Das, D.: Voltage stability analysis of radial distribution. Int. J. Electr. Power Energy Syst. **23**, 129–135 (2001)

16. Eberhart, R., Kennedy, J.: A new optimizer using particle swarm theory. In: Proceedings of the Sixth International Symposium on Micro Machine and Human Science (1995)

Online Data-Driven Surrogate-Assisted Particle Swarm Optimization for Traffic Flow Optimization

Shuo-wei Cai, Shi-cheng Zha, and Wei-neng Chen[⊠]

South China University of Technology, Guangzhou, China
cwnraul634@aliyun.com

Abstract. Traffic flow optimization is an important and challenging problem in handling traffic congestion issues in intelligent transportation systems (ITS). As the simulation and prediction of traffic flows are time-consuming, it is inefficient to apply evolution algorithms (EAs) as the optimizer for this problem. To address this problem, this paper aims to introduce surrogate-assisted EAs (SAEAs) to solve the traffic flow optimization problem. We build a traffic flow model based on cellular automata to simulate the real-world traffic and a surrogate-assisted particle swarm algorithm (SA-PSO) is presented to optimize this time-consuming problem. In the proposed algorithm, a surrogate model based on generalized regression neural network (GRNN) is constructed and local search particle swarm algorithm is applied to select best solutions according to the surrogate model. Then candidate solutions are evaluated using the original traffic flow model, and the surrogate model is updated. This search process iterates until the limited number of function evaluations (FEs) are exhausted. Experimental results show that this method is able to maintain a good performance even with only 600 FEs needed.

Keywords: Online data-driven · Traffic flow problem · Surrogate · Particle Swarm Optimization

1 Introduction

With the rapid development of urbanization as well as the revolution of automobiles, the number of vehicles in cities grows explosively. Traffic congestion has been a key concern in almost all big cities in the world. Compared to the growing number of automobiles, the development of traffic facilities is much slower, which leads to the limited traffic capacity and vehicle speed. How to efficiently

This work was in part by the National Natural Science Foundation of China under Grants 61976093, in part by the Science and Technology Plan Project of Guangdong Province 2018B050502006, and in part by Guangdong Natural Science Foundation Research Team 2018B030312003.

M. Han et al. (Eds.): ISNN 2020, LNCS 12557, pp. 47–58, 2020.
https://doi.org/10.1007/978-3-030-64221-1_5

manage the traffic flows in road network so that the utilization of road network facilities can be optimized has become a crucial problem in ITS.

In general, there are two steps in traffic flow management. The first step is to simulate traffic conditions in road networks by modeling. To build traffic simulation models, researchers have developed mechanistic models like the petri net [1], data-driven models like Kalman filters [2], autoregressive integrated moving average (ARIMA) [3] and neural networks [4]. Traffic simulation software like VISSIM [5] have also been developed. These models and tools can provide a good simulation of the real-world traffic flow under the given condition even with limited traffic data. Then, the second step is to optimize the traffic flow based on the model constructed. Many global search algorithms, like genetic algorithm [6], water flow algorithm [7] and particle swarm optimization algorithms (PSO) [8] have applied to optimize the traffic flow problem. The popularity of evolution algorithm (EAs) in this area is probably due to the NP-hard characteristic of the traffic flow problem. EAs, which are inspired by natural phenomenon and biological behaviors, are able to handle multi-modal problems even without an analytical model.

Despite the popularity of EAs, there remains some challenges for them to handle real-world traffic control problems. EAs assume that a mathematical analytical evaluation function is provided for evaluating and assessing the candidate solutions generated during the process of the algorithms. However, in many applications, the evaluations of candidate solutions are time-consuming. In the traffic flow problem considered in this paper, the prediction and simulation of traffic flows in existing models and tools are usually computational expensive. Since EAs require a population of individuals to evolve iteratively, time-consuming FEs will make the execution of EAs become unaffordable.

To overcome this challenge, surrogate-assisted EAs (SAEAs) have been introduced in the literature [9,20], where a model is trained as an approximation of the original model to provide extra information for the searching algorithm. In this method, a proportion of the expensive FEs performed by the original model is replaced by the approximation model often known as the surrogate model. Therefore, the computational expense for solving computational-intensive optimization problems can be significantly reduced. In an online data-driven surrogate-assisted optimization process, FEs are conducted during the process and the surrogate model are updated to gain higher accuracy [10]. Usually, machine learning methods like polynomial regression [11], radio basis function neural networks [12], Kriging models [13], are commonly used in SAEAs as surrogate models.

This paper aims to introduce SAEAs into the optimization of the classic traffic flow problem to overcome the problem of large computation expense during the optimization. A traffic flow model based on cellular automata is constructed to simulate real-world traffic according to the problem. Then, an online surrogate-assisted particle swarm optimization algorithm (SA-PSO) is proposed to optimize the average vehicle velocity based on this model. During the optimization process, local-search PSO (LS-PSO) will be applied on the surrogate model repeatedly and FEs provided by the original traffic flow model will update the surrogate. By bringing SAEAs into this question, we significantly reduce the computational expense and can still gain a good performance during the optimization and find a good configuration of the road net.

The rest of this paper is organized as follows. In Sect. 2, a traffic flow model based on cellular automata is introduced. In Sect. 3, the structure and detailed processing of the surrogate-assisted particle swarm optimization algorithm is proposed. Section 4 shows the experiment results of the algorithm on the traffic flow model. Section 5 concludes the paper.

2 Definition of the Traffic Flow Problem

2.1 Problem Description

With the rapid development of urbanization, urban traffic pressure has been increasing. How to optimize urban road net management, improve the utilization of transportation facilities, and ultimately improve the efficiency of travel are the core problems in urban traffic management.

Traditionally, the timing scheme of traffic lights was fixed. However, in recent years, with the development of intelligent urban traffic management, the use of urban road net resources becomes more flexible. This enables people to optimize the road net configuration in an unconventional way, like adjusting the signal timing scheme. Our goal is to obtain the optimal traffic signal timing scheme which maximizes the average vehicle speed. As a result, a traffic flow model based on a specific road net which can accurately reflect the relationship between the signal timing scheme and the average vehicle speeds should be constructed.

2.2 Basic Structure

A traffic flow model based on cellular automata was established by converting an urban road network into square grids by means of topological equivalence [14]. This cellular automata model can represent a local road network with entrances and intersections. Cellular automata, which is used to construct the traffic flow model, can be viewed as a basic cluster consists of various individual segments (cells). The interaction between particles (vehicles) in the cells is restricted to its nearby neighbors. Every individual cell is in a particular state, which updates after some time subject to the conditions of its nearby neighbors [15]. The vehicle's motion algorithm is based on the CA-184 rule [15]. The CA-184 rule can reasonably represent the real situation of vehicles moving in road network.

2.3 Road Network Establishment

We consider an urban area of $l \times l$ meters divided into two-dimensional grids and each cell (grid) is given a label C_{ij}. C_{ij} can be a number from 0 to 8, where 0 represents a unreachable cell, 1 means a cell in a lane with direction eastbound driveway, 2 means a cell in a lane with direction westbound driveway, 3 represents north and 4 represents south, and $5, 6, 7, 8$ are special cells that make up a complete intersection. Take Fig. 1 as an example, in the i–j coordinate system, the cell highlighted at the bottom-left is $(0, 0)$ and C_{00} is 0, which represents an unreachable cell. In the middle of the figure, cell with label $5, 6, 7, 8$ form an intersection. $1, 2, 3, 4$ are four lanes with different directions, the lanes and intersections form a simple road net.

In the traffic flow model, we set l as 3020 and each cell as a $5 \times 5\,\mathrm{m}^2$ grid, so there are in total 364816 cells in this model. The road net inside this model has 3 bidirectional single-lanes from south to north, 3 bidirectional single-lanes from east to west. Each lane is 200 cells away from the nearest lanes with the same direction and in total there are twelve entrances (exits) in this road net. Those 6 lanes form 9 intersections in the road net and the distance from an entrance to the nearest intersections is also 200 cells. Two variables are assigned for the traffic lights in each intersection, one for the cycle time of the signal and one for the proportion of the time of green light in east-west direction. If the second parameter is negative, the traffic light in east-west direction will be initialized with red (in default). All traffic lights are initialized at the same time, and they start their cycle at the beginning of the simulation.

0	0	0	0	4	3	0	0	0	0
0	0	0	0	4	3	0	0	0	0
0	0	0	0	4	3	0	0	0	0
0	0	0	0	4	3	0	0	0	0
2	2	2	2	5	6	2	2	2	2
1	1	1	1	7	8	1	1	1	1
0	0	0	0	4	3	0	0	0	0
0	0	0	0	4	3	0	0	0	0
0	0	0	0	4	3	0	0	0	0
0	0	0	0	4	3	0	0	0	0

Fig. 1. Example of a two-dimensional grid of road net

2.4   Motion Algorithm

We set the time unit of this model to be seconds and the maximum speed to be $72\,\mathrm{km/h}$, which means each particle (vehicle) can move at most 4 cells per time unit. Each particle will stop and wait if there is a particle in the next cell or the cell in front is with label $5, 6, 7, 8$ if the traffic light is red. Once the red light becomes green, the particles that waiting will start to move one after one (in different time unit) and their speed will increase to maximum in only one time unit. There is no acceleration and deceleration in this model for simplicity, and no accidental events to affect the traffic flow inside the road net.

The probability of a vehicle to turn left or right randomly at the intersection is 20% and the probability of going straight is 60%. A random vehicle will enter the road network from a random entrance every time unit and will be removed when it reaches an exit. The random number in this model is fixed as a representation of traffic conditions outside the road net, and it also keeps the output unchanged with the same configuration of signal timing.

2.5   Problem Definition

The fitness function of the model is:

$$f(x) = \frac{\sum_{i=t_0}^{t} \bar{v}_i}{t - t_0} \tag{1}$$

which is the average vehicle velocity in the road net after t_0. In the formula, x represents signal timing settings with 18 dimensions (2 for each of the 9 intersections), $\bar{v}_i$ represents the average vehicle speed in the ith time unit, t represents the total counting time of the model, t_0 means the time that the recording starts. Because the road net is empty in initial, we set $t_0 = 50$ to let vehicles fill the road net. The total counting time is $t = 350$, a 5-min simulation of the whole road net. Because the average speed is determined by the given traffic light cycle, therefore, a higher fitness value represents a better traffic signal configuration.

3   Surrogate-Assisted Particle Swarm Optimization

Because the simulation process of the model is time-consuming, conventional EAs are unsuited for optimizing this problem. As a result, an online surrogate-assisted particle swarm optimization algorithm is proposed.

3.1   Overall Framework

Initially, a dataset D_0 is created to store the FEs generated by the expensive original model, where the subscript of D represents the number of FEs is performed by the original traffic flow model. Then, Latin hypercube sampling (LHS) [16] is applied to initialize the data set with M FEs, these data point with configuration x_i, fitness value y_i and default particle velocity $v_i = 0$ are stored into the dataset as the initial data to construct the surrogate model.

$$y = f(x) \tag{2}$$

$$D_M = \{< x_1, v_1, y_1 >, < x_2, v_2, y_2 >, ..., < x_M, v_M, y_M >\} \tag{3}$$

Then, the algorithms switch to the surrogate model management part to generate a surrogate model $\hat{f}$ based on the dataset. After that, local search PSO (LS-PSO) is conducted to find candidate solutions based on the surrogate model. Once a candidate solution x_{t+1} is returned, it will be evaluated by the original traffic flow model to get y_{t+1} and added into the dataset accompany with the particle velocity v_{t+1} in the LS-PSO and the solution itself.

$$y = \hat{f}(x) \tag{4}$$

$$D_{t+1} = D_t \cup \{< x_{t+1}, v_{t+1}, y_{t+1} >\} \tag{5}$$

This process will continue iterating until the maximum number of FEs N during the whole process is reached. In final, the best solution is obtained by:

$$x_{best} = argmax_{x_i} \ y_i \tag{6}$$

where x_i represents an arbitrary data in the dataset D_N (Fig. 2).

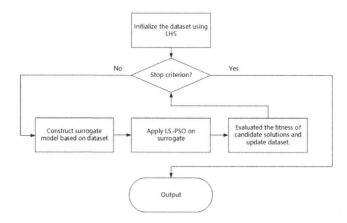

Fig. 2. Generic diagram of surrogate-assisted particle swarm optimization algorithm.

3.2 Surrogate Model Management

The surrogate model management consists of two parts, data selection and model construction. In the data selection part, several data points from the dataset will be selected according to their fitness value. Firstly, the best P solutions will be extracted from the dataset into the training set D_{train}, then $M - P$ solutions will be extracted with the same interval in their rank of the fitness in the candidate solutions remain. With proper parameters setting, the surrogate model constructed will have a good approximation on the peak part of the solution space and still maintain a good overview of the global of the function. It is a balance of exploration and exploitation.

In the model construction part, a surrogate model is generated by generalized regression neural network (GRNN), which is a kind of radio-based function (RBF) network. The data selected by the strategy above will be used and a multimodal surrogate will be constructed as an approximation of the original model to provide information to the search algorithm. For further exploitation, a local model management strategy is able to be involved when the size of the dataset becomes larger in the second half process of the algorithm.

3.3 Local Search Particle Swarm Algorithm

Conventional particle swarm optimization algorithms are unsuitable for the search on the surrogate model because the model is inaccurate, a strong exploitation search will result in huge uncertain. Therefore, local search particle swarm optimization algorithm adapted from social learning PSO [17] is applied on the surrogate model.

The particles are initialized at the positions of the training data with their velocity given. This provides a high certainty of the initial data. The update function of the LS-PSO is:

$$v_{new} = c_1 r_1 v + c_2 r_2 (x_l - x) + c_3 r_3 (x_m - x) \tag{7}$$

$$x_{new} = x + v \tag{8}$$

where x and v represent the old position and velocity of the particle, x_{new} and v_{new} represent the new position and the velocity of the particle; x_l is the position of a random particle that has a better fitness value than the particle itself and x_m is the mean position of all the particles. The particle with the best fitness value will not be updated. r_1 and r_2 are two random numbers in $[0,1]$, c_1 and c_2 are set to 1. There is a quadratic adaptation on c_3, which is increased from 0 to 0.4 with generation g, where g_{max} is the maximal number of the iteration.

$$c_3 = 0.4 \times (\frac{g}{g_{max}})^2 \tag{9}$$

Different from conventional PSOs, LS-PSO performs only one local search, updates only once during the one LS-PSO search. After the update is finished, the best 2 solutions (in which 2 is set to improve fault tolerance) are selected according to the fitness of the surrogate model are returned with their position and velocity.

$$x_{best} = argmax_{x_t} \hat{f}(x_t) \tag{10}$$

According to this strategy, LS-PSO can make full use of the surrogate model but not highly rely on it according to its uncertainty.

3.4 Local Model Management Strategy

Because of the complicated structure of the traffic flow model, it is difficult to get an accurate approximation by a single surrogate model. Therefore, we introduce a hybrid model, a global model is trained using the strategy above and local models are trained with niching strategy [18]. Then we can get a comprehensive general surrogate model for global and accurate surrogate models for local.

In local surrogate construction, we split existing solutions into multiple sets using K-means, establish an independent local model for each sets and use the local model to compute the fitness which has the shortest Euclidean distance between the position of solution x and cluster center c_j. The formula of local model is shown as follow:

$$\hat{f}_2(x) = g_j(x) \tag{11}$$

$$j = argmin_{i=1...K} \, ||x - c_j||_2 \tag{12}$$

where, K represents the cluster centers of K-means and $g_j(x)$ represents the local model trained by a separated data set clustered by K-means. Both the global and the local model are trained by GRNN, and the formula of the total surrogate model is defined as follow:

$$\hat{f}(x) = w_1 \times \hat{f}_1(x) + w_1 \times \hat{f}_2(x) \tag{13}$$

where $\hat{f}_1$, $\hat{f}_2$ represent the global and local surrogate model; w_1, w_2 are weights that change with FEs that have already performed. Due to the fact that the construction of local model requires more data, we set w_1, w_2 to 1, 0 in initial, and switch their value to 0, 1 after L FEs have already performed during the optimization. Figure 3 shows the process of surrogate model construction after involving local surrogate model.

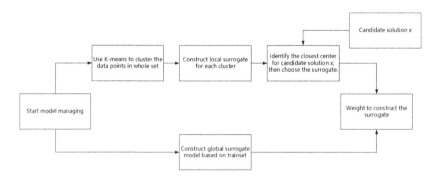

Fig. 3. Model management after involving local model

4 Experimental Results

In this section, we empirically analyze the performance of SA-PSO on the traffic flow model. We select classic method as well as some surrogate-assisted methods to optimize the traffic flow model and compare the performance between them. In total 4 algorithms are tested in experiment: Classic PSO, 2 variants of SA-PSO without local model (SA-PSO-I and SA-PSO-II), and SA-PSO with local model (SA-PSO-L). Through the comparison of these methods we can see the significance of modifications and some characteristics of SA-PSO.

4.1 Experimental Settings

The maximum number of FEs, parameter N is set to 600 for all four algorithms, which is based on the real-world situation of the time constraint because the majority computational expense of the algorithm are the FEs of the original model.

For classic PSO, both the cognitive and social parameters are set to 1.49445 as suggested in [19]. The number of particle and iteration are adjusted to 20 and 29 according to the constraint of 600 FEs. PSO is selected as a simple contrast to evaluate the effect of involving surrogate model into traffic flow optimization.

As for the 3 kinds of SA-PSO, the training size M is set to 100, which enable the surrogate model to construct a fine global surrogate of the original model, then further search can perform on it. The two variants of SA-PSO have no local model constructed. In SA-PSO-I, the parameter P, which is the best candidate solutions that extracted into the training set, is set to 100, which is same as the value of M. In SA-PSO-II, P is set to 80, and $M - P$ is 20. These two tests is held to examine the importance of maintaining a balance between exploration and exploitation with only the global model constructed. As for the case with local surrogate constructed, SA-PSO-L has the parameter P set to 80; cluster number K set to 4, parameter L (the numbers of FEs performed before the weight w_1, w_2 changes) is set to $M + 160$ (260).

In order to make fair comparison and for better assessment, SA-PSO-I, SA-PSO-II and SA-PSO-L are initialized with the same M data extracted by LHS.

4.2   Result Comparisons

The statistic of the result obtained by algorithms on 20 independent runs are shown in the Table 1.

Table 1. Results of 4 algorithms

Algorithm	Best	Worst	Average	Std
PSO	67.41991206	64.8815123	66.3521579	0.70478528
SA-PSO-I	68.02472452	65.9117196	66.8293796	0.48916306
SA-PSO-II	67.52465329	66.31099015	67.02550605	0.30397467
SA-PSO-L	67.69284148	65.87936601	66.88648851	0.440627011

PSO behaves poorly on this problem. This unsurprising result is due to the characteristic of EAs. Without enough FEs, it is hard for EAs to explore and exploit. Moreover, the std of PSO is the highest among the four algorithms because of the relative few particle number. If particles are initialized with bad fitness value, they will do a series of pointless search and waste their FEs.

Compared to PSO, SA-PSO-I has a significant increase in final outcome in average contribute to the surrogate model constructed in the algorithm. Its standard deviation is lower, and its best and worst case are better. The surrogate model brings stability to the result of this optimization problem.

Figure 4 is a typical case of convergence of SA-PSO-I, where the orange line represents the fitness value of the better candidate solution returned by LS-PSO and the blue line represents the best fitness value of all the data point been examined. Initially, surrogate is constructed and some solutions around fitness value 64 are evaluated. Then a better surrogate is constructed according to these newly added data, which results in a candidate solution with a better fitness value (65.5) been evaluated. This newly added data point updates the surrogate model significantly and then new points around the better fitness value are evaluated. This process repeatedly continues until the number of FEs runs out. In contrast to PSO, SA-PSO-I significantly reduce the amount of FEs with a low fitness value. In this way, the information that facilitates the optimization is provided in a more effective way. That's why SA-PSO-I still maintains a good performance under the strict limitation of FEs.

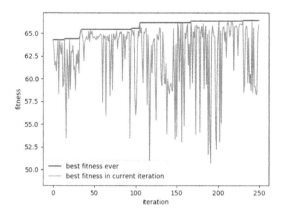

Fig. 4. A typical case of convergence of SA-PSO-I

However, there is still limitations of SA-PSO-I. We can see that in Fig. 4, around iteration 140, the fitness values of candidate solutions are significantly lower. This phenomenon is result from the error of the surrogate model. Because SA-PSO-I selects only the best solutions in the data set and the points selected may gather around some peaks in the solution spaces, which leads to a high uncertainty of the region with spare data. The error of the surrogate in these regions leads to a bad performance of LS-PSO, and the algorithm might use the bad surrogate for several iterations because its training set could remain unchanged when the outcomes of LS-PSO are not the best P solutions.

Therefore, we propose SA-PSO-II, where $M - P \neq 0$. This strikes a balance between exploitation at local and approximation of global. According to Table 1, SA-PSO-II gains the best average outcome and the smallest standard deviation by eliminating the abrupt decrease happens in SA-PSO-I.

In SA-PSO-L, a local surrogate model is involved to separate global approximation and local exploitation. According to Table 1, SA-PSO-L outperforms SA-PSO-I, but underperforms SA-PSO-II. This is probably due to the incompatible elements in this search method. LS-PSO has a trend of converging together and exploiting a single peak in the solution spaces, which is a conflict to the assumption of the local model based: several peaks are in the solution spaces.

From the above results, we can conclude that the performance of SA-PSO-II is the best among the algorithms given on the traffic flow optimization problem with a strict FEs constraint.

5   Conclusion

This paper introduces a surrogate-assisted EA for solving the traffic flow problem. In the proposed SA-PSO, a surrogate model is constructed and LS-PSO is used to search the best candidate solution on the surrogate model. The experimental result shows that under the same constraints, SA-PSO performs the best compared to conventional EAs. In further work, it is interesting to consider using other swarm intelligence techniques like ant colony optimization [21,25] for the traffic flow optimization problems. It is also promising to apply cooperative co-evolutionary approaches [22–24] to solve large-scale traffic flow optimization problems.

References

1. Qu, Y., Li, L., Liu, Y., Chen, Y., Dai, Y.: Travel routes estimation in transportation systems modeled by Petri Nets. In: Proceedings of 2010 IEEE International Conference on Vehicular Electronics and Safety, QingDao, China, pp. 73–77 (2010)
2. Okutani, I., Stephanedes, Y.J.: Dynamic prediction of traffic volume through Kalman filtering theory. Transp. Res. Part B Methodol. **18**(1), 1–11 (1984)
3. Van Der Voort, M., Dougherty, M., Watson, S.: Combining Kohonen maps with ARIMA time series models to forecast traffic flow. Transp. Res. Part C Emerg. Technol. **4**(5), 307–318 (1996)
4. Hui, S., Liu, Z.G., Li, C.J.: Research on traffic flow forecasting design based on BP neural network. J. Southwest Univ. Sci. Technol. **23**(2), 72–75 (2008)
5. Yang, Y., Lu, Y., Jia, L., Qin, Y., Dong, H.: Optimized simulation on the intersection traffic control and organization based on combined application of simulation softwares. In: Proceedings of the 24th Chinese Control and Decision Conference (CCDC 2012), Taiyuan, pp. 3787–3792 (2012)
6. Dezani, H., Marranghello, N., Damiani, F.: Genetic algorithm-based traffic lights timing optimization and routes definition using Petri net model of urban traffic flow. In: Proceedings of the 19th World Congress, The International Federation of Automatic Control, pp. 11326–11331 (2014)
7. Utama, D.N., Zaki, F.A., Munjeri, I.J., Putri, N.U.: A water flow algorithm based optimization model for road traffic engineering. In: Proceedings of the International Conference on Advanced Computer Science and Information Systems (ICACSIS 2016), Malang, pp. 591–596 (2016)
8. Qian, Y., Wang, C., Wang, H., Wang, Z.: The optimization design of urban traffic signal control based on three swarms cooperative-particle swarm optimization. In: Proceedings of the 2007 IEEE International Conference on Automation and Logistics, Jinan, pp. 512–515 (2007)
9. Jin, Y.: Surrogate-assisted evolutionary computation: recent advances and future challenges. Swarm Evol. Comput. **1**(2), 61–70 (2011)

10. Wang, H., Jin, Y., Doherty, J.: Committee-based active learning for surrogate-assisted particle swarm optimization of expensive problems. IEEE Trans. Cybern. **47**(9), 2664–2677 (2017)
11. Zhou, Z., Ong, Y.S., Nguyen, M.H., Lim, D.: A study on polynomial regression and Gaussian process global surrogate model in hierarchical surrogate-assisted evolutionary algorithm. In: Proceedings of the IEEE Congress on Evolutionary Computation (CEC), Edinburgh, U.K., vol. 3, pp. 2832–2839 (2005)
12. Sun, C., Jin, Y., Zeng, J., Yu, Y.: A two-layer surrogate-assisted particle swarm optimization algorithm. Soft Comput. **19**(6), 1461–1475 (2014). https://doi.org/10.1007/s00500-014-1283-z
13. Chugh, T., Jin, Y., Miettinen, K., Hakanen, J., Sindhya, K.: A surrogate-assisted reference vector guided evolutionary algorithm for computationally expensive many-objective optimization. IEEE Trans. Evol. Comput. **22**(1), 129–142 (2018)
14. Pang, H., Yang, X.: Simulation of urban macro-traffic flow based on cellular automata. In: Proceedings of the Chinese Control and Decision Conference (CCDC 2019), Nanchang, China, pp. 520–524 (2019)
15. Angeline, L., Choong, M.Y., Chua, B.L., Chin, R.K.Y., Teo, K.T.K.: A traffic cellular automaton model with optimised speed. In: Proceedings of the IEEE International Conference on Consumer Electronics-Asia (ICCE-Asia 2016), Seoul, pp. 1–4 (2016)
16. Stein, M.: Large sample properties of simulations using Latin hypercube sampling. Technometrics **29**(2), 143–151 (1987)
17. Cheng, R., Jin, Y.: A social learning particle swarm optimization algorithm for scalable optimization. Inf. Sci. **291**, 43–60 (2015)
18. Magele, C., Köstinger, A., Jaindl, M., Renhart, W., Cranganu-Cretu, B., Smajic, J.: Niching evolution strategies for simultaneously finding global and pareto optimal solutions. IEEE Trans. Magn. **46**(8), 2743–2746 (2010)
19. Eberhart, R.C., Shi, Y.: Comparing inertia weights and constriction factors in particle swarm optimization. In: Proceedings of the IEEE Congress on Evolutionary Computation (CEC), vol. 1, pp. 84–88 (2000)
20. Wei, F.-F., et al.: A classifier-assisted level-based learning swarm optimizer for expensive optimization. IEEE Trans. Evol. Comput. Accepted in 2020
21. Huang, Z.-M., et al.: An ant colony system with incremental flow assignment for multi-path crowd evacuation. IEEE Trans. Cybern. Accepted in 2020
22. Chen, W.-N., et al.: A cooperative co-evolutionary approach to large-scale multi-source water distribution network optimization. IEEE Trans. Evol. Comput. **23**(5), 188–202 (2019)
23. Zhao, T.-F., et al.: Evolutionary divide-and-conquer algorithm for virus spreading control over networks. IEEE Trans. Cybern. (2020, in press)
24. Jia, Y.-H., et al.: Distributed cooperative co-evolution with adaptive computing resource allocation for large scale optimization. IEEE Trans. Evol. Comput. **23**(2), 188–202 (2019)
25. Yang, Q., et al.: Adaptive multimodal continuous ant colony optimization. IEEE Trans. Evol. Comput. **21**(2), 191–205 (2017)

Neurodynamics, Complex Systems, and Chaos

Complex Dynamic Behaviors in a Discrete Chialvo Neuron Model Induced by Switching Mechanism

Yi Yang[1,2(✉)], Changcheng Xiang[3], Xiangguang Dai[1,2], Liyuan Qi[1,2], and Tao Dong[4(✉)]

[1] College of Computer Science and Engineering, Chongqing Three Gorges University, Chongqing, China
yang1595@126.com
[2] Key Laboratory of Intelligent Information Processing and Control of Chongqing Municipal Institutions of Higher Education, Chongqing Three Gorges University, Chongqing, China
[3] School of Mathematics and Statistics, Hubei Minzu University, Enshi, China
[4] College of Electronics and Information Engineering, Southwest University, Chongqing, China
david_312@126.com

Abstract. Switching policy has been considered in many biological systems and can exhibit rich dynamical behaviors which include different types of the bifurcations and deterministic chaos. The Chialvo neuron model analyzed in this article illustrates how bifurcations and multiple attractors can arise from the combination of the switching mechanism acting on membrane potential. The elementary dynamics of the system without the switching policy are analyzed firstly using phase plane methods. The comparisons of the bifurcation analysis with or without switching mechanism near the fixed points are provided. It can be concluded that the switching policy can be prone to give rise to the coexistence of multiple periodic attractors, which indicates there exist abundant firing modes in the switching system with the same system parameters and different initial values. More complex bifurcation and dynamical behaviors can be observed since applying the switching policy.

Keywords: Chialvo neuron model · Switching mechanism · Bifurcation analysis · Multiple attractors coexistence

1 Introduction

Dynamics of neuronal excitability have been extensively studied since Hodgkin and Huxley introduced the Hodgkin-Huxley model (H-H) in 1952 [9–13]. In these articles, Hodgkin and Huxley have proved the existence of ionic movements by experiments, revealed the excitation mechanism, the form and the rate of action potential. However, owing to the higher dimension of H-H model, scholars have

© Springer Nature Switzerland AG 2020
M. Han et al. (Eds.): ISNN 2020, LNCS 12557, pp. 61–73, 2020.
https://doi.org/10.1007/978-3-030-64221-1_6

been committed to simplify the neuronal model, which can not only retain the most characteristics of the biological neurons, but also make the calculation and simulation easier. Consequently, many simplified models were put forward in the following decades, such as FitzHugh-Nagumo model [6], Morris-Lecar model [19], Hindmarsh-Rose model [8], Chialvo model [2] and etc. The Chialvo model is a two-dimensional discrete-time nonlinear equation [2], the map-based model is as follows

$$x(n + 1) = x^2(n) \exp(y(n) - x(n)) + k,$$
$$y(n + 1) = ay(n) - bx(n) + c,$$

(1)

where $x(n)$ and $y(n)$ represent the membrane potential variable and the recovery variable, respectively. Symbol n represents the iteration step of the discrete equation (1) and the parameter k act as a constant bias or as a perturbation. Other three positive parameters a, b, and offset c can determine the fixed point of the recovery variable. Dynamical behaviors such as fixed point, bifurcation and chaos of the Chialvo model have been analyzed in relevant literatures [2,17,23].

Recently, the spiking neuron models with state variables threshold resetting have been proposed [14–16], they use these simple models to build networks of spiking neurons capable of exhibiting collective dynamics and rhythms similar to those of the mammalian cortex. One of the discrete-time Izhikevich model [16] used Eulier method take the form of

$$x(n + 1) = 0.04x^2(n) + 6x(n) + 140 - y(n) + I(n),$$
$$y(n + 1) = 0.004x(n) + 0.98y(n),$$

(2)

with the auxiliary after-spike resetting, if $x(n) \geq x_{TH} = 30mv$

$$x(n + 1) = c,$$
$$y(n + 1) = y(n) + d,$$

(3)

where x_{TH} is the resetting membrane potential threshold, c and d are constants, $I(n)$ is the external excitation current. After the spike reaches its apex (+30 mV), the membrane voltage and the recovery variable are reset according to the rules (3). The mapping can produce a variety of bursting patterns and most fundamental neuro-computational properties. It also provided a useful classification scheme for studying the dynamical mechanisms of most possible bursters. Subsequently, this type of state variables reset issue has been discussed in literatures [7,20,21,24]. Gang Zheng and Arnaud Tonnelier [7] consider an adaptive quadratic integrate-and-fire model with the modified reset rules based on Izhikevich's work, which can exhibit a chaotic behavior under a constant current which significantly differs from those in previous literatures that report chaos in IF models with periodic forcing and nonlinear term. Sou Nobukawa et al. [21] have examined the chaotic behaviors of Izhikevich neuron model incorporating the resetting process by using Lyapunov exponent with saltation matrix and Poincare section methods in two representative parameter regions. That is, the chaotic state appears through the period-doubling bifurcation route or through the intermittent route. In [20] they further reveal the emergence of the chaotic states through tangent bifurcation in continuous FHN neuron model with the resetting process. Yang et al. [24] considered two-dimensional Hindmarsh–Rose model with state variable resetting process, the period-adding bifurcation phenomena can be induced by the bifurcation parameter of auxiliary resetting process, the emergence of irregular limit cycle manifests the spiking and bursting both in abnormal discharge mode with the truncated firing peak.

Inspired by Izhikevich and the followers' work, in order to improve the pattern that the membrane potential varies dramatically in the resetting process in the literatures [7,14–16,20,21] and to study how the threshold x_{TH} in 1 influences the firing mode of the system, we change the membrane potential indirectly by adjusting the coefficient of recovery variable in the resetting process. For this purpose, we consider a discrete Chialvo neuron model guiding by the following switching policy based on model (1).

$$\begin{cases} x(n+1) = x^2(n)\exp\left(y(n) - x(n)\right) + k, \\ y(n+1) = \alpha y(n) - bx(n) + c, \end{cases} if\ x(n) < x_{TH}, \\ \begin{cases} x(n+1) = x^2(n)\exp\left(y(n) - x(n)\right) + k, \\ y(n+1) = \beta y(n) - bx(n) + c, \end{cases} if\ x(n) \geq x_{TH}, \tag{4}$$

where b, c, k are control parameters, $\alpha < \beta$ account for the resetting strategy that when the membrane potential reaches and exceeds the x_{TH}, the corresponding recovery variable take the larger value than the recovery variable in the subsystem when $x(n) < x_{TH}$. The membrane potential is often reset to a certain constant value below the resetting threshold in the previous resetting process. In our resetting process, the membrane potential change indirectly through recovery variable. The threshold control policy is also referred to as an on-off control introduced by Filippov [5] and applied in mechanic system [1], biological systems [3,4,18], electro-mechanical system [22], neuronal system [25]. The main purpose of this paper is to formulate complex dynamical behaviors such as bifurcation, multiple attractors and chaos of the Filippov Chialvo system in which membrane potential threshold control is implemented.

The rest of the paper is structured as follows. After providing the existence and stability of fixed points of system (1) in Sect. 2, the bifurcation analysis of system (4) is introduced in Sect. 3 and we focus on the issue of multiple attractors induced by switching mechanism in the fourth part. Finally, the conclusion remarks are illustrated in Sect. 5.

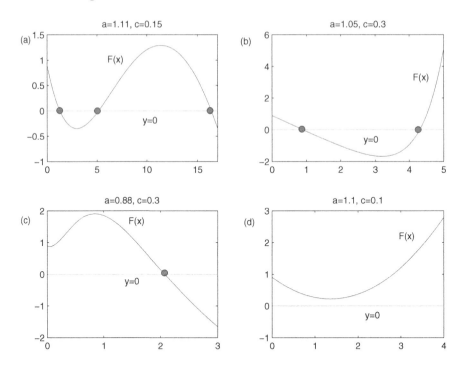

Fig. 1. Graphs of the roots of $F(x)$, (a) three fixed points, where $a = 1.11$, $b = 0.1$, $c = 0.15$, $k = 0.9$, (b) two fixed points, the parameters are $a = 1.05$, $b = 0.1$, $c = 0.3$, $k = 0.9$, (c) one fixed point, where $a = 0.88$, $b = 0.1$, $c = 0.3$, $k = 0.9$, (d) without fixed point, $a = 1.1$, $b = 0.1$ $c = 0.1$, $k = 0.9$.

2   Fixed Point and Its Stability

2.1   Existence of Fixed Points

The fixed point (x_*, y_*) of system (1) should satisfy the following equations

$$
\begin{aligned}
x_* &= x_*^2 \exp\left(y_* - x_*\right) + k \triangleq F(x_*, y_*), \\
y_* &= ay_* - bx_* + c \triangleq G(x_*, y_*),
\end{aligned}
\tag{5}
$$

that is to say, x_* should satisfy the equation $x = x^2 \exp\left(\frac{-bx+c}{1-a} - x\right) + k$. Let $F(x) = x^2 \exp\left(\frac{-bx+c}{1-a} - x\right) + k - x$, $y = ay - bx + c$, the fixed points of system (1) have four cases. The two nullclines and their intersections are plotted in Fig. 1 by fixing $b = 0.1$, $k = 0.9$. There exists three red dots where $a = 1.11, c = 0.15$ and each dot represents a fixed point in Fig. 1(a). Figure 1(b) shows two fixed points when $a = 1.05, c = 0.3$, it is clear that when $a = 0.88, c = 0.3$ there is only one fixed point at Fig. 1(c). Notice that subtle changes of a and c result in a change in the number of fixed points, the two nullclines of Fig. 1(d) have no intersection when $a = 1.1, c = 0.1$. In a neuronal system, a fixed point means that the system

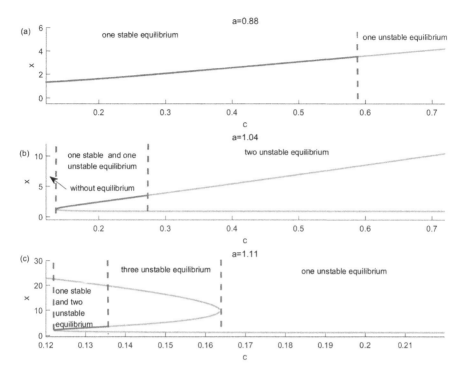

Fig. 2. Bifurcation curve of the fixed points about parameter c, the other parameters are fixed as $b = 0.1, k = 0.9$. (a) one fixed point for $a = 0.88$, (b) two fixed points for $a = 1.04$, (c) three fixed points for $a = 1.11$. (Color figure online)

state is in a relative resting state. A small disturbance may produce bifurcation at the fixed point if it is unstable, but if the fixed point is stable, then when the disturbance disappears, the system will enter the resting state again.

2.2 Stability of Fixed Points

The Jacobian matrix J of system (5) at the fixed point $P = (x_*, y_*)$ is presented as follows:

$$J = \begin{pmatrix} \frac{\partial F(x_*,y_*)}{\partial x_*} & \frac{\partial F(x_*,y_*)}{\partial y_*} \\ \frac{\partial G(x_*,y_*)}{\partial x_*} & \frac{\partial G(x_*,y_*)}{\partial y_*} \end{pmatrix} = \begin{pmatrix} \left(2x_* - x_*^2\right)\exp\left(y_* - x_*\right) & x_*^2 \exp\left(y_* - x_*\right) \\ -b & a \end{pmatrix},$$

(6)

the eigenvalues of system (5) can be expressed by trace and determinant

$$\lambda_{1,2} = \frac{\mathrm{Tr}(J) \pm \sqrt{[\mathrm{Tr}(J)]^2 - 4\,\mathrm{Det}(J)}}{2},$$

where Tr and Det represent the trace and the determinant of the Jacobian matrix J, respectively. The fixed point P is stable if the eigenvalues $|\lambda_{1,2}| < 1$.

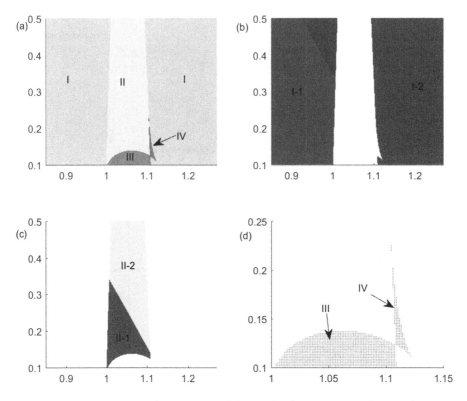

Fig. 3. Parameter regions about a, c for different fixed points cases, $b = 0.1, k = 0.95$, (a) parameter regions for I, II, III, IV, (b) parameter regions for I-1, I-2, (c) parameter regions for II-1, II-2, (d) parameter regions for III, IV

Stability range of fixed points corresponding to Fig. 1 about parameter c is illustrated in Fig. 2. According to Fig. 1, fixed $b = 0.1, k = 0.9$, when $a = 0.88$, system (5) has only one fixed point. The fixed point is stable when $c \in (0, 0.585)$ but unstable if $c > 0.585$, the red curve indicates that the fixed point is stable while the green curve means unstable and two areas separated by dashed lines which is plotted in Fig. 2(a). The portrait provided in Fig. 2(b) are the cases of two fixed points and no fixed point where $a = 1.04$. There is no fixed point when c is in a small range at the left side of the first dashed line. There exists one stable fixed point and one unstable fixed point when c is located between the first dashed line and the second dashed line. Similarly, we use the red curve represents the stable fixed point and the green curve represents the unstable fixed point. But once c is over 0.27, the stable fixed point becomes unstable. Compared with the above three cases, there exists three fixed points when a increases at 1.11 which is shown in Fig. 2(c). When c locates in the first dotted line and the second dotted line of $(0.12, 0.14)$, one fixed point is stable and two fixed points are unstable, and till c cross the second dotted line, the stable fixed point turns to an unstable fixed point and all the fixed points are under

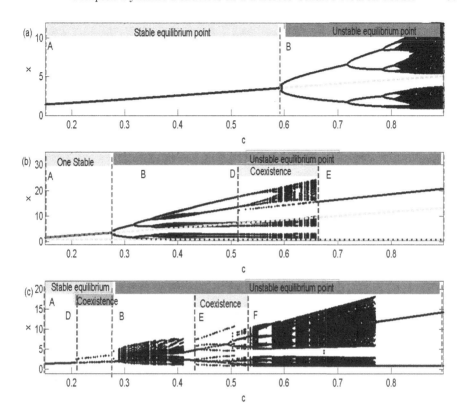

Fig. 4. Bifurcation diagrams of system (4). The parameters are fixed as $b = 0.1, k = 0.9$, (a) bifurcation diagram of the subsystem of system (4) where $\alpha = 0.88$, (b) bifurcation diagram of the subsystem of system (4) where $\beta = 1.04$, (c) bifurcation diagram of system (4) where $\alpha = 0.88, \beta = 1.04, x_{TH} = 2.5$. (Color figure online)

unstability. Once c crosses the third dotted line, two unstable fixed points disappear and there's only one unstable fixed point left. It can be concluded from Fig. 2 that with the increase of the number of fixed points and a, the range of offset parameter c with stable fixed points becomes smaller.

In order to further explain the relationship between fixed points and parameters. Parameter regions about a, c for the existence of fixed points are elaborated in Fig. 3. Here $b = 0.1$ and $k = 0.95$, we use Roman alphabet I, II, III, IV represent the parameter regions of one fixed point, two fixed points, three fixed points and without fixed point at the region of $a \in (0.85, 1.25)$ and $c \in (0.1, 0.5)$. From Fig. 3(a) it is obviously that the region of one fixed point is the largest (region I), followed by the region of two fixed points (region II). The region of three fixed points is very small (region III), and the region without a fixed point is almost negligible (region IV). In Fig. 3(b), I-1 and I-2 indicate the stable region of one fixed point and the unstable region of one fixed point, respectively. Similarly, II-1 denotes the coexistence region of a stable fixed point and an unstable fixed

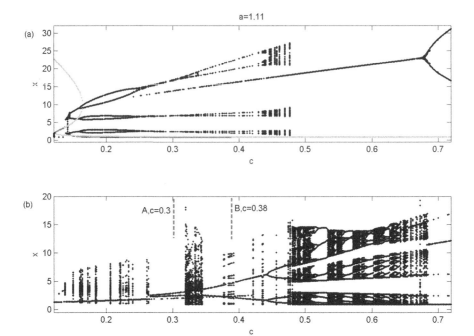

Fig. 5. Bifurcation diagrams of system (4).The parameters are fixed as $b = 0.1, k = 0.9$, (a) bifurcation diagram of the subsystem of system (4) where $\beta = 1.11$, (b) bifurcation diagram of system (4) where $\alpha = 0.85, \beta = 1.11, x_{TH} = 2.5$. (Color figure online)

point while II-2 reflects the region of two unstable fixed points. Figure 3(d) is an enlargement of the region of three fixed points and the region of no fixed point in Fig. 3(a), the stable and unstable regions are not subdivided due to their small sizes.

3   Bifurcation Analysis

The second part analyzes the existence and stability of the fixed points of system (1) in detail. The switching dynamical behaviors of system (4) will be discussed in this section. Bifurcation structures involved in the firing activities of neurons are viewed in the subsystem of system (4) and the whole system. In Fig. 4(a) when $\alpha = 0.88$, and c locates between 0 and 0.59, there exists one stable fixed point which represented by the red dotted line. At the same time, we use solid black lines to denote the system's solutions(just membrane potential x), it is obvious that the solution of subsystem is coincident with the fixed point when the fixed point is stable. That is, the subsystem trajectory will be attracted to the fixed point eventually. When $c \in (0.59, 0.9)$, the fixed point is unstable and denoted by green dotted line. Accordingly, the subsystem of system (4) undergoes a series of bifurcations including period doubling bifurcation, and

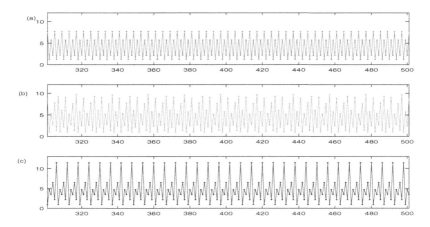

Fig. 6. Time series of membrane potential x corresponding to Fig. 4(c) when $c = 0.52$. The other parameters are fixed as $\alpha = 0.88, \beta = 1.04, b = 0.1, k = 0.9$. (Color figure online)

chaotic states. From Fig. 4(b) there exists two fixed points when $\beta = 1.04$ and $c \in (0.15, 0.275)$, where one fixed point represented by red line is stable and the other represented by green line is unstable, the subsystem trajectory and the stable fixed point overlap together. However, when $c > 0.275$, the stable fixed point becomes an unstable fixed point, period doubling bifurcation is generated instead of the stable solution. In addition to the period doubling bifurcation, we also note that multiple attractors may coexist between the two blue dashed lines D and E where c belong to the range of $(0.5, 0.7)$. Panel (a) and (b) of Fig. 4 show the bifurcation diagrams of the subsystems of system (4) where there exists one fixed point and two fixed points, respectively. The switching dynamical behaviors of system (4) is depicted in Fig. 4(c) supposing $\alpha = 0.88$ and $\beta = 1.04$, $x_{TH} = 2.5$. For Fig. 4(a), (b), (c), the first 1101 simulated values are omitted to remove the initial transients and only the next 100 values are plotted. Note that the membrane potential are different for the same parameter c compared with Fig. 4(a), (b), (c). For instance, the coexistences of multiple attractors are emerged when c is between dotted lines D and B, E and F, the number and the time series diagram of attractors will be introduced later. Switching strategy induce the system produce more multiple attractors coexistence regions.

Analogously, if the subsystems of one fixed point and three fixed points are placed in the same switching system, the qualitative properties of switching system will change greatly. As shown in Fig. 5(a) where $\beta = 1.11$, there exists three fixed points in the subsystem of system (4) and the subsystem undergoes a series of bifurcations including period doubling bifurcation and period-halving phenomena. Besides bifurcation it is also observed that several attractors may coexist as c increases beyond 0.3 and 0.4. Instead, it appears that the attractors of the switching system shown in Fig. 5(b) are relatively scattered and the bifurcation and chaos become more complex, where the parameters are fixed as $\alpha = 0.85$, $\beta = 1.11$ and $x_{TH} = 2.5$.

4 Multiple Attractors

The coexistence of multiple attractors is a phenomenon that with the same parameters the trajectory of the system will be attracted to distinct attractors if the initial value is in different attraction domain. It is noted from Fig. 4(c) when $c = 0.52$ there exists three periodic attractors, the time series diagrams of these periodic attractors are shown in Fig. 6(a), (b), (c), respectively. In other words, if the initial values of membrane potential and recovery variable are different within a certain range, the trajectory of the system will be stable on one of three periodic

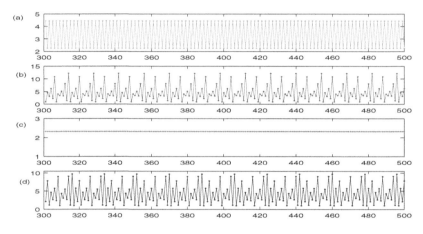

Fig. 7. Time series of membrane potential x corresponding to Fig. 5(b) when $c = 0.38$. The other parameters are fixed as $\alpha = 0.85, \beta = 1.04, b = 0.1, k = 0.9$. (Color figure online)

attractors which denoted by magenta, green and blue color curves. Similarly, when $c = 0.38$, the time series diagrams of four attractors of the switching system in Fig. 5(b) are shown in Fig. 7(a), (b), (c), (d), respectively, the four attractors are denoted by green, red, magenta, blue color curves. It can be seen from Fig. 4 and Fig. 5 that the system may have multiple attractors coexistence by choosing some suitable parameters, nevertheless it does not signify that there exists multiple attractors under any initial values, only when the initial values in a certain range this phenomenon can occur. The attraction basins are defined as the set of the initial conditions whose trajectories asymptotically approach that attractor as time evolves. The attraction basin shown in Fig. 8(a) indicates that the two attractors of the subsystem of system (4) emerged in Fig. 4(b) when $c = 0.52$ can coexist if the initial values $(x(0), y(0)) \in (1, 3.5)$. For example, when $(x(0), y(0)) = (2, 3)$ locates in the cyan color area of Fig. 8(a), the system trajectory will be attracted to the one of two attractors(time series diagrams are omitted). Suppose the initial values $(x(0), y(0)) = (2, 2.5)$ locates in the magenta color area of Fig. 8(b), the system trajectory will be attracted to the one of three

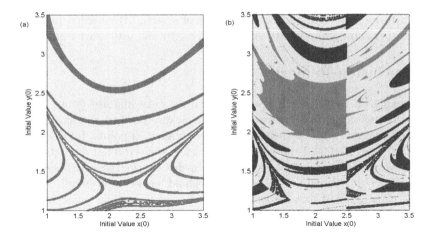

Fig. 8. Attraction basins of multiple attractors, (a) attraction basins of attractors shown in Fig. 4(b) when $c = 0.52$, (b) attraction basins of attractors shown in Fig. 4(c) when $c = 0.52$.

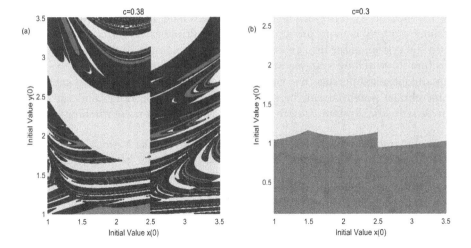

Fig. 9. Attraction basins of multiple attractors, (a) attraction basins of attractors shown in Fig. 5(b) when $c = 0.38$, (b) attraction basins of attractors shown in Fig. 5(b) when $c = 0.3$.

attractors corresponding to Fig. 6(a). When the parameters of the subsystem and the switching system are the same except for the switching threshold x_{TH}, the number and attraction regions of attractors are distinct. In the switching system, the attraction regions of multiple attractors undergo dramatic changes near the switching line. The attraction basins shown in Fig. 9(a) indicates that the four attractors of system (4) emerged in Fig. 5(b) when $c = 0.38$ can coexist if the initial values $(x(0), y(0))$ start from $(1, 3.5)$. The attraction basins shown in Fig. 9(b) indicate that the two attractors of system (4) emerged in Fig. 5(b) when $c = 0.3$ can coexist if $x(0)$ start from $(1, 3.5)$ and $y(0)$ start from $(1, 2.5)$.

5   Conclusion Remarks

The research on hybrid neuron models such as state impulse (state resetting process) or switching neuron model have been discussed in recent years. The dynamic behaviors of these neuron models such as fixed points, bifurcation and chaos have been studied. However, most of such systems did not pay attention to the issue of multiple attractors coexistence. In this article, the switching Chialvo neuron model with membrane potential threshold has been proposed. There are at most three fixed points for the system, the corresponding parameter areas of the fixed points are plotted which show the area of one fixed point is much larger than that of two fixed points and three fixed points. If the trajectory of the system will be stable at the fixed point eventually, that is to say, the membrane potential stay in a resting state. If the fixed point is unstable, a small disturbance may cause the system to deviate from the fixed point and then bifurcate or enter into chaos. The bifurcation diagrams about switching neuron system compare with the neuron system without applying switching policy indicate that the switching strategy is more prone to generate multiple attractor coexistence and induce the system enter into chaos. Moreover, the number and area of multiple attractors in switching systems are more than those in non-switching systems. From a biophysical perspective, the switching control can be designed such that the membrane potential can be stabilized at a desirable range and the firing mode of switching system is more abundant than that of non-switching system.

Acknowledgments. This work was supported in part by the National Natural Science Foundation of China under Grant 11961024, in part by Youth Project of Scientific and Technological Research Program of Chongqing Education Commission (KJQN201901203, KJQN201901218), in part by the Chongqing Technological Innovation and Application Project under Grant cstc2018jszx-cyzdX0171, in part by Chongqing Basic and Frontier Research Project under Grant cstc2019jcyj-msxm2105, in part by the Science and Technology Research Program of Chongqing Municipal Education Commission under Grant KJQN201900816, in part by Chongqing Social Science Planning Project under Grant 2019BS053.

References

1. Brogliato, B.: Erratum to: nonsmooth mechanics. Nonsmooth Mechanics. CCE, pp. E1–E11. Springer, Cham (2016). https://doi.org/10.1007/978-3-319-28664-8_9
2. Chialvo, D.R.: Generic excitable dynamics on a two-dimensional map. Chaos, Solitons Fractals **5**(3–4), 461–479 (1995)
3. da Silveira Costa, M.I.: Harvesting induced fluctuations: insights from a threshold management policy. Math. Biosci. **205**(1), 77–82 (2007)
4. da Silveira Costa, M.I., Meza, M.E.M.: Application of a threshold policy in the management of multispecies fisheries and predator culling. Math. Med. Biol. **23**(1), 63–75 (2006)

5. Filippov, A.F.: Differential Equations with Discontinuous Right-Hand Side. Kluwer Academic Publishers, Boston (1988)
6. Fitzhugh, R.: Impulses and physiological states in theoretical models of nerve membrane. Biophys. J. **1**(6), 445–466 (1961)
7. Gang, Z., Tonnelier, A.: Chaotic solutions in the quadratic integrate-and-fire neuron with adaptation. Cogn. Neurodyn. **3**(3), 197–204 (2009)
8. Hindmarsh, J.L., Rose, R.M.: A model of neuronal bursting using three coupled first order differential equations. Proc. Roy. Soc. London B Biol. Sci. **221**, 87–102 (1984)
9. Hodgkin, A.L., Huxley, A.F.: The components of membrane conductance in the giant axon of Loligo. J. Physiol. **116**(4), 473–496 (1952)
10. Hodgkin, A.L., Huxley, A.F.: Currents carried by sodium and potassium ions through the membrane of the giant axon of Loligo. J. Physiol. **116**(4), 449–472 (1952)
11. Hodgkin, A.L., Huxley, A.F.: The dual effect of membrane potential on sodium conductance in the giant axon of Loligo. J. Physiol. **116**(4), 497–506 (1952)
12. Hodgkin, A.L., Huxley, A.F.: A quantitative description of membrane current and its application to conduction and excitation in nerve. J. Physiol. **117**(4), 500–544 (1952)
13. Hodgkin, A.L., Huxley, A.F., Katz, B.: Measurement of current-voltage relations in the membrane of the giant axon of Loligo. J. Physiol. **116**(4), 424–448 (1952)
14. Izhikevich, E.M.: Simple model of spiking neurons. IEEE Trans. Neural Networks **14**(6), 1569–1572 (2003)
15. Izhikevich, E.M.: Which model to use for cortical spiking neurons? IEEE Trans. Neural Networks **15**(5), 1063–1070 (2004)
16. Izhikevich, E.M., Hoppensteadt, F.: Classification of bursting mappings. Int. J. Bifurcat. Chaos **14**(11), 3847–3854 (2004)
17. Jing, Z., Yang, J., Feng, W.: Bifurcation and chaos in neural excitable system. Chaos Solitons Fractals **27**(1), 197–215 (2006)
18. Meza, M.E.M., Bhaya, A., Kaszkurewicz, E., da Silveira Costa, M.I.: Threshold policies control for predator-prey systems using a control Liapunov function approach. Theor. Popul. Biol. **67**(4), 273–284 (2005)
19. Morris, C., Lecar, H.: Voltage oscillations in the barnacle giant muscle fiber. Biophys. J. **35**(1), 193–213 (1981)
20. Nobukawa, S., Nishimura, H., Yamanishi, T.: Routes to chaos induced by a discontinuous resetting process in a hybrid spiking neuron model. Sci. Rep. **8**(1), 379 (2018). https://doi.org/10.1038/s41598-017-18783-z
21. Nobukawa, S., Nishimura, H., Yamanishi, T., Liu, J.Q.: Chaotic states induced by resetting process in Izhikevich neuron model. J. Artif. Intell. Soft Comput. Res. **5**(2), 109–119 (2015)
22. Utkin, V., Guldner, J., Shi, J.: Sliding Mode Control in Electro-mechanical Systems. CRC Press, Boca Raton (2009)
23. Wang, F., Cao, H.: Mode locking and quasiperiodicity in a discrete-time Chialvo neuron model. Commun. Nonlinear Sci. Numer. Simul. **56**, 481–489 (2017)
24. Yang, Y., Liao, X., Dong, T.: Period-adding bifurcation and chaos in a hybrid Hindmarsh-Rose model. Neural Netw. **105**, 26–35 (2018)
25. Yang, Y., Xiaofeng, L.: Filippov Hindmarsh-Rose neuronal model with threshold policy control. IEEE Trans. Neural Netw. Learn. Syst. **30**(1), 306–311 (2019)

Multi-resolution Statistical Shape Models for Multi-organ Shape Modelling

Zhonghua Chen[1,2], Tapani Ristaniemi[2], Fengyu Cong[1,2,3,4], and Hongkai Wang[1(✉)]

[1] School of Biomedical Engineering, Faculty of Electronic Information and Electrical Engineering, Dalian University of Technology, Dalian 116024, China
chenzh4693@foxmail.com, {cong,wang.hongkai}@dlut.edu.cn
[2] Faculty of Information Technology, University of Jyväskylä, Jyväskylä 40100, Finland
tapani.e.ristaniemi@jyu.fi
[3] School of Artificial Intelligence, Faculty of Electronic Information and Electrical Engineering, Dalian University of Technology, Dalian, China
[4] Key Laboratory of Integrated Circuit and Biomedical Electronic System, Dalian University of Technology, Dalian, Liaoning, China

Abstract. Statistical shape models (SSMs) are widely used in medical image segmentation. However, traditional SSM methods suffer from the High-Dimension-Low-Sample-Size (HDLSS) problem in modelling. In this work, we extend the state-of-the-art multi-resolution SSM approach from two dimension (2D) to three dimension (3D) and from single organ to multiple organs. Then we proposed a multi-resolution multi-organ 3D SSM method that uses a downsampling-and-interpolation strategy to overcome HDLSS problem. We also use an inter-surface-point distance thresholding scheme to achieve multi-resolution modelling effect. Our method is tested on the modelling of multiple mouse abdominal organs from mouse micro-CT images in three different resolution levels, including global level, single organ level and local organ level. The minimum specificity error and generalization error of this method are less than 0.3 mm, which are close to the pixel resolution of mouse micro-CT images (0.2 mm) and better than the modelling results of traditional principal component analysis (PCA) method.

Keywords: Multi-resolution multi-organ SSM · PCA · HDLSS · Mouse micro-CT image

1 Introduction

Over the last three decades, SSM modelling approaches, as one of the most important methods, have been widely used to segment and register organs for medical image analysis [1]. The applications of SSMs includes but not limit to the following fields: 1) Medical image segmentation and registration [2–4]. 2) Clinical diagnosis and treatment [5, 6]. 3) Analysis of organ contraction [7].

Due to the complexity of medical images, SSMs of 3D organs are playing an increasingly important role in medical image segmentation. To represent 3D organ shapes, landmarks are sampled from organ surfaces. However, a challenging problem of the

© Springer Nature Switzerland AG 2020
M. Han et al. (Eds.): ISNN 2020, LNCS 12557, pp. 74–84, 2020.
https://doi.org/10.1007/978-3-030-64221-1_7

construction for a 3D SSM is that the number of training samples is small while the number of landmarks is large [8]. To fully capture the great variability of a 3D shape, traditional PCA modelling methods need to provide a large number of representative training samples to achieve a good modelling effect, which is usually labor-intensive and is even impossible to complete. This problem is later called HDLSS problem, which leads to insufficient and inaccurate expression of the model.

In order to solve the HDLSS problem, Wilms et al. [9] proposed a multi-resolution SSM modelling method with the traditional PCA method based on local distance constraints in 2017, and they used this method to construct 2D multi-resolution SSMs of human hand shapes and cardiopulmonary shapes respectively. The method is an important extension of traditional SSM methods, which can be used to obtain variation modes of SSMs at different resolution levels. In addition, it makes the resulting models achieve better generalization and specificity based on fewer training samples. However, one limitation of this method is that it takes up a lot of computer memory and is not suitable for cases where there are many sampling vertices. Therefore, Wilms et al. only modelled simple 2D shapes in their study. Unfortunately, the 3D shape vectors of multiple organs usually contain thousands or even tens of thousands of sampling vertices. Moreover, since the modelling method usually requires more sampling vertices for multiple organs than a single organ, this method is not applicable for the modelling of multiple organs as well. These drawbacks limit the application of this method in multi-organ 3D shape modelling.

In this study, we propose a solution to extend the multi-resolution SSM approach to 3D shape modelling of multiple organs with large number of surface vertices. Our method combines down-sampled 3D object surface with a Laplacian smoothing function to construct a multi-resolution multi-organ 3D SSM. We obtain deformation components in three resolution levels, which are the "global level", "single organ level" and "local organ level". The models obtained from the above three resolution levels are compared quantitatively with the traditional PCA modelling methods in terms of model generalization and specificity performances, and we obtain better modelling performances than the traditional methods for "global level".

2   Materials and Methods

2.1   Description of Mouse Micro-CT Data

The multi-organ training samples of mouse micro-CT images are taken from the Molecular Imaging Centre of the University of California, Los Angeles [10, 11]. During the imaging process, mice are injected with liver contrast agent Fenestra LC (ART, Montreal, QC, Canada) for clear imaging of abdominal organs. The weights of tested mice range from 15 to 30 grams, and the data are selected according to the following principles for modelling: (1) Boundaries of the abdominal organs (i.e., liver, spleen, left kidney and right kidney) of each mouse micro-CT image are clear. (2) There are no motion artefacts in each mouse micro-CT image. (3) There are no cases where the liver, spleen and kidneys of a mouse deviate from the normal shape. Mice are imaged in a multi-mode indoor prone position that provides anesthesia and heating [12]. Although the imaging room limits the possible postures of the mice, these postures are not strictly normalized. The random body bending postures in the left, right, and backward directions are included in the data set. The imaging system is MicroCAT II Small Animal CT (Siemens Preclinical Solutions, Knoxville, TN). Equipment acquisition parameters for imaging: exposure setting 70 kVp, 500 mAs, 500 ms and 360 step rotation, 2.0 mm aluminum filter. In the image acquisition process, an improved Field Kamp process is used to reconstruct the image so that the isotropic voxel size is 0.2 mm, the image matrix size is 256 × 256 × 496, and the pixel resolution is 0.2 mm.

In this study, 98 mouse micro-CT images are collected as training samples for model construction. Small animal imaging experts are invited to segment the 3D regions of livers, spleens, left kidneys, and right kidneys from the images, and then use the moving cube algorithm [13] to convert the segmented label maps to meshes. According to these principles, first, one of the 98 sample meshes is selected as the template. Second, the point cloud registration algorithm [14, 15] is used to register the template to all other training samples, so that different meshes have the same number of vertices, and each vertex corresponds to the same anatomical position in different meshes, thus completing the preparation of all training data. Figure 1 illustrates the entire training data preparation process.

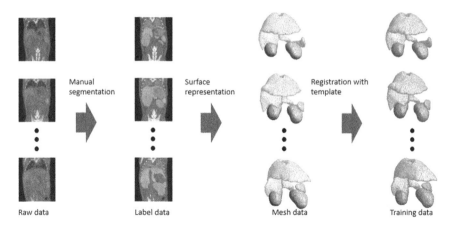

Fig. 1. The construction process of training data

2.2 Description of Algorithms

The Construction of Multi-resolution SSM

1. Given $Nd(d = 3)$ dimensional mouse abdominal multi-organ training data $\{S_i\}_1^N$, where $S_i = \{X_{1,i}, \ldots, X_{j,i}, \ldots, X_{M,i}\}$ contains M vertices $X_{j,i} = (x_{j,i}, y_{j,i}, z_{j,i})^T$ which are distributed on the surface of the training data. Then calculate the mean model of these training data, the calculation formula is shown as follows:

$$\bar{\mu} = \frac{1}{N}\sum_{i=1}^{N} S_i \tag{1}$$

2. After normalization, the covariance matrix of the coordinates of different dimensional vertices is calculated, and the calculation formula is shown in formula (2):

$$C = \frac{1}{N-1} \sum_{i=1}^{N} (S_i - \vec{\mu})(S_i - \vec{\mu})^T \qquad (2)$$

3. Calculate the point-to-point geodesic distance $d_{geo}(X_i, X_j)$ on mean model $\vec{\mu}$, and use rule (3) to set the values of the two sides of the covariance matrix C symmetrical to 0:

$$\rho_{i,j} = \begin{cases} \frac{cov(X_i, X_j)}{\sigma_i \sigma_j} & \text{if } d_{geo}(X_i, X_j) \leq \tau \\ 0 & \text{else} \end{cases} \qquad (3)$$

where σ_i and σ_j are the standard deviations of the i-th and j-th dimensions, respectively, and $\tau \left(0 \leq \tau \leq max_{i,j} d_{geo}(X_i, X_j)\right)$ is the threshold given in the experiment, so that a simplified symmetric matrix $R_1 = \begin{pmatrix} \rho_{1,1} & \cdots & \rho_{1,M} \\ \vdots & \ddots & \vdots \\ \rho_{M,1} & \cdots & \rho_{M,M} \end{pmatrix}$.

4. Since R_1 is not positive semi-definite and cannot be implemented eigenvalue decomposition, it is necessary to use the approximation method [16] to find an approximate positive semi-definite matrix R_2 replacing R_1 with formula (4):

$$R_2 = \min_A A - R_{1F}$$
$$det(A) \geq 0$$
$$diag(A) = 1 \qquad (4)$$

5. Compute the eigenvector matrix U_τ and the corresponding eigenvalue matrix Λ_τ, as shown in formula (5):

$$\begin{pmatrix} \sigma_1 & \cdots & 0 \\ \vdots & \ddots & \vdots \\ 0 & \cdots & \sigma_{dM} \end{pmatrix} R_2 \begin{pmatrix} \sigma_1 & \cdots & 0 \\ \vdots & \ddots & \vdots \\ 0 & \cdots & \sigma_{dM} \end{pmatrix} = U_\tau \Lambda_\tau U_\tau^T \qquad (5)$$

where the eigenvector set included in U_τ is represented as $\mathbf{P}$, and the corresponding eigenvalue vector on the diagonal of Λ_τ is represented as $\vec{\lambda}$.

6. When different values of distance threshold τ are selected in Eq. (3), the model will show different deformation capabilities locally. When $\tau \geq max_{i,j} d_{geo}(\vec{x}_i, \vec{y}_i)$, the constructed model is a traditional SSM; when $\tau = 0$, the vertice coordinates on all training data lose their relevance, and the constructed model is useless. By defining a series of thresholds $\tau_1 > \tau_2 > \ldots > \tau_L$, a multi-resolution scheme is defined to obtain a set of shape models $\{\vec{\mu}, P_1, \ldots, P_L, \vec{\lambda}_1, \ldots, \vec{\lambda}_L\}$ that vary from global

to local. However, these models are highly dependent and redundant, and do not form a single shape space. Therefore, it is necessary to combine these models into a subspace, so that the deformation components/eigenvectors provided by local SSMs can optimally represent deformation information. Based on step 1 to 6, the algorithm for constructing a multi-resolution shape model is derived as follows:

Suppose that there are N training data matrices $X = \left(\vec{s_1} \middle| \vec{s_2} \ldots \vec{s_N} \right) \in R_{m \times N}$, the thresholds of geodesic distance on each model surface are $\tau_1 > \tau_2 > \ldots > \tau_L$. Compute the mean model $\vec{\mu} = \frac{1}{N} \sum_{i=1}^{NN} \vec{s_i}$ of the training data. And define the distance matrix d_{geo} on the mean model. Assuming that the iteration index r ranges from 1 to L in the calculation process, where r represents the number of models, the deformation coefficient of the local SSM is defined as $\overrightarrow{\lambda_{\tau_r}}$, and the deformation components corresponding to the coefficients are defined as P_{τ_r}.

When $r = 1$, it means that there is only one shape model space, and the global multi-organ SSM can be obtained by directly using the traditional PCA method; when $r > 1$, it means that there are multiple shape spaces, and these spaces need to be combined for singular value decomposition. The decomposition process is as follows:

$$U(\cos\theta)V^T \leftarrow svd\left(P_{MR}^T P_{\tau_r}\right) \tag{6}$$

$$S = \begin{pmatrix} \cos\theta_{k \times k} \cdots & 0 \\ \vdots & \ddots & \vdots \\ 0 & \cdots I_{(l-k) \times (l-k)} \end{pmatrix} \tag{7}$$

where $svd()$ represents singular value decomposition and the transform base $\hat{B}$ is calculated (i.e., $\hat{B} = P_{\tau_r} VS^T$). Then, calculate the covariance matrix after spatial transformation:

$$\hat{\Sigma}_{\tau_{MR}} = \left(\sigma_{i,j}^{\tau_{MR}}\right) = U^T \begin{pmatrix} \lambda_{\tau_{MR},1} \cdots & 0 \\ \vdots & \ddots & \vdots \\ 0 & \cdots \lambda_{\tau_{MR},k} \end{pmatrix} U \tag{8}$$

$$\hat{\Sigma}_{\tau_r} = \left(\sigma_{i,j}^{\tau_r}\right) = V^T \begin{pmatrix} \lambda_{\tau_r,1} \cdots & 0 \\ \vdots & \ddots & \vdots \\ 0 & \cdots \lambda_{\tau_r,l} \end{pmatrix} V \tag{9}$$

$\hat{\Sigma}_{MR,r} = \left(\sigma_{i,j}^{\tau_{MR},\tau_r}\right)$, where $\sigma_{i,j}^{\tau_{MR},\tau_r} = \begin{cases} \sigma_{i,j}^{\tau_{MR}} & if \ i,j \in [1,k] \\ \sigma_{i,j}^{\tau_r} & if \ i,j \in [k+1,l] \\ 0 & else \end{cases}$, finally calculate

the uncorrelated eigenvectors and the corresponding eigenvalues:

$$\left[P_{MR}, \hat{\lambda}_{MR}\right] \leftarrow eig\left(\hat{B}\hat{\Sigma}_{MR,r}\hat{B}^T\right) \tag{10}$$

where $eig()$ represents eigenvalue decomposition. And the multi-resolution SSM μ is represented as follows:

$$\mu = \vec{\mu} + P_{MR}\hat{\lambda}_{MR} \tag{11}$$

Multi-resolution Multi-organ SSMs

Since the original approach of Wilms et al. [9] did not specifically consider the problem of modelling multiple organs, but only imposed geodesic distance constraints on the range of local deformation to generate multi-resolution models. However, in the case of multi-organ modelling, a simple geodesic distance cannot properly describe the distance relationship between two vertices belonging to different organs. For example, if the Euclidean distance between a vertex at the bottom of a lung and another vertex at the top of the liver is very close, these two vertices should have strong correlation in terms of common deformation. But these two vertices belong to different organs, and the geodesic distance will be farther, so that the correlation between them in a model becomes smaller and does not meet the deformation regulation of adjacent organs according to the methods of Wilms et al. In this study, we use Euclidean distance instead of geodesic distance as a constraint.

Based on the locally deformed multi-resolution SSM constructed in steps 1 to 6, the distances of surface vertices are computed by combining the modified Euclidean distance between different organs with the geodesic distance expressed in Eq. (3). This article specifies that the geodesic distance of vertices on different organ models is infinite, and the Euclidean distance of vertices on different organ models can be calculated. Given a model vector $\vec{s}$ containing O target calibration vertices, and $\vec{s} = \left(\vec{x}_{1,1}^T, \ldots, \vec{x}_{1,M_1}^T, \vec{x}_{2,1}^T, \ldots, \vec{x}_{2,M_2}^T, \ldots, \vec{x}_{O,1}^T, \ldots, \vec{x}_{O,M_O}^T\right)^T$, where M_i represents the number of landmarks of the i-th model, $i \in \{1, \ldots, O\}$. Define the undirected graph $G_g(V, E_g), V = \{\vec{x}_{i,j} | i \in \{1, \ldots, O\}, j \in \{1, \ldots M_i\}\}$ represents the vertices of the undirected graph, $E_g = \{(\vec{x}_{i,j}, \vec{x}_{i,k}) | i \in \{1, \ldots, O\}, j \in \{1, \ldots, M_i\}, k \in N(\vec{x}_{i,j})\}$, $N(\vec{x}_{i,j})$ is the direct neighbourhood of vertex $\vec{x}_{i,j}$ on target i. The weight $w_{i,j,k}^g$ of edge $(\vec{x}_{i,j}, \vec{x}_{i,k})$ is represented by the Euclidean distance between two vertices and the distance d between two vertices is computed by:

$$d = \begin{cases} w_{i,j,k}^g = \|\vec{x}_{i,j} - \vec{x}_{l,k}\| & \text{if } i \neq l, \\ d_{geo}(\vec{x}_{i,j}, \vec{x}_{l,k}) & \text{if } i = l. \end{cases} \tag{12}$$

There is no connection between different targets in G_g, so the geodesic distance between vertices on different targets is infinite and the Euclidean distance is $w_{i,j,k}^g$. If two vertices are on the same target, their distance is represented by geodesic distance $d_{geo}(\vec{x}_{i,j}, \vec{x}_{l,k})$

In order to extend the modelling method [9] to the construction of 3D multi-organ models with a large number of vertices, the idea adopted in this study is to first down-sample the vertices on the training data. Then we train the down-sampled data to obtain an initial multi-resolution multi-organ SSM. Finally, we interpolate the eigenvectors of the down-sampled vertices to generate eigenvectors of all vertices on the mean model and obtain the final multi-resolution multi-organ SSM. Laplacian smoothing function realizes the interpolation process:

$$\vec{X}_{i_{(n+1)}} = \vec{X}_{i_n} + \frac{\lambda}{M} \sum_{j=0}^{M} \left(\vec{X}_{j_n} - \vec{X}_{i_{\bar{n}}} \right) \tag{13}$$

where n represents the number of iterations, i is the vertex index, $\vec{x}_{i_n}$ represents the eigenvector of the i-th vertex coordinate at the n-th iteration, M is the number of vertices that share a common edge with vertex i, λ represents the smooth intensity coefficient. For $n = 0$, the eigenvector $\vec{x}_{i_0}$ of a down-sampled vertex is set as a vector P_{MR}, and the eigenvector of the other vertices is 0. In the iteration process, keep the eigenvectors of down-sampled vertices always in P_{MR}, and the eigenvectors of other vertices are computed by Eq. (13). In order to obtain the desirable interpolation effect, through repeated testing, the maximum number of iterations is set to 1000, and the value of λ is set to 0.8.

There are 3759 vertices in four kinds of organs (liver, spleen, left kidney and right kidney) in this study, and 375 vertices are obtained after 10 times down sampling, which can be used to construct a model in a computer with 16 G memory. Figure 2 shows the idea of this improved process, where Fig. 2-a shows the down-sampled vertices (marked in red) on the surface of the model. Figure 2-b shows the deformation components (represented by black arrows) on the down-sampled vertices of the model, and Fig. 2-c shows the deformation components on all vertices by interpolating the eigenvectors of down-sampled vertices over the entire model surface. From Fig. 2-c, we know that after interpolation, the deformation components on a small number of down-sampled vertices smoothly spread to the entire model surface. Although the deformation components are obtained by interpolation instead of by training all vertices, in the case of limited memory, this method can obtain reasonable deformation components for a large number of vertices of multiple organs. After observation and quantitative measurement (see this in Results section), the modelling results are better than the traditional global shape model.

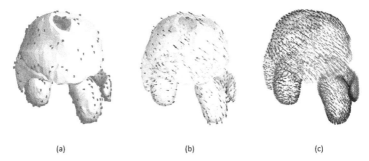

(a)   (b)   (c)

Fig. 2. Schematic diagram of improved method based on down-sampling training and eigenvectors interpolation. (a) Down-sampled vertices; (b) Deformation components of the down-sampled vertices; (c) The interpolated deformation components of all vertices on the mean model.

From above analysis, we set the ratio σ of the distance threshold τ to $1, 0.99$ and 0.5 respectively. A multi-resolution multi-organ SSM with traditional global level model ($\sigma = 1$, constructed by the traditional PCA modelling method), the single organ level model ($\sigma = 0.99$) and the local organ level model ($\sigma = 0.5$) is constructed. This modelling method well reflects the different levels of deformation in a multi-organ combination system and describes the deformation of multiple organs better than traditional global models.

3   Results

Figure 3 shows the modelling effect of the deformation components of the multi-resolution multi-organ SSM constructed in this work. Due to the limited space of this article, the model of each resolution level only shows the results of the first three deformation components on the mean model. Deformation components are denoted by PC_1, PC_2 and PC_3, respectively. $\lambda_1, \lambda_2, \lambda_3$ are the corresponding eigenvalues, and $\alpha_1, \alpha_2, \alpha_3$, are the shape coefficients of the multi-resolution multi-organ SSM. For each resolution level, the first row shows the mean models (the mean models of the three resolution levels are the same), and the second to fourth rows show the variation modes of the mean model through the first three deformation components. Figure 3-a, -b and -c show different shapes when shape coefficients of a deformation component takes different values, and the parts with obvious deformation are circled in the right column. From Fig. 3-a, we know that the organ deformation reflected by different deformation components all occur together among multiple organs. For example, PC_1 reflects the change in the distance between the left lower lobe of the liver and the spleen, which is most likely caused by the size change of the stomach between them. PC_2 reflects the closeness between the anterior half of the spleen and the left kidney, and PC_3 reflects the change in the distance between the liver and the two kidneys. Figure 3-b reflects the deformation of a single organ level, in which PC_1, PC_2 and PC_3 correspond to the deformation of the livers, left kidneys and spleens, respectively. Figure 3-c Reflects the local deformation of each organ, such as PC_1 reflects the deformation of the left lower lobe of the liver, PC_2 reflects the deformation of the anterior half of the spleen, PC_3 reflects changes in the anterior curvature of the right kidney. When local deformation is performed, other parts of the same organ keep unchanged. These results show that the method in this study can effectively model the deformation of organs at different resolution levels. We also employ generalization and specificity [3] to evaluate the accuracy of model construction in the study.

The generalization of the model is used to measure the model's ability to represent new shapes (that is, shapes not included in the training samples). Generalization is measured by using Leave-One-Out (LOO) method: assuming there are N training samples, one sample is left as the test sample S_j, and the other $N - 1$ samples $\{S_i | i = 1, 2, \ldots, N, i \neq j\}$ are used to train the model M^*. Then fit S_j through the deformation of M^*, and compute the average distance between the fitting result and S_j as the fitting error e_j. This process is repeated N times (i.e. $j = 1, 2, \ldots, N$), and set the average error $e_g = \frac{\sum_{j=1}^{N} e_j}{N}$ of N times as a measure of model generalization ability, the smaller the value of e_g is, the better the model generalization ability is.

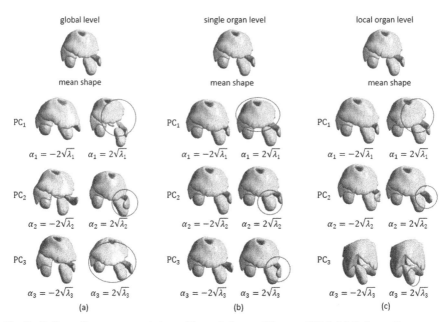

Fig. 3. Deformation components in multi-resolution multi-organ SSM. (a) Deformation components at the global resolution level; (b) Deformation components at the single organ resolution level; (c) Deformation components at the local organ level.

The specificity of the model is used to measure the model's ability to represent its own training samples. The specificity of the SSM is tested by randomly generating shape samples: when we get the model based on N training samples, the shape coefficient vectors $\{\vec{\alpha}_j | j = 1, 2, \ldots, K\}$ of K group models is randomly generated based on the normal distribution. The mean value of the normal distribution is 0, and the standard deviation is obtained by eigenvalue decomposition. Based on each randomly generated coefficient $\vec{\alpha}_j$, obtain its corresponding 3D shape, and find a sample whose surface distance is closest to this shape in the training sample set. Set this surface distance as the error e_j of the j-th random sample. Then calculate the average error $e_s = \frac{\sum_{j=1}^{K} e_j}{K}$ of K random samples as a measure of model specificity, the smaller the value of e_s is, the better the model specificity ability is.

In order to reflect the improvement effect of the multi-resolution multi-organ SSM on the generalization error e_g and the specificity error e_s, this experiment computes the results of e_g and e_s at different resolution levels, as shown in Fig. 4. Both for generalization error and specificity error, the mean value and standard deviation of the three model errors from global resolution level to local organ resolution level are within 1.0 mm. When the model changes from global level to local organ level, the mean value and variance of the errors are gradually decreasing, which indicates that this SSM with local organ resolution level is more accurate for boundary registration. This means that the multi-resolution multi-organ SSM constructed in this study has better generalization and specificity than the global model with global level constructed by traditional PCA method. Encouragingly, even for generalization errors, the minimum mean value of the multi-resolution multi-organ SSM has reached about 0.3 mm, which is close to the minimum mean value of the specificity error of 0.25 mm. More importantly, it is also close to the pixel resolution of mouse micro-CT images of 0.2 mm, and is lower than the average specificity error of the traditional global model of 0.31 mm.

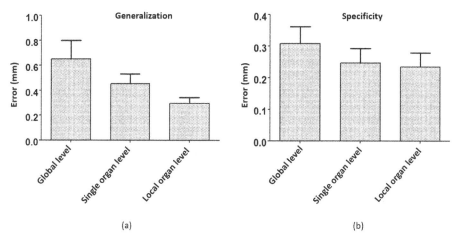

Fig. 4. Quantitative evaluation for modelling effect of the multi-resolution multi-organ SSM (a) Generalization error; (b) Specificity error

4   Conclusion

This paper proposes a multi-resolution multi-organ SSM construction method and we use it to model multiple abdominal organs of mouse micro-CT images. Compared to the recently state-of-the-art 2D multi-resolution SSM method proposed by Wilms et al., our method solves the shortcomings of memory occupation and thus extend the method to 3D space. On the other hand, this work extend the method to multi-organ modelling and can be used for modelling the inter-subject shape changes of multiple organs, single organ and local organ levels. This method surpasses the traditional PCA modelling method in terms of both generalization and specificity. It should be pointed out that although this work builds a model based on the abdominal organs of mouse, the method in this study is also applicable to the multi-organ modelling of human or other animal bodies. The model constructed in this work lays the foundation of shape prior knowledge for further multi-organ image segmentation.

Since the model constructed by the method in this paper has better deformation effects, we will use this method to construct a multi-resolution multi-organ model of human abdomen in the future, and further segment the organs in human CT images. The current results of the preliminary segmentation of human internal organs have gradually shown better results, and these results need to be further validate by our future work.

Acknowledgements. This study was funded by the general program of the National Natural Science Fund of China (No. 81971693, 81401475), the Science and Technology Innovation Fund of Dalian City (2018J12GX042) and the Fundamental Research Funds for the Central Universities (DUT19JC01). We thank the Molecular Imaging Centre of the University of California, Los Angeles for providing 98 mouse CT images and the scholarships from China Scholarship Council (No. 201806060163).

References

1. BMVA homepage. http://www.bmva.org/theses/2002/2002-davies.pdf. Accessed 15 Sep 2020
2. Cootes, T., Edwards, G., Taylor, C.: Active appearance models. IEEE Trans. Pattern Anal. Mach. Intell. **23**(6), 681–685 (2001)
3. Cootes, T., Taylor, C., Cooper, D., Graham, J.: Active shape models-their training and application. Comput. Vis. Image Underst. **61**(1), 38–59 (1995)
4. Heimann, T., Meinzer, H.: 3D statistical shape models for medical image segmentation. In: Proceedings of the 2nd International Conference on 3-D Digital Imaging and Modeling, pp. 414–423. IEEE, Ottawa (1999)
5. Bondiau, P.-Y., Malandain, G.: Eye reconstruction and CT-retinography fusion for proton treatment planning of ocular diseases. In: Troccaz, J., Grimson, E., Mösges, R. (eds.) CVRMed/MRCAS -1997. LNCS, vol. 1205, pp. 705–714. Springer, Heidelberg (1997). https://doi.org/10.1007/BFb0029296
6. Sandor, S., Leahy, R.: Surface-based labeling of cortical anatomy using a deformable atlas. IEEE Trans. Med. Imaging **16**(1), 41–54 (1997)
7. Rueckert, D., Burger, P.: Shape-based segmentation and tracking in 4D cardiac MR images. In: Troccaz, J., Grimson, E., Mösges, R. (eds.) CVRMed/MRCAS -1997. LNCS, vol. 1205, pp. 43–52. Springer, Heidelberg (1997). https://doi.org/10.1007/BFb0029223
8. Stegmann, M., Gomez, D.: A brief introduction to statistical shape analysis (2002)
9. Matthias, W., Heinz, H., Jan, E.: Multi-resolution multi-object statistical shape models based on the locality assumption. Med. Image Anal. **38**(2), 17–29 (2017)
10. Suckow, C., Stout, D.: MicroCT liver contrast agent enhancement over time, dose, and mouse strain. Mol. Imaging Biol. **10**(2), 114–120 (2008)
11. Wang, H., Stout, D., Chatziioannou, A.: Estimation of mouse organ locations through registration of a statistical mouse atlas with micro-CT images. IEEE Trans. Med. Imaging **31**(1), 88–102 (2012)
12. Suckow, C., Kuntner, C., Chow, P.: Multimodality rodent imaging chambers for use under barrier conditions with gas anesthesia. Mol. Imaging Biol. **11**(2), 100–106 (2009)
13. Lorensen, W., Cline, H.: Marching cubes: a high resolution 3D surface construction algorithm. ACM SIGGRAPH Comput. Graph. **21**(8), 163 (1987)
14. Roberto, M., Vito, R., Massimiliano, N., Tiziana, D., Ettore, S.: A modified iterative closest point algorithm for 3D point cloud registration. Comput. Aided Civ. Infrastruct. Eng. **31**(7), 515–534 (2016)
15. Park, S., Lim, S.: Template-based reconstruction of surface mesh animation from point cloud animation. ETRI J. **36**(12), 1008–1015 (2014)
16. Highm, N.: Computing the nearest correlation matrix—a problem from finance. IMA J. Numer. Anal. **22**(3), 329–343 (2002)

A Neural Network for Distributed Optimization over Multiagent Networks

Jiazhen Wei[1], Sitian Qin[1(✉)], and Wei Bian[2(✉)]

[1] Department of Mathematics, Harbin Institute of Technology at Weihai,
Weihai 264209, People's Republic of China
`qinsitian@163.com`

[2] Department of Mathematics, Harbin Institute of Technology, Harbin 150001,
People's Republic of China
`bianweilvse520@163.com`

Abstract. This paper is concerned with a distributed optimization problem with inequality constraint over a multiagent network. The objective function is the sum of multiple local convex functions, which can be nonsmooth. Based on graph theory and nonsmooth analysis, we propose a neural network with a time-varying auxiliary function. The boundedness of the state solution is demonstrated by using the properties of the auxiliary function. Moreover, it is proved that the designed neural network with any initial conditions reaches a consensus and converges to the global optimal solution. Finally, a numerical simulation is discussed to verify the theoretical results.

Keywords: Distributed optimization · Multiagent system · Neurodynamic approach

1 Introduction

In recent years, the distributed optimization has been widely used in science and engineering problems, including sensor networks, source location and distributed data regression (see [3,5,15]). Taking advantage of the properties of multiagent networks, many scholars begin to study the distributed optimization problem over multiagent networks.

Generally speaking, the main algorithms for distributed optimization can be divided into discrete-time and continuous-time algorithm. In the mid-1980s, for linear programming problems, Tank and Hopfield [6] proposed a novel optimization method. Then Kennedy and Chua [8] proposed a nonlinear optimization method based on recurrent neural networks, which is called neurodynamic optimization approach. Since then, neurodynamic approach has been widely researched (see [11,13,14]).

In recent years, in order to solve various distributed optimization problems, scholars have designed many neural networks. Liu *et al.* [12] proposed a collective neurodynamic network to solve distributed convex optimization problem.

© Springer Nature Switzerland AG 2020
M. Han et al. (Eds.): ISNN 2020, LNCS 12557, pp. 85–95, 2020.
https://doi.org/10.1007/978-3-030-64221-1_8

Jiang *et al.* [7] designed a neural network based on a penalty-like method for a class of nonsmooth distributed optimization problem.

Drived by the above researches, a neural network to solve the nonsmooth distributed optimization problem with inequality constraint is presented. Different from references [9,14], this paper does not depend on the additional assumption that the objective function is bounded from below. And we avoid utilizing the projection operator for distributed optimization problem. Besides, the existence and boundedness of the state solution of the neural network are guaranteed, which have not written about in [12].

2   Preliminaries

In this paper, denote the set of non-negative real numbers by $\mathbb{R}_+$. Denote $n+m$-dimensional vector $(y^{\mathrm{T}}, z^{\mathrm{T}})^{\mathrm{T}} \in \mathbb{R}^{n+m}$ by $\mathrm{col}(y, z) \in \mathbb{R}^{n+m}$. $\|\cdot\|$ indicates the Euclidean norm, $\mathbf{1}_n = (1, \ldots, 1)^{\mathrm{T}} \in \mathbb{R}^n$, $\mathbf{0}_n = (0, \ldots, 0)^{\mathrm{T}} \in \mathbb{R}^n$, I_n is the identity matrix in $\mathbb{R}^{n \times n}$. Let $\mathrm{diag}\{A_1, A_2, \ldots, A_k\}$ be the block-diagonal matrix of $A_1, A_2, \ldots, A_k$. For $\tilde{y} \in \mathbb{R}^n$ and $r > 0$, $B(\tilde{y}, r) = \{y \in \mathbb{R}^n : \|y - \tilde{y}\| < r\}$. $P_\Omega(x)$ denotes the projection of x on Ω. Let $\otimes$ be Kronecker product.

2.1   Graph Theory

Let $\mathcal{G} = (V, E, \mathcal{A})$ be a weighted undirected graph. $V = (v_1, v_2, \cdots, v_n)$ is a vertex set and $E \subseteq V \times V$ is an edge set. $\mathcal{A} = (a_{ij}) \in \mathbb{R}^{n \times n}$ is the adjacent matrix, where $a_{ij} > 0$ if $(v_i, v_j) \in E$ and $a_{ij} = 0$ otherwise. The weighted degree of vertex v_i is defined as $d^w(v_i) = \sum_{j=1}^n a_{ij}$. The weighted degree matrix $\mathcal{D} = \mathrm{diag}\{d_{11}, \ldots, d_{nn}\} \in \mathbb{R}_+^{n \times n}$ with $d_{ii} = d^w(v_i)$. The Laplacian matrix is $L = \mathcal{D} - \mathcal{A}$ and $L\mathbf{1}_n = \mathbf{0}_n$.

2.2   Nonsmooth Analysis

Assume that $\Psi : [0, T) \times \mathbb{R}^n \to \mathbb{R}^n$ is a set-valued mapping. Consider the following differential inclusion:

$$\begin{cases} \dot{x}(t) \in \Psi(t, x(t)) \\ x(0) = x_0 \end{cases} \tag{1}$$

Definition 1 [1]. *$x : [0, T) \to \mathbb{R}^n$ is a state solution of differential inclusion (1) on $[0, T)$, if x is an absolutely continuous function and satisfies (1) almost everywhere on $[0, T)$.*

Definition 2 [2]. *If $\varphi : \mathbb{R}^n \to \mathbb{R}$ is a convex function, then the subdifferential of φ at x is defined by*

$$\partial\varphi(x) = \{\xi \in \mathbb{R}^n : \varphi(x) - \varphi(y) \le \langle \xi, x - y \rangle, \forall y \in \mathbb{R}^n\}$$

According to [2], we get that $\partial\varphi(\cdot)$ is upper semicoutinuous and monotone, i.e., $\langle y - y_0, x - x_0 \rangle \ge 0$, for any $y \in \partial\varphi(x)$ and $y_0 \in \partial\varphi(x_0)$. In addition, $\partial\varphi(x)$ is a nonempty convex compact set of $\mathbb{R}^n$.

Lemma 1 [2]. *Assume that $\mathcal{F} : \mathbb{R}^n \rightarrow \mathbb{R}$ is regular at $y(t)$, $y : \mathbb{R} \rightarrow \mathbb{R}^n$ is differentiable at t and Lipschitz at t, then for a.e. $t \in [0, +\infty)$, we have*

$$\frac{\mathrm{d}}{\mathrm{d}t} \mathcal{F}(y(t)) = \langle \xi, \dot{y}(t) \rangle, \ \forall \xi \in \partial \mathcal{F}(y(t))$$

3 Problem Description and Optimization Algorithm

Consider the following distributed optimization problem:

$$\begin{aligned} &\min \ \sum_{i=1}^{k} f_i(x) \\ &\text{s.t.} \ \ g_i(x) \leq \mathbf{0}, \ i \in \{1, 2, \dots, k\} \end{aligned} \tag{2}$$

where $x \in \mathbb{R}^n$, $f_i : \mathbb{R}^n \rightarrow \mathbb{R}$ is the objective function of the ith agent, and $g_i = (g_{i1}, \dots, g_{im_i}) : \mathbb{R}^n \rightarrow \mathbb{R}^{m_i}$ is the inequality constraint of the ith agent.

Assumptions:

- (i) Communication graph $\mathcal{G}$ is undirected and connected.
- (ii) Function f_i is strictly convex, but not necessarily differential. All components of function g_i are convex and not necessarily differential.
- (iii) There exists $x^* \in \mathbb{R}^n$, such that $g_i(x^*) < 0$ $(i = 1, 2, \dots, k)$.
- (iv) For $i = 1, 2, \dots, k$, inequality constraint set $\{x \in \mathbb{R}^n : g_i(x) \leq 0\}$ is bounded, i.e., there is $\hat{r}$ satisfying $\{x \in \mathbb{R}^n : g_i(x) \leq 0\} \subset B(0, \hat{r})$.

Denote $\boldsymbol{x} = \mathrm{col}(x^1, x^2, \dots, x^k)$, $\boldsymbol{g}(\boldsymbol{x}) = \mathrm{col}(g_1(x^1), g_2(x^2), \dots, g_k(x^k))$. Let L_k be the Laplacian matrix of graph $\mathcal{G}$ and $\boldsymbol{L} = L_k \otimes I_n \in \mathbb{R}^{kn \times kn}$. By the connectedness of undirected graph, we know $\boldsymbol{Lx} = \mathbf{0}$ is equivalent to $x^1 = \dots = x^k$. According to [4], we obtain the following lemma.

Lemma 2 [4]. *If (i) in **Assumptions** is satisfied, problem (2) is equivalent to the following problem,*

$$\begin{aligned} &\min \ f(\boldsymbol{x}) = \sum_{i=1}^{k} f_i(x^i) \\ &\text{s.t.} \ \ \boldsymbol{g}(\boldsymbol{x}) \leq \mathbf{0}, \ \boldsymbol{Lx} = \mathbf{0} \end{aligned} \tag{3}$$

Let $\mathcal{I} = \{\boldsymbol{x} \in \mathbb{R}^{kn} : \boldsymbol{g}(\boldsymbol{x}) \leq \mathbf{0}\}$ be the inequality constraint set of (3).

Denote

$$\tilde{g}_{ij}(x^i) = g_{ij}(x^i) + \max\{\|x^i\|, \hat{r}\} - \hat{r}$$
$$G_i(x^i) = \sum_{j=1}^{m_i} \max\{\tilde{g}_{ij}(x^i), 0\}, \ G(\boldsymbol{x}) = \sum_{i=1}^{k} G_i(x^i)$$

where $\hat{r}$ is from (iv) in **Assumptions**. It is obvious that $G(\boldsymbol{x})$ is convex.

Lemma 3 [10]. *If (ii)-(iv) in **Assumptions** hold, then $\lim\limits_{\|x^i\| \rightarrow \infty} G_i(x^i) = +\infty$.*

Lemma 4 [16]. *Assume that (ii)-(iv) in **Assumptions** hold, there is a constant $\omega > 0$ such that for any $\boldsymbol{x} \notin \mathcal{I}$, one has*

$$\langle \xi,\ \boldsymbol{x} - \boldsymbol{x}^* \rangle > \omega,\ \forall \xi \in \partial G(\boldsymbol{x})$$

*where $\boldsymbol{x}^* = 1_k \otimes x^*$ and x^* is (iii) in **Assumptions**.*

Based on the above analysis, we propose a neural network:

$$\dot{\boldsymbol{x}}(t) \in -E(\boldsymbol{x}(t))\big(\boldsymbol{L}\boldsymbol{x}(t) + \gamma(t)\partial f(\boldsymbol{x}(t))\big) - (t+1)\partial G(\boldsymbol{x}(t)) \tag{4}$$

where $E(\boldsymbol{x}) = \mathrm{diag}\{e(G_1(x^1)), e(G_2(x^2)), \ldots, e(G_k(x^k))\} \otimes I_n \in \mathbb{R}^{kn \times kn}$. And $e(y)$ is a time-varying auxiliary function which is defined as

$$e(y) = \begin{cases} 1 - y, & 0 \le y \le 1 \\ 0, & y > 1 \end{cases}$$

Furthermore, $\gamma : \mathbb{R}_+ \to \mathbb{R}_+$ is defined with the following properties:

$$\lim_{t \to +\infty} \gamma(t) = 0,\ \dot{\gamma}(t) \le 0,\ \int_0^\infty \gamma(t)\mathrm{d}t = \infty$$

For example, $\gamma(t)$ can be defined as $\gamma(t) = \frac{\gamma_0}{(t+1)^r}$, where $\gamma_0 > 0$ and $r \in (0, 1]$.

The distributed implementation of neural network (4) can be described as:

$$\begin{aligned} \frac{\mathrm{d}}{\mathrm{d}t}x^i(t) \in -e\big(x^i(t)\big)\Big(\sum_{j \in N_i} a_{ij}\big(x^i(t) - x^j(t)\big) + \gamma(t)\partial f_i\big(x^i(t)\big) \Big) \\ -(t+1)\partial G_i\big(x^i(t)\big),\ i = 1, 2, \ldots, k \end{aligned} \tag{5}$$

where a_{ij} indicates the connection weight between the agents $\{i, j\}$ and N_i represents the index set of neighbors of the ith node.

4   Main Results

In this section, the consensus and convergence of the neural network in (4) is analyzed. Firstly, we prove the boundedness of the state solution.

Theorem 1. *Under (i)-(iv) in **Assumptions** described above, for any initial point $\boldsymbol{x}(0) \in \mathbb{R}^{kn}$, the state solution of neural network (4) is bounded.*

Proof. Since the r.h.s of differential inclusion (4) is nonempty multifunction with compact convex values, according to [4], we know for any initial point $\boldsymbol{x}(0)$, there exist $T > 0$ and a local solution $\boldsymbol{x}(t)$ such that (4) holds for *a.e.* $t \in [0, T)$. Therefore, there are $\eta_i(t) \in \partial f_i(x^i(t))$ and $\xi_i(t) \in \partial G_i(x^i(t))$ satisfying

$$\frac{\mathrm{d}}{\mathrm{d}t}x^i(t) = -e\big(x^i(t)\big)\Big(\sum_{j \in N_i} a_{ij}\big(x^i(t) - x^j(t)\big) + \gamma(t)\eta_i(t) \Big) - (t+1)\xi_i(t)$$

for *a.e.* $t \in [0, T)$. By Lemma 3, it implies that there is $r_i > 0$ satisfying

$$\{x^i(0)\} \cup \{x^i \in \mathbb{R}^n : 0 \leq G_i(x^i) \leq 1\} \subset B(x^*, r_i) \tag{6}$$

where x^* is from (iii) in **Assumptions**.

Next, we will prove that for any $i = 1, 2, \ldots, k$, $x^i(t) \in B(x^*, r_i)$, $\forall t \in [0, T)$. If not, then there exist t_1 and $\rho > 0$, such that

$$\|x^i(t_1) - x^*\| = r_i, \ \|x^i(t) - x^*\| > r_i, \ \forall t \in (t_1, t_1 + \rho] \tag{7}$$

From (6) and (7), we have $G_i(x^i(t)) > 1$, $e(G_i(x^i(t))) = 0$, $\forall t \in (t_1, t_1 + \rho]$. Then for *a.e.* $t \in (t_1, t_1 + \rho]$, the neural network (5) can be described as

$$\frac{\mathrm{d}}{\mathrm{d}t} x^i(t) = -(t+1)\xi_i(t) \tag{8}$$

Differentiating $\frac{1}{2}\|x^i(t) - x^*\|^2$ along the solution of (8), for *a.e.* $t \in (t_1, t_1 + \rho]$, we have

$$\frac{1}{2}\frac{\mathrm{d}}{\mathrm{d}t}\|x^i(t) - x^*\|^2 = -(t+1)(x^i(t) - x^*)^{\mathrm{T}}\xi_i(t)$$
$$\leq -(t+1)G_i(x^i(t)) \leq 0$$

Then, it implies that $\|x^i(t_1 + \rho) - x^*\| \leq \|x^i(t_1) - x^*\| \leq r_i$, which contradicts with (7). Thus, $x^i(t) \in B(x^*, r_i)$ for any $i = 1, 2, \ldots, k$. The proof is complete.

Remark 1. From Theorem 1, we have the local state solution of neural network (4) is bounded. Thus, by the extension theorem of solution in [1], the global solution $x(t)$ of neural network (4) with any initial point $x(0) \in \mathbb{R}^{kn}$ exists.

Theorem 2. *Under (i)-(iv) in **Assumptions**, for the state solution $x(t)$ of neural network (4) starting from any initial point $x(0) \in \mathbb{R}^{kn}$, there exists $\tilde{T} \geq 0$ such that $x(t) \in \mathcal{I}$, $\forall t \in [\tilde{T}, +\infty)$.*

Proof. For the state solution of neural network (4), there are $\eta(t) \in \partial f(x(t))$ and $\xi(t) \in \partial G(x(t))$ satisfying

$$\dot{x}(t) = -E(x(t))\big(\boldsymbol{L}x(t) + \gamma(t)\eta(t)\big) - (t+1)\xi(t), \ \textit{a.e. } t \geq 0 \tag{9}$$

Taking the derivation of $G(x(t))$ along the state solution of (4), we have

$$\frac{\mathrm{d}}{\mathrm{d}t}G(x(t)) = -\big\langle \xi(t), E(x(t))\big(\boldsymbol{L}x(t) + \gamma(t)\eta(t)\big) + (t+1)\xi(t)\big\rangle, \ \textit{a.e. } t \geq 0$$

Based on Lemma 4, for any $x(t) \notin \mathcal{I}$, one has $\|\xi(t)\| \geq \frac{\omega}{R}$.

In the following content, we show that there is a $\tilde{T} > 0$ such that $x(\tilde{T}) \in \mathcal{I}$. If not, we assume that $x(t) \notin \mathcal{I}$, $\forall t > 0$. Noting that $\partial f(\cdot)$ is upper semicontinuous, then $\eta(t)$ is bounded. Then, for *a.e.* $t \geq 0$, there are constant $M_1, M_2 > 0$ such that

$$\frac{\mathrm{d}}{\mathrm{d}t}G(x(t)) \leq \|\xi(t)\|\big(M_1 + M_2 - (t+1)\|\xi(t)\|\big)$$
$$\leq \frac{\omega}{R}\Big(M_1 + M_2 - (t+1)\frac{\omega}{R}\Big) \tag{10}$$

Choosing $T_0 = \frac{1}{\omega}(M_1 + M_2)R - 1$, then there exists constant $\tau > 0$ satisfying

$$\frac{\mathrm{d}}{\mathrm{d}t}G(\boldsymbol{x}(t)) \leq -\tau \tag{11}$$

for *a.e.* $t \geq T_0$. By integrating (11) from T_0 to $t(\geq T_0)$, it gives that

$$G(\boldsymbol{x}(t)) \leq G(\boldsymbol{x}(T_0)) - \tau(t - T_0)$$

When $t \to \infty$, we have $G(\boldsymbol{x}(t)) \to -\infty$, which is a contradiction with the fact that $G(\boldsymbol{x}(t)) \geq 0$. Hence, there exists $\tilde{T} > 0$ satisfying $\boldsymbol{x}(\tilde{T}) \in \mathcal{I}$.

Finally, we claim that $\boldsymbol{x}(t) \in \mathcal{I}$, $\forall t \in [\tilde{T}, \infty)$. If it does not hold, then there are t_1, t_2 such that $t_2 \geq t_2 \geq \tilde{T}$ and $\boldsymbol{x}(t_1) \in \mathcal{I}$, $\boldsymbol{x}(t) \notin \mathcal{I}$, $\forall t \in (t_1, t_2]$, which deduces that $G(\boldsymbol{x}(t_1)) = 0$. Then based on the analysis above, we know that (11) holds for $\forall t \in (t_1, t_2]$. Taking the integral of (11) from t_1 to t_2, one has

$$G(\boldsymbol{x}(t_2)) \leq G(\boldsymbol{x}(t_1)) - \tau(t_2 - t_1) = -\tau(t_2 - t_1) < 0$$

which contradicts with $G(\boldsymbol{x}(t_2)) \geq 0$. The proof is complete.

Definition 3 [2]. *The neural network achieves consensus, if for any initial conditions, one has* $\lim\limits_{t \to \infty} \|x^i(t) - x^j(t)\| = 0$, $\forall i, j \in 1, 2, \ldots, k$.

Theorem 3. *Under* (i)–(iv) *in* **Assumptions**, *the state solution* $\boldsymbol{x}(t)$ *of neural network* (4) *with any initial point* $\boldsymbol{x}(0) \in \mathbb{R}^{kn}$ *converges to the feasible region of problem* (3), *which means that the state reaches consensus ultimately.*

Proof. By Theorem 1, there is $r > 0$ satisfying $\boldsymbol{x}(t) \in B(\boldsymbol{x}^*, r)$, $\forall t \geq 0$. Let $m_f = \inf_{\boldsymbol{x} \in B(\boldsymbol{x}^*, r)} f$. From Theorem 2, we can assume that $\boldsymbol{x}(t) \in \mathcal{I}$ for $t \geq 0$, consequently $E(\boldsymbol{x}(t)) = \boldsymbol{I}_{kn \times kn}$. Thus, for *a.e.* $t \geq 0$, (9) can be rewritten as

$$\dot{\boldsymbol{x}}(t) = -\boldsymbol{L}\boldsymbol{x}(t) - \gamma(t)\eta(t) - (t + 1)\xi(t) \tag{12}$$

Next, we construct the following Lyapunov function:

$$V(\boldsymbol{x}, t) = \frac{1}{2}\boldsymbol{x}(t)^{\mathrm{T}}\boldsymbol{L}\boldsymbol{x}(t) + (t + 1)G(\boldsymbol{x}(t)) + \gamma(t)\big(f(\boldsymbol{x}(t)) - m_f\big) \tag{13}$$

Taking the derivation of $V(\boldsymbol{x}, t)$ along the state solution of (4), one has

$$\frac{\mathrm{d}}{\mathrm{d}t}V(\boldsymbol{x}(t), t) = \big\langle \boldsymbol{L}\boldsymbol{x}(t) + (t + 1)\xi(t) + \gamma(t)\eta(t), \dot{\boldsymbol{x}}(t) \big\rangle + \dot{\gamma}(t)\big(f(\boldsymbol{x}(t)) - m_f\big) \tag{14}$$

Combining (12) with (14), we have

$$\frac{\mathrm{d}}{\mathrm{d}t}V(\boldsymbol{x}(t), t) = -\|\dot{\boldsymbol{x}}(t)\|^2 + \dot{\gamma}(t)\big(f(\boldsymbol{x}(t)) - m_f\big) \leq -\|\dot{\boldsymbol{x}}(t)\|^2 \leq 0$$

which means that $V(\boldsymbol{x}(t), t)$ is nonincreasing. Since $V(\boldsymbol{x}(t), t) \geq 0$, we get $\lim\limits_{t \to +\infty} \frac{1}{2}\boldsymbol{x}(t)^{\mathrm{T}}\boldsymbol{L}\boldsymbol{x}(t) = \lim\limits_{t \to +\infty} W(\boldsymbol{x}(t), t)$ exists.

Next, we will prove that $\lim\limits_{t \to +\infty} \frac{1}{2}\boldsymbol{x}(t)^{\mathrm{T}}\boldsymbol{L}\boldsymbol{x}(t) = 0$. If not, suppose that there is $t > T_1$ satisfying $\frac{1}{2}\boldsymbol{x}(t)^{\mathrm{T}}\boldsymbol{L}\boldsymbol{x}(t) \geq \frac{\varepsilon}{2}$. Let $\mathcal{X}$ be the optimal solution set of (3), consider the following function

$$V_1(\boldsymbol{x}, t) = \frac{1}{2}\|\boldsymbol{x} - P_{\mathcal{X}}(\boldsymbol{x})\|^2 + V(\boldsymbol{x}, t)$$

Differentiating $V_1(\boldsymbol{x}, t)$ along the state solution of (4), for $a.e.\ t \geq 0$, one obtains

$$\begin{aligned}
\frac{\mathrm{d}}{\mathrm{d}t}V_1(\boldsymbol{x}(t), t) &= \big\langle \boldsymbol{x}(t) - P_{\mathcal{X}}(\boldsymbol{x}(t)) + \boldsymbol{L}\boldsymbol{x}(t) + (t+1)\xi(t) + \gamma(t)\eta(t), \dot{\boldsymbol{x}}(t) \big\rangle \\
&\quad + \dot{\gamma}(t)\big(f(\boldsymbol{x}(t)) - m_f\big) + G(\boldsymbol{x}(t)) \\
&\leq \big\langle \boldsymbol{L}\boldsymbol{x}(t) + (t+1)\xi(t) + \gamma(t)\eta(t), \dot{\boldsymbol{x}}(t) \big\rangle + \big\langle \dot{\boldsymbol{x}}(t), \boldsymbol{x}(t) - P_{\mathcal{X}}(\boldsymbol{x}(t)) \big\rangle \\
&= -\|\dot{\boldsymbol{x}}(t)\|^2 - \gamma(t)\big\langle \eta(t), \boldsymbol{x}(t) - P_{\mathcal{X}}(\boldsymbol{x}(t)) \big\rangle \\
&\quad + \big\langle \boldsymbol{L}\boldsymbol{x}(t) + (t+1)\xi(t), \boldsymbol{x}(t) - P_{\mathcal{X}}(\boldsymbol{x}(t)) \big\rangle \\
&\leq -\gamma(t)\big\langle \eta(t), \boldsymbol{x}(t) - P_{\mathcal{X}}(\boldsymbol{x}(t)) \big\rangle - \frac{1}{2}\boldsymbol{x}(t)^{\mathrm{T}}\boldsymbol{L}\boldsymbol{x}(t)
\end{aligned}$$

Based on $\lim\limits_{t \to +\infty} \gamma(t) = 0$, we have $\lim\limits_{t \to +\infty} \gamma(t)\big\langle \eta(t), \boldsymbol{x}(t) - P_{\mathcal{X}}(\boldsymbol{x}(t)) \big\rangle = 0$. That is, there exists $T_2 > T_1$, such that $\big|\gamma(t)\big\langle \eta(t), \boldsymbol{x}(t) - P_{\mathcal{X}}(\boldsymbol{x}(t)) \big\rangle\big| \leq \frac{\varepsilon}{4}$, $\forall t > T_2$. Hence

$$\frac{\mathrm{d}}{\mathrm{d}t}V_1(\boldsymbol{x}(t), t) \leq \frac{\varepsilon}{4} - \frac{\varepsilon}{2} = -\frac{\varepsilon}{4}, \ \forall t \in [T_2, +\infty) \tag{15}$$

Taking the integral of (15) from T_2 to t, we obtain

$$V_1(\boldsymbol{x}(t), t) - V_1(\boldsymbol{x}(T_2), T_2) \leq -\frac{\varepsilon}{4}(t - T_2), \ \forall t \in [T_2, +\infty)$$

As $t \to \infty$, $V_1(\boldsymbol{x}(t), t) \to -\infty$, which leads to the contradiction with $V_1(\boldsymbol{x}(t), t) > 0$. Then, we have $\lim\limits_{t \to +\infty} \frac{1}{2}\boldsymbol{x}(t)^{\mathrm{T}}\boldsymbol{L}\boldsymbol{x}(t) = 0$. Therefore, $\lim\limits_{t \to +\infty} \boldsymbol{L}\boldsymbol{x}(t) = 0$ holds, that is, $\boldsymbol{x}(t)$ converges to the feasible region. It follows that $x^1(t) = \ldots = x^k(t)$ as $t \to \infty$, which means that the state solution reach consensus ultimately.

Theorem 4. *Suppose (i)–(iv) in **Assumptions** are satisfied, the state solution $\boldsymbol{x}(t)$ of neural network (4) with any initial point $\boldsymbol{x}(0) \in \mathbb{R}^{kn}$ converges to the global optimal solution of optimization problem (2) $(i = 1, 2, \ldots, k)$.*

Proof. From Theorem 3, we have

$$\frac{\mathrm{d}}{\mathrm{d}t}V_1(\boldsymbol{x}(t), t) \leq -\gamma(t)\big\langle \eta(t), \boldsymbol{x}(t) - P_{\mathcal{X}}(\boldsymbol{x}(t)) \big\rangle \tag{16}$$

In the following content, we will consider three cases to prove that $\boldsymbol{x}(t)$ converges to the optimal solution set $\mathcal{X}$ of (3).

Case 1: There is $T \geq 0$, such that $f(\boldsymbol{x}(t)) \leq f(\bar{\boldsymbol{x}})$, $\forall t \geq T$, where $\bar{\boldsymbol{x}} \in \mathcal{X}$.

Based on Theorem 3, we know that the state solution $\boldsymbol{x}(t)$ converges to the feasible region of (3). Combined with $f(\boldsymbol{x}(t)) \leq f(\bar{\boldsymbol{x}})$, $\forall t \geq T$, it follows that $\boldsymbol{x}(t)$ converges to $\mathcal{X}$.

Case 2: There is $T \geq 0$, such that $f(\boldsymbol{x}(t)) > f(\bar{\boldsymbol{x}})$, $\forall t \geq T$, where $\bar{\boldsymbol{x}} \in \mathcal{X}$.

By the convexity of f, we have

$$\langle \eta_x(t), \boldsymbol{x}(t) - \bar{\boldsymbol{x}} \rangle \geq f(\boldsymbol{x}(t)) - f(\bar{\boldsymbol{x}}) > 0, \ \forall \eta_x(t) \in \partial f(\boldsymbol{x}(t)) \tag{17}$$

Letting $\eta_x(t) := \eta(t)$ in (17) and combining with (16), one has

$$\frac{\mathrm{d}}{\mathrm{d}t} V_1(\boldsymbol{x}(t), t) \leq -\gamma(t)\big(f(\boldsymbol{x}(t)) - f(P_{\mathcal{X}}(\boldsymbol{x}(t)))\big) = -\gamma(t)\big(f(\boldsymbol{x}(t)) - f(\bar{\boldsymbol{x}})\big) < 0$$

which means that $\lim_{t\to\infty} V_1(\boldsymbol{x}(t), t)$ exists. Then, we know $\lim_{t\to\infty} \|\boldsymbol{x}(t) - P_{\mathcal{X}}(\boldsymbol{x}(t))\|$ exists. By (17), we get $\liminf_{t\to\infty} \langle \eta(t), \boldsymbol{x}(t) - P_{\mathcal{X}}(\boldsymbol{x}(t)) \rangle \geq 0$.

Next, we will prove that

$$\liminf_{t\to\infty} \langle \eta(t), \boldsymbol{x}(t) - P_{\mathcal{X}}(\boldsymbol{x}(t)) \rangle = 0 \tag{18}$$

If not, assume that $\liminf_{t\to\infty} \langle \eta(t), \boldsymbol{x}(t) - P_{\mathcal{X}}(\boldsymbol{x}(t)) \rangle = \varepsilon > 0$. Then, there exists $T_3 > T$ satisfying $\langle \eta(t), \boldsymbol{x}(t) - P_{\mathcal{X}}(\boldsymbol{x}(t)) \rangle \geq \frac{\varepsilon}{2}, \ \forall t > T_3$. By (16), one has

$$\frac{\mathrm{d}}{\mathrm{d}t} V_1(\boldsymbol{x}(t), t) \leq -\frac{\varepsilon}{2} \gamma(t), \ \forall t > T_3 \tag{19}$$

By integrating from T_3 to t, we have

$$V_1(\boldsymbol{x}(t), t) - V_1(\boldsymbol{x}(T_3), T_3) \leq -\int_{T_3}^{t} \frac{\varepsilon}{2} \gamma(s)\mathrm{d}s$$

Base on the definition of $\gamma(t)$, we get $\lim_{t\to\infty} V_1(\boldsymbol{x}(t), t) = -\infty$, which contradicts with $V_1(\boldsymbol{x}(t), t) \geq 0$. Therefore, (18) holds. It follows that there exists a subsequence $\{t_n\}$, such that $\lim_{t\to\infty} t_n = \infty$ and $\lim_{t\to\infty} \langle \eta(t_n), \boldsymbol{x}(t_n) - P_{\mathcal{X}}(\boldsymbol{x}(t_n)) \rangle = 0$.

From the boundedness of $\{\boldsymbol{x}(t_n)\}$, there is a subsequence $\{t_{n_k}\} \subseteq \{t_n\}$, such that $\boldsymbol{x}(t_{n_k}) \to \hat{\boldsymbol{x}}$ as $k \to \infty$. Owing to the upper semi-continuity of ∂f and the continuity of projection operator, there exists $\hat{\eta} \in \partial f(\hat{\boldsymbol{x}})$ satisfying $\langle \hat{\eta}, \hat{\boldsymbol{x}} - P_{\mathcal{X}}(\hat{\boldsymbol{x}}) \rangle = 0$. By the convexity of f, we have

$$f(\hat{\boldsymbol{x}}) - f(P_{\mathcal{X}}(\hat{\boldsymbol{x}})) \leq \langle \hat{\eta}, \hat{\boldsymbol{x}} - P_{\mathcal{X}}(\hat{\boldsymbol{x}}) \rangle = 0$$

By Theorem 3, we know $\hat{\boldsymbol{x}} \in \mathcal{X}$. That is, $\lim_{t\to\infty} \|\boldsymbol{x}(t_{n_k}) - P_{\mathcal{X}}(\boldsymbol{x}(t_{n_k}))\| = 0$. Since $\lim_{t\to\infty} \|\boldsymbol{x}(t_n) - P_{\mathcal{X}}(\boldsymbol{x}(t_n))\|$ exists, we have $\lim_{t\to\infty} \|\boldsymbol{x}(t_n) - P_{\mathcal{X}}(\boldsymbol{x}(t_n))\| = 0$. Hence, for case 2, $\boldsymbol{x}(t)$ converges to $\mathcal{X}$.

Case 3: Both $\Delta_1 = \{t \in [0, +\infty) : f(\boldsymbol{x}(t)) \leq f(\bar{\boldsymbol{x}})\}$ and $\Delta_2 = \{t \in [0, +\infty) : f(\boldsymbol{x}(t)) > f(\bar{\boldsymbol{x}})\}$ are unbounded.

Since Δ_2 is an open set, we have $\Delta_2 = \bigcup_{k\in\mathbb{N}}(t_{2k+1}, t_{2k+2})$. Therefore, $\Delta_1 = [0, +\infty)\backslash\Delta_2 = \bigcup_{k\in\mathbb{N}}[t_{2k}, t_{2k+1}]$.

For any $t \in [t_{2k}, t_{2k+1}] \subseteq \Delta_1$, after a similar analysis in case 1, we get

$$\lim_{\substack{t\to\infty \\ t\in\Delta_1}} \|\boldsymbol{x}(t) - P_{\mathcal{X}}(\boldsymbol{x}(t))\| = 0$$

For any $t \in (t_{2k+1}, t_{2k+2}) \subseteq \Delta_2$, we set $\rho(t) = \sup_{\substack{s \leq t \\ s \in [t_{2k}, t_{2k+1}]}} s$. Then, $\rho(t) \in [t_{2k}, t_{2k+1}]$. Since the unboundedness of Δ_2, it follows that $\rho(t) \to \infty$ as $t \to \infty$. Since $V_1(\boldsymbol{x}(t), t)$ is nonincreasing, it follows

$$\limsup_{t \to \infty} V_1(\boldsymbol{x}(t), t) \leq \limsup_{t \to \infty} V_1(\boldsymbol{x}(\rho(t)), \rho(t)) = 0$$

Then $\limsup_{\substack{t \to \infty \\ t \in \Delta_2}} V_1(\boldsymbol{x}(t), t) = 0$, which means that $\lim_{\substack{t \to \infty \\ t \in \Delta_2}} \|\boldsymbol{x}(t) - P_{\mathcal{X}}(\boldsymbol{x}(t))\| = 0$.

Therefore, for case 3, $\boldsymbol{x}(t)$ converges to the global optimal solution set $\mathcal{X}$.

Finally, from Lemma 2, we know $\boldsymbol{x}(t)$ also converges to the global optimal solution set of problem (2). According to (ii) in **Assumptions**, we know f is strictly convex, which has a unique minimizer. Hence, the state solution $\boldsymbol{x}(t)$ converges to the global optimal solution of problem (2).

5 Numerical Examples

Example 1. Consider the following distributed optimization problem:

$$\min \sum_{i=1}^{4} f_i(x) \tag{20}$$
$$\text{s.t.} \quad g_i(x) \leq \mathbf{0}, \ i \in \{1, 2, 3, 4\}$$

where $x = (x_1, x_2)^{\mathrm{T}}$, $f_1(x) = |x_1|$, $f_2(x) = |x_1 + x_2|$, $f_3(x) = \frac{1}{2}x_1^2 + \frac{1}{2}x_2^2$, $f_4(x) = x_1^2 + x_2$, $g_1(x) = x_1^2 + (x_2 - 2)^2 - 4$, $g_2(x) = x_1 - \frac{1}{2}$, $g_3(x) = (x_1 - 1)^2 + x_2^2 - 1$, $g_4(x) = -x_2$.

The communication graph $\mathcal{G}$ is shown in Fig. 1, where we can know local constraints of each agent. We can verify that (i)-(iv) in **Assumptions** hold. Hence, we can apply neural network (4) to solve problem (20). Figure 2 displays the state solution of the neural network (4). It shows the state solution reaches consensus and converges to the optimal solution $\bar{x} = (0.471, 0.055)^{\mathrm{T}}$.

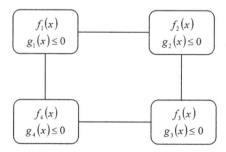

Fig. 1. Communication graph with 4 agents in Example 1.

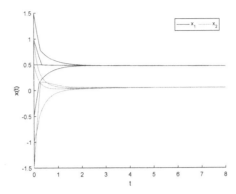

Fig. 2. The state trajectories of neural network (4) in Example 1.

6   Conclusions

For a distributed convex optimization problem with inequality constraint, this paper presents a neural network over multiagent systems. Then it is proved that the state solution of the neural network is bounded and reaches a consensus output at the global optimal solution. We also give a numerical example to show the efficiency of the proposed neural network.

References

1. Aubin, J.P., Cellina, A.: Differential Inclusions: Set-valued Maps and Viability Theory. Springer-Verlag, Berlin (1984)
2. Clarke, F.H.: Optimization and Nonsmooth Analysis. SIAM, Philadelphia (1990)
3. Deng, Z., Liang, S., Hong, Y.: Distributed continuous-time algorithms for resource allocation problems over weight-balanced digraphs. IEEE Trans. Cybern. **48**(11), 3116–3125 (2018)
4. Gharesifard, B., Corts, J.: Distributed continuous-time convex optimization on weight-balanced digraphs. IEEE Trans. Autom. Control **59**(3), 781–786 (2014)
5. Gravina, R., Alinia, P., Ghasemzadeh, H., Fortino, G.: Multi-sensor fusion in body sensor networks: state-of-the-art and research challenges. Inf. Fusion **35**, 68–80 (2016)
6. Hopefiled, J., Tank, D.: Neural computation of decisions in optimization problems. Biol. Cybern. **52**(3), 141–152 (1985)
7. Jiang, X., Qin, S., Xue, X.: A penalty-like neurodynamic approach to constrained nonsmooth distributed convex optimization. Neurocomputing **377**(15), 225–233 (2020)
8. Kennedy, M., Chua, L.: Neural networks for nonlinear programming. IEEE Trans. Circ. Syst. **35**(5), 554–562 (1988)
9. Li, Q., Liu, Y., Zhu, L.: Neural network for nonsmooth pseudoconvex optimization with general constraints. Neurocomputing **131**, 336–347 (2014)
10. Liu, N., Qin, S.: A neurodynamic approach to nonlinear optimization problems with affine equality and convex inequality constraints. Neural Netw. **109**, 147–158 (2019)

11. Liu, N., Qin, S.: A novel neurodynamic approach to constrained complex-variable pseudoconvex optimization. IEEE Trans. Cybern. **49**(11), 3946–3956 (2019). (Regular paper)
12. Liu, Q., Yang, S., Wang, J.: A collective neurodynamic approach to distributed constrained optimization. IEEE Trans. Neural Netw. Learn. Syst. **28**(8), 1747–1758 (2017)
13. Qin, S., Xue, X.: A two-layer recurrent neural network for non-smooth convex optimization problems. IEEE Trans. Neural Netw. Learn. Syst. **26**(6), 1149–1160 (2015). (Regular paper)
14. Qin, S., Yang, X., Xue, X., Song, J.: A one-layer recurrent neural network for pseudoconvex optimization problems with equality and inequality constraints. IEEE Trans. Cybern. **47**(10), 3063–3074 (2017)
15. Wu, Y., Sun, X., Zhao, X., Shen, T.: Optimal control of Boolean control networks with average cost: a policy iteration approach. Automatica **100**, 378–387 (2019)
16. Xue, X., Bian, W.: Subgradient-based neural networks for nonsmooth convex optimization problems. IEEE Trans. Circ. Syst. **55**(8), 2378–2391 (2008)

Dynamically Weighted Model Predictive Control of Affine Nonlinear Systems Based on Two-Timescale Neurodynamic Optimization

Jiasen Wang[1,3(✉)], Jun Wang[1,2,3(✉)], and Dongbin Zhao[4]

[1] Department of Computer Science, City University of Hong Kong,
Kowloon, Hong Kong
jiasewang2-c@my.cityu.edu.hk, jwang.cs@cityu.edu.hk
[2] School of Data Science, City University of Hong Kong, Kowloon, Hong Kong
[3] Shenzhen Research Institute, City University of Hong Kong, Shenzhen, China
[4] Institute of Automation, Chinese Academy of Sciences, Beijing, China
dongbin.zhao@ia.ac.cn

Abstract. This paper discusses dynamically weighted model predictive control based on two-timescale neurodynamic optimization. Minimax optimization problems with dynamic weights in objective functions are used in the model predictive control. The minimax optimization problems are solved by using a two-timescale neurodynamic optimization approach. Examples on controlling HVAC (heating, ventilation, and air-conditioning) and CSTR (cooling continuous stirred tank reactor) systems are elaborated to substantiate the efficacy of the control approach.

Keywords: Model predictive control · Neurodynamic optimization · HVAC · CSTR

1 Introduction

Model predictive control (MPC), also known as receding horizon control, refers to control via solving rolling horizon optimization problems with predictions on finite horizons. It is an optimal control approach with online adaptation and prediction capability. It is widely used in process control [18], vehicle control [4], CSTR control [42], and HVAC control [1].

A key step in MPC is solving optimization problems with various constraints effectively and efficiently. Neurodynamic approaches based on recurrent neural networks are suitable for optimization. For example, many neurodynamic models are available for solving various optimization problems [2, 5, 7, 8, 12, 12–17, 23,

This work was supported in part by the National Natural Science Foundation of China under Grant 61673330, by International Partnership Program of Chinese Academy of Sciences under Grant GJHZ1849, and the Research Grants Council of the Hong Kong Special Administrative Region of China, under Grants 11208517 and 11202318.

© Springer Nature Switzerland AG 2020
M. Han et al. (Eds.): ISNN 2020, LNCS 12557, pp. 96–105, 2020.
https://doi.org/10.1007/978-3-030-64221-1_9

27, 29–35, 37, 41, 45]. In addition, many neurodynamics-based control approaches are available [3, 6, 9, 10, 19–22, 24, 36, 38–40, 42–44, 46]. For example, a nonlinear MPC method based on neurodynamic optimization is proposed in [44]. A robust nonlinear MPC approach based on collective neurodynamic optimization is presented in [43].

In this paper, MPC of affine systems is formulated as sequential minimax optimization problems with dynamic weights in the objective function. A two-timescale neurodynamic approach is used for solving the minimax optimization problem sequentially. Simulation results on HVAC and CSTR systems are reported to substantiate the efficacy of the neurodynamics-based dynamically-weighted MPC approach.

2 Neurodynamics-Based MPC

2.1 Problem Formulation

Consider a discrete-time affine system:

$$x(k + 1) = a(x(k)) + b(x(k))u(k), \tag{1}$$

where k is a time step, x and u are system state and input vectors, respectively, a and b are differentiable functions. The control objective is to regulate $x(k)$ such that it tracks a given target x_d by using the system input $u(k)$.

To achieve the control objective, MPC of system (1) is formulated as a sequential minimax optimization problem as follows for $k = 0, 1, \dots$:

$$\min_{u(\cdot|k)} \max_{\lambda \in [0,1]} \lambda \sum_{t=1}^{T} \|x(t + k|k) - x_d\|_Q^2 + (1 - \lambda) \sum_{t=0}^{T-1} \|\Delta u(t + k|k)\|_R^2 \tag{2a}$$

$$\text{s.t. } x(t + k + 1|k) = a(x(t + k|k - 1))$$
$$+ b(x(t + k|k - 1))u(t + k|k), t = 0, ..., T - 1 \tag{2b}$$

$$\underline{u} \le u(t + k|k) \le \bar{u}, \ t = 0, ..., T - 1 \tag{2c}$$

$$\underline{x} \le x(t + k|k) \le \bar{x}, \ t = 1, ..., T \tag{2d}$$

$$x(k|k - 1) = x(k), \tag{2e}$$

where $u(t|k)$ and $x(t|k)$ denote respectively the predictions of u and x at step t with known information at step k, $x(k)$ is the state at step k, $\Delta u(t + k|k) = u(t + k|k) - u(t - 1 + k|k)$, T is the prediction horizon, Q and R are positive definite matrices, $\|x\|_Q^2 = x^T Q x$, $\underline{u}$, $\bar{u}$, $\underline{x}$, and $\bar{x}$ are bound vectors satisfying $\underline{u} \le \bar{u}$ and $\underline{x} \le \bar{x}$, λ is a dynamic weight.

The objective function in (2a) is a dynamically weighted sum of the tracking errors and the variations of control variables with a dynamic weight λ. Equation (2b) is the system dynamic equation in a predicted form same as that in [39]. Equations (2c) and (2d) are control and state constraints, respectively. Equation (2e) is the initial state condition of the system.

As usual in MPC, $u(k) = u^*(k|k)$ is used to control system (1) at time k ($k = 0, 1, 2, ...$).

2.2   Two-Timescale Neurodynamic Optimization

Let $\mathbf{u} = [u(k|k)^T, u(k+1|k)^T, ..., u(k+T-1|k)^T]^T$, $J(\mathbf{u}, \lambda) = \lambda \sum_{t=1}^{T} \|x(t + k|k) - x_d\|_Q^2 + (1-\lambda) \sum_{t=1}^{T-1} \|\Delta u(t+k|k)\|_R^2$, $\underline{\mathbf{u}} = \mathrm{col}(\underline{u}, ..., \underline{u})$, $\bar{\mathbf{u}} = \mathrm{col}(\bar{u}, ..., \bar{u})$, and

$$g(\mathbf{u}) = \begin{bmatrix} a(x(k|k-1) + b(x(k|k-1))u(k|k) - \bar{x} \\ \underline{x} - a(x(k|k-1) - b(x(k|k-1))u(k|k) \\ a(x(k+1|k-1) + b(x(k+1|k-1))u(k+1|k) - \bar{x} \\ \underline{x} - a(x(k+1|k-1) - b(x(k+1|k-1))u(k+1|k) \\ \cdots \\ a(x(k+T-1|k-1) + b(x(k+T-1|k-1))u(k+T-1|k) - \bar{x} \\ \underline{x} - a(x(k+T-1|k-1) - b(x(k+T-1|k-1))u(k+T-1|k) \end{bmatrix}.$$

Problem (2) is rewritten as follows:

$$\min_{\mathbf{u}} \max_{\lambda} \ J(\mathbf{u}, \lambda)$$
$$\text{s.t. } g(\mathbf{u}) \leq 0 \tag{3}$$
$$\mathbf{u} \in [\underline{\mathbf{u}}, \bar{\mathbf{u}}], \lambda \in [0,1],$$

where $\mathbf{u}$ and λ are decision variables.

A few neurodynamic models can be used for solving problem (3). In particular, a two-timescale neurodynamic model is proposed in [11] with the following dynamic equation:

$$\epsilon_1 \frac{d\mathbf{u}}{dt} = -\mathbf{u} + P_U(\mathbf{u} - (\nabla_{\mathbf{u}} J(\mathbf{u}, \lambda) + \nabla_{\mathbf{u}} g(\mathbf{u})\mu)$$
$$\epsilon_2 \frac{d\lambda}{dt} = -\lambda + P_\Lambda(\lambda - \nabla_\lambda J(\mathbf{u}, \lambda)) \tag{4}$$
$$\epsilon_2 \frac{d\mu}{dt} = -\mu + (\mu + g(\mathbf{u}))^+,$$

where ϵ_1 and ϵ_2 are time constants, $\mathbf{u}$, λ, and μ are neuronal states, $\nabla_{\mathbf{u}} g(\mathbf{u}) = [\nabla_{\mathbf{u}} g_1(\mathbf{u}), \nabla_{\mathbf{u}} g_2(\mathbf{u}), ...]$, $\mu_i^+ = \max(\mu_i, 0)$, $U = [\underline{\mathbf{u}}, \bar{\mathbf{u}}]$, $\Lambda = [0,1]$,

$$P_U(\mathbf{u}_i) = \begin{cases} \bar{u}_i & u_i > \bar{u}_i, \\ u_i & \underline{u}_i \leq u_i \leq \bar{u}_i, \\ \underline{u}_i & u_i < \underline{u}_i, \end{cases} \qquad P_\Lambda(\lambda) = \begin{cases} 1 & \lambda > 1, \\ \lambda & 0 \leq \lambda \leq 1, \\ 0 & \lambda < 0. \end{cases}$$

3   Simulation Results

3.1   Neurodynamics-Based HVAC Control

Consider a discrete-time HVAC system in (1) with its dynamic equation being defined as follows [25, 28]:

$$a(x) = \begin{bmatrix} x_1 \\ x_2 + \frac{\tau}{V_{ts}} \frac{q_l}{\rho c_p} \end{bmatrix}, \quad b(x) = \begin{bmatrix} \frac{\tau}{V_{he}\rho c_p} & \frac{\tau}{V_{he}}(x_2 - x_1) & \frac{\tau}{V_{he}}(T_0 - x_2) \\ 0 & \frac{\tau}{V_{ts}}(x_1 - x_2) & 0 \end{bmatrix},$$

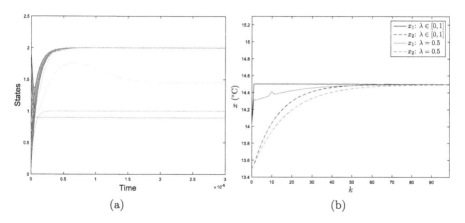

(a)   (b)

Fig. 1. Neurodynamics-based HVAC control: (a) a snapshot of transient neuronal states of the neurodynamic model with $\epsilon_1 = 10^{-6}$ and $\epsilon_2 = 0.5 \times 10^{-6}$ at $k = 0$; (b) HVAC states with $T = 20$. (Color figure online)

where x_1 and x_2 are temperature states, u_1, u_2, and u_3 are control variables, $1/\tau$ is a sampling frequency, V_{he}, T_0, V_{ts}, q_l, ρ, and c_p are HVAC system parameters.

The HVAC system parameters in [25] are adopted. $Q = R = I$, $T = 20$, $\bar{x} = [14.5, 30]$, $\underline{x} = [5, 10]$, $\bar{u} = [4000, 2, 2]$, $\underline{u} = [-4000, 0.0354, 0.0354]$. Same as [28], the sampling rate is set to 0.1Hz. $x_d = [14.5, 14.5]^T$, $x(0) = [14, 13.5]^T$.

Figures 1, 2, 3 and 4 depict the simulation results on the HVAC system. Specifically, Fig. 1(a) depicts a snapshot of transient neuronal states of the two-timescale neurodynamic model at $k = 0$, where the green line depicts the transient behavior of λ converging to one. Figure 1(b) depicts the controlled HVAC system states for dynamic and fixed weights. It shows that x_1 and x_2 are regulated to the reference temperature 14.5°C. In addition, it shows that MPC with dynamic weights reaches steady state faster than that with fixed weight. Figure 2 depicts the computed control inputs to the HVAC system for dynamic and fixed weights. Figure 3(a)–(b) depict the states of the neurodynamics-based HVAC MPC system with prediction horizons $T = 10$ and 100, where little differences appear in the state variables. Figure 4(c)–(d) depict respectively λ in HVAC control system with $T = 20$, $T = 5$ and $T = 100$.

3.2   Neurodynamics-Based CSTR Control

Consider a discrete-time CSTR system with a and b in (1) being defined as follows [26, 42]:

$$a(x) = \begin{bmatrix} x_1 + \tau[-x_1 + D_a(1 - x_1)\exp(\frac{x_2}{1+0.05x_2})] \\ x_2 + \tau[-x_2 + BD_a(1 - x_1)\exp(\frac{x_2}{1+0.05x_2}) - 0.3x_2] \end{bmatrix}, \quad b(x) = \begin{bmatrix} 0 \\ 0.3\tau \end{bmatrix},$$

where x_1 and x_2 are the reactant concentration and reactor temperature, respectively, B and D_a are CSTR system parameters.

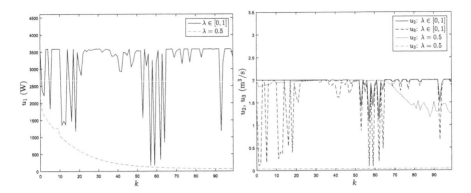

Fig. 2. Control inputs to the HVAC system with $T = 20$.

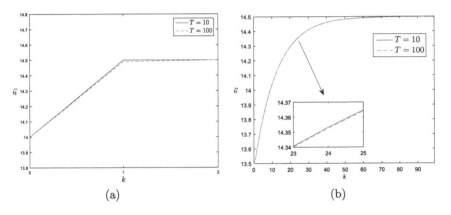

(a)   (b)

Fig. 3. States of the HVAC control system with prediction horizons $T = 10$ and 100: (a) x_1, (b) x_2.

Same as [42], $B = 1$, $D_a = 0.072$, $x_d = [0.4472, 2.7520]^T$, the sampling rate is set to 10Hz. $Q = R = I$, $T = 20$, $\bar{x} = [0.85, 5]$, $\underline{x} = [0, 0]$, $\bar{u} = 18$, $\underline{u} = 7$. The initial state is $x(0) = [0, 0]^T$.

Figures 5, 6 and 7 depict the simulation results on the CSTR system. Specifically, Fig. 5(a) depicts a snapshot of neuronal states of the two-timescale neurodynamic model, where the red line depicts the transient behaviors of λ converging to one. Figure 5(b) depicts the dynamic weights and fixed weight in the CSTR control. Figure 6(a) depicts the controlled CSTR states for dynamic and fixed weights, which shows that states x_1 and x_2 are regulated to the references 0.4472 and 2.7520, respectively. In addition, it shows that dynamically weighted MPC CSTR system reaches steady state faster than that with fixed weight. Figure 6(b) depicts the control inputs to the CSTR MPC system for dynamic and fixed weights. Figure 7 depicts the states and weights in the neurodynamics-based CSTR MPC system with predction horizons of 10 and 50, where a little improvements appear in the state variables with a shorter prediction horizon.

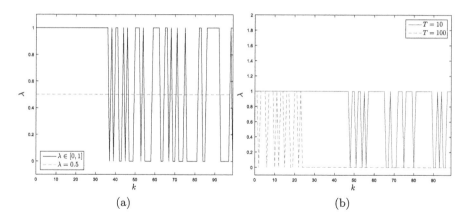

Fig. 4. Dynamic weight λ in the HVAC control system: (a) prediction horizon $T = 20$, (b) $T = 10$ and 100.

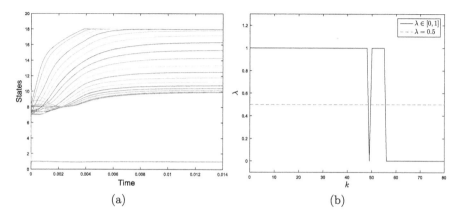

Fig. 5. Neurodynamics-based in CSTR control: (a) A snapshot of transient neuronal states of the neurodynamic model with $\epsilon_1 = 10^{-6}$ and $\epsilon_2 = 0.5 \times 10^{-6}$ at $k = 0$, (b) λ.

Simulation results on a two-mass-spring linear system [11] show that the dynamically weighted MPC approach achieves comparable performances compared to that with fixed weight. However, no obvious advantage is observed on controlling the two-mass-spring system.

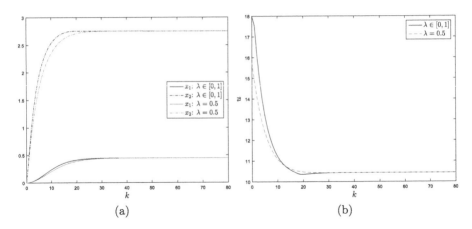

Fig. 6. (a) System states of CSTR with $T = 20$. (b) Control inputs to the CSTR system with $T = 20$.

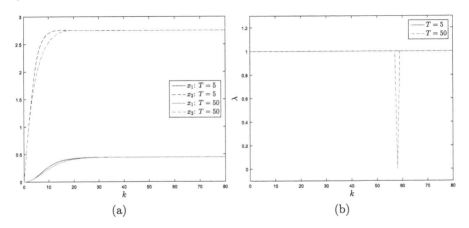

Fig. 7. States and dynamic weights in neurodynamics-based CSTR control with prediction horizons $T = 5$ and 50: (a) x; (b) λ.

4   Conclusions

In this paper, a two-timescale neurodynamic model is used for dynamically weighted model predictive control (MPC) of affine nonlinear systems. The MPC problem is formulated as a sequential minimax optimization problem, and solved by using a two-timescale neurodynamic model. Simulation results on setpoint regulation for HVAC and CSTR MPC systems are presented to substantiate the efficacy of the neurodynamics-based MPC approach.

References

1. Aswani, A., Gonzalez, H., Sastry, S.S., Tomlin, C.: Provably safe and robust learning-based model predictive control. Automatica **49**(5), 1216–1226 (2013)
2. Che, H., Wang, J.: A collaborative neurodynamic approach to global and combinatorial optimization. Neural Netw. **114**, 15–27 (2019)
3. Ding, H., Wang, J.: Recurrent neural networks for minimum infinity-norm kinematic control of redundant manipulators. IEEE Trans. Syst. Man Cybern. - Part A **29**(3), 269–276 (1999)
4. Dunbar, W.B., Murray, R.M.: Distributed receding horizon control for multi-vehicle formation stabilization. Automatica **42**(4), 549–558 (2006)
5. Guo, Z., Liu, Q., Wang, J.: A one-layer recurrent neural network for pseudoconvex optimization subject to linear equality constraints. IEEE Trans. Neural Netw. **22**(12), 1892–1900 (2011)
6. Han, M., Fan, J., Wang, J.: A dynamic feedforward neural network based on Gaussian particle swarm optimization and its application for predictive control. IEEE Trans. Neural Netw. **22**(9), 1457–1468 (2011)
7. Hu, X., Wang, J.: Design of general projection neural networks for solving monotone linear variational inequalities and linear and quadratic optimization problems. IEEE Trans. Syst. Man Cybern. - Part B: Cybern. **37**(5), 1414–1421 (2007)
8. Hu, X., Wang, J.: An improved dual neural network for solving a class of quadratic programming problems and its k-winners-take-all application. IEEE Trans. Neural Netw. **19**(12), 2022–2031 (2008)
9. Le, X., Wang, J.: Robust pole assignment for synthesizing feedback control systems using recurrent neural networks. IEEE Trans. Neural Netw. Learn. Syst. **25**(2), 383–393 (2014)
10. Le, X., Wang, J.: Neurodynamics-based robust pole assignment for high-order descriptor systems. IEEE Trans. Neural Netw. Learn. Syst. **26**(11), 2962–2971 (2015)
11. Le, X., Wang, J.: A two-time-scale neurodynamic approach to constrained minimax optimization. IEEE Trans. Neural Netw. Learn. Syst. **28**(3), 620–629 (2017)
12. Li, G., Yan, Z., Wang, J.: A one-layer recurrent neural network for constrained nonconvex optimization. Neural Netw. **61**, 10–21 (2015)
13. Liang, X., Wang, J.: A recurrent neural network for nonlinear optimization with a continuously differentiable objective function and bound constraints. IEEE Trans. Neural Netw. **11**(6), 1251–1262 (2000)
14. Liu, Q., Wang, J.: A one-layer recurrent neural network with a discontinuous hard-limiting activation function for quadratic programming. IEEE Trans. Neural Netw. **19**(4), 558–570 (2008)
15. Liu, Q., Wang, J.: A one-layer recurrent neural network for constrained nonsmooth optimization. IEEE Trans. Syst. Man Cybern. - Part B: Cybern. **40**(5), 1323–1333 (2011)
16. Liu, Q., Yang, S., Wang, J.: A collective neurodynamic approach to distributed constrained optimization. IEEE Trans. Neural Netw. Learn. Syst. **28**(8), 1747–1758 (2017)
17. Liu, S., Wang, J.: A simplified dual neural network for quadratic programming with its KWTA application. IEEE Trans. Neural Netw. **17**(6), 1500–1510 (2006)
18. Mayne, D., Rawlings, J., Rao, C., Scokaert, P.: Constrained model predictive control: stability and optimality. Automatica **36**(6), 789–814 (2000)

19. Pan, Y., Wang, J.: Model predictive control for nonlinear affine systems based on the simplified dual neural network. In: Proceedings of IEEE International Symposium on Intelligent Control, pp. 683–688. IEEE (2009)
20. Pan, Y., Wang, J.: Model predictive control of unknown nonlinear dynamical systems based on recurrent neural networks. IEEE Trans. Ind. Electron. **59**(8), 3089–3101 (2012)
21. Peng, Z., Wang, J., Han, Q.: Path-following control of autonomous underwater vehicles subject to velocity and input constraints via neurodynamic optimization. IEEE Trans. Ind. Electron. **66**(11), 8724–8732 (2019)
22. Peng, Z., Wang, D., Wang, J.: Predictor-based neural dynamic surface control for uncertain nonlinear systems in strict-feedback form. IEEE Trans. Neural Netw. Learn. Syst. **28**(9), 2156–2167 (2017)
23. Qin, S., Le, X., Wang, J.: A neurodynamic optimization approach to bilevel quadratic programming. IEEE Trans. Neural Netw. Learn. Syst. **28**(11), 2580–2591 (2017)
24. Tang, W.S., Wang, J.: A recurrent neural network for minimum infinity-norm kinematic control of redundant manipulators with an improved problem formulation and reduced architectural complexity. IEEE Trans. Syst. Man Cybern. - Part B **31**(1), 98–105 (2001)
25. Teeter, J., Chow, M.Y.: Application of functional link neural network to HVAC thermal dynamic system identification. IEEE Trans. Ind. Electron. **45**(1), 170–176 (1998)
26. Uppal, A., Ray, W., Poore, A.: On the dynamic behavior of continuous stirred tanks. Chem. Eng. Sci. **29**, 957–985 (1974)
27. Wang, J.: A deterministic annealing neural network for convex programming. Neural Netw. **7**(4), 629–641 (1994)
28. Wang, J.S., Wang, J., Gu, S.: Neurodynamics-based receding horizon control of an HVAC system. In: International Symposium on Neural Networks, vol. 2, pp. 120–128 (2019)
29. Xia, Y., Feng, G., Wang, J.: A recurrent neural network with exponential convergence for solving convex quadratic program and related linear piecewise equations. Neural Netw. **17**(7), 1003–1015 (2004)
30. Xia, Y., Feng, G., Wang, J.: A novel neural network for solving nonlinear optimization problems with inequality constraints. IEEE Trans. Neural Netw. **19**(8), 1340–1353 (2008)
31. Xia, Y., Leung, H., Wang, J.: A projection neural network and its application to constrained optimization problems. IEEE Trans. Circ. Syst. Part I **49**(4), 447–458 (2002)
32. Xia, Y., Wang, J.: A general methodology for designing globally convergent optimization neural networks. IEEE Trans. Neural Netw. **9**(6), 1331–1343 (1998)
33. Xia, Y., Wang, J.: Global exponential stability of recurrent neural networks for solving optimization and related problems. IEEE Trans. Neural Netw. **11**(4), 1017–1022 (2000)
34. Xia, Y., Wang, J.: A general projection neural network for solving monotone variational inequalities and related optimization problems. IEEE Trans. Neural Netw. **15**(2), 318–328 (2004)
35. Xia, Y., Wang, J.: A recurrent neural network for nonlinear convex optimization subject to nonlinear inequality constraints. IEEE Trans. Circ. Syst. - Part I **51**(7), 1385–1394 (2004)

36. Xia, Y., Wang, J., Fok, L.M.: Grasping force optimization of multi-fingered robotic hands using a recurrent neural network. IEEE Trans. Robot. Autom. **20**(3), 549–554 (2004)
37. Xia, Y., Wang, J.: A bi-projection neural network for solving constrained quadratic optimization problems. IEEE Trans. Neural Netw. Learn. Syst. **27**(2), 214–224 (2016)
38. Yan, Z., Le, X., Wang, J.: Tube-based robust model predictive control of nonlinear systems via collective neurodynamic optimization. IEEE Trans. Ind. Electron. **63**(7), 4377–4386 (2016)
39. Yan, Z., Wang, J.: Model predictive control of nonlinear affine systems based on the general projection neural network and its application to a continuous stirred tank reactor. In: Proceedings of International Conference on Information Science and Technology, pp. 1011–1015. IEEE (2011)
40. Yan, Z., Wang, J.: Model predictive control of nonlinear systems with unmodeled dynamics based on feedforward and recurrent neural networks. IEEE Trans. Ind. Inform. **8**(4), 746–756 (2012)
41. Yan, Z., Fan, J., Wang, J.: A collective neurodynamic approach to constrained global optimization. IEEE Trans. Neural Netw. Learn. Syst. **28**(5), 1206–1215 (2017)
42. Yan, Z., Wang, J.: A neurodynamic approach to bicriteria model predictive control of nonlinear affine systems based on a goal programming formulation. In: Proceedings of International Joint Conference on Neural Networks (IJCNN), pp. 1–7. IEEE (2012)
43. Yan, Z., Wang, J.: Robust model predictive control of nonlinear systems with unmodeled dynamics and bounded uncertainties based on neural networks. IEEE Trans. Neural Netw. Learn. Syst. **25**(3), 457–469 (2014)
44. Yan, Z., Wang, J.: Nonlinear model predictive control based on collective neurodynamic optimization. IEEE Trans. Neural Netw. Learn. Syst. **26**(4), 840–850 (2015)
45. Yan, Z., Wang, J., Li, G.: A collective neurodynamic optimization approach to bound-constrained nonconvex optimization. Neural Netw. **55**, 20–29 (2014)
46. Zhang, Y., Wang, J.: Recurrent neural networks for nonlinear output regulation. Automatica **37**(8), 1161–1173 (2001)

An Efficient Method of Advertising on Online Social Networks

Xitao Zou[1], Huan Liu[1], Xiangguang Dai[1], Jiang Xiong[1], and Nian Zhang[2(✉)]

[1] Key Laboratory of Intelligent Information Processing and Control of Chongqing Municipal Institutions of Higher Education, Chongqing Three Gorges University, Chongqing 40044, China
xiaotao1009@sina.cn, 429321043@qq.com, daixiangguang@163.com, xjcq123@126.com
[2] Department of Electrical and Computer Engineering, University of the District of Columbia, Washington, D.C. 20008, USA
nzhang@udc.edu

Abstract. How to advertise in online social networks is a hot and open research topic. In short, the main goal is to post the advertisement on a few of most influential users' profiles to spread the advertising information to the potential suitable recipients. Typical research works about this topic are limitedly involved with two aspects: one is Spreading Maximization problem and the other is Centrality measures. The Spreading Maximization is proven to be an NP-hard problem, which means the corresponding method is mostly inefficient to exactly find the most influential spreaders. Second, traditional centrality measures, such as degree centrality, closeness centrality etc., roughly take the geometric information (degree or distance) to calculate the potentially most influential users, rather than considering online users' personal interests or preference which are more likely to determine the set of people whether read/accept the advertising contents or not. In this paper, we put closed-related labels for each individual's profile in the online social network and assign particularly set-up scores to these attribute labels. Based on these labels, we apply the weighted k-shell decomposition method to identify the core users in the networks, which is also regarded as the most influential users in this paper. The experimental results show that the proposed method is sufficient to identify the most influential users in some artificial networks. More importantly, the proposed method shows good discrimination degree of influence ranking.

Keywords: Advertising · Online social networks · K-Shell decomposition · Susceptible-infected model · Centrality measure

1 Introduction

With the advent of popular Web destinations such as Facebook and Twitter, a new kind of online community now occupy the center stage in e-commerce, that

© Springer Nature Switzerland AG 2020
M. Han et al. (Eds.): ISNN 2020, LNCS 12557, pp. 106–116, 2020.
https://doi.org/10.1007/978-3-030-64221-1_10

is, social networking communities. The rapid growth of online social networking communities has caught the attention of advertisers that hope to find new ways to harness these communities for their advertising purposes. The effectiveness of targeting a small portion of customers for advertising has long been recognized by businesses [1]. Traditional approach to targeted advertising is to analyze a historical database of previous transactions and the features associated with the customers, possibly with the help of some statistical tools and identify a list of customers most likely to respond to the advertisement of the product. With the advent of new online social networks, it is advocated that mining more data to identify potential customers [2]. Hence, many recommender systems, whose basic idea is to advertise products based on users' preferences which can be obtained by ratings either explicitly stated by the users or implicitly inferred from previous transaction records, Web logs, or cookies, have emerged over the past few years. One of the recommendation technique was called the *content-based approach* [3]. The other is *collaborative approach* [4]. Both recommendation systems consider individuals' preference to recommend products to them. However, in this paper, we take another way to advertise on online social networks. First, we focus on how to quickly and efficiently identify potential users who might show interest in the advertising information, and based on these potential users, we want to find out the most influential users who can be considered as initial spreaders of the advertisement in order to save resources such as money and time.

Intuitively, the topic about advertising in online social networks is closely related to the problem of maximizing the spreading of influence through a social network. Both want to maximize the spreading range of information in a short time. In the work [5,6], Domingos and Richardson posed a fundamental algorithmic problem for such systems [5,6]. Suppose that we have data on a social network, with estimates for the extent to which individuals influence one another, and we would like to market a new product that we hope will be adopted by a large fraction of the network. The premise of viral marketing is that by initially targeting a few *influential* members of the network – say, giving them free samples of the product – we can trigger a cascade of influence by which friends will recommend the product to other friends, and many individuals will ultimately try it. But how should we choose the few key individuals to use for seeding this process? In Refs [5,6], this question was considered in a probabilistic model of interaction; heuristics were given for choosing customers with a large overall effect on the network, and methods were also developed to infer the influence data necessary for posing these types of problems. In this paper, we consider the issue of choosing influential sets of individuals as a problem in discrete optimization. The optimal solution is NP-hard for most models that have been studied, including the model of [5]. However, we are fortunate and don't need to solve this NP-hard problem mathematically. Instead, if we can empirically figure out a way to identify several potentially most influential online social network users and then spread the advertising information through them. We can definitely save a lot of time and money. This is one of main purpose why our research works are given in this paper.

On the other hand, how to identify most influential nodes in networks is closely related to the topic of this paper. Various centrality measures have been proposed over the years to capture network individuals meanings – influence [7], importance [8–10], popularity [11], recommendation [12–14], controllability [15], spreading efficiency [16] and so on [17–19], according to their degree and weight strength of each node and topological importance in the network structure [20].

The methods commonly used in binary networks are *Degree centrality* (DC), *Betweenness centrality* (BC) [21] and *Closeness centrality* (CC) [22]. The DC method is very simple but of little relevance, since it does not take into account the global structure of complex network. BC and CC are well-known metrics which can better capture the global structure of a network, but are difficult to be applied in large-scale networks due to the computational complexity. Meanwhile, another limitation of CC is the lack of adjustability to networks with disconnected components: two nodes that belong to different components but do not have a finite distance between them. These three centrality measures have already been extended to be applied in weighted networks [21]. In 2011, Chen et al. [7] proposed a effective *Semi-local Centrality* which can give better result in low computational complexity than other method, such as BC and CC, but is incapable to be applied in weighted networks. Several spectral centrality measures are also available, such as eigenvector centrality (EC) [23], Katz's centrality [24], subgraph centrality [25], PageRank [26], LeaderRank [27]. Some centrality measures have been extended to weighted networks [21,28–30]. In a word, the design of an effective ranking method to identify influential nodes is still an open issue.

To sum up, in this paper we avoid treating the problem as the same with the maximization spreading problem. Instead, by the weighted k-shell decomposition method, we efficiently find a set of core users in the network which can be regarded as the most influential users. Before applying this method, each user in the networks is put on labels with values to measure its individual interests or preference on the content of the advertisement so as to filter those who have no interest. The rest parts of this paper are organized as follows. Some basic knowledge about k-shell decomposition and degree centrality are introduced. Then, we introduce the proposed method in Sect. 3. In Sect. 4, the performance of the proposed method is shown through several numerical examples. Section 5 concludes the paper.

2   Basic Knowledge

In this section, we introduce some basic knowledge about the original K-shell decomposition for binary networks and the extended K-shell decomposition for weighted networks, and also degree centrality and closeness centrality.

2.1 K-Shell Decomposition for Binary Networks

As shown in Fig. 1, nodes are assigned to k shells according to their remaining degree, which is obtained by successive pruning of nodes with degree smaller than the k_S value of the current layer. We start by removing all nodes with degree $k = 1$. After removing all the nodes with $k = 1$, some nodes may be left with one link, so we continue pruning the system iteratively until there is no node left with $k = 1$ in the network. The removed nodes, along with the corresponding links, form a k shell with index $k_S = 1$. In a similar fashion, we iteratively remove the next k shell, $k_S = 2$, and continue removing higher-k shells until all nodes are removed. As a result, each node is associated with one k_S index, and the network can be viewed as the union of all k shells. The resulting classification of a node can be very different than when the degree k is used.

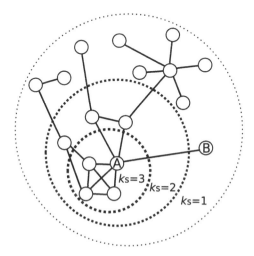

Fig. 1. Illustration of the layered structure of a network, obtained using the k-shell decomposition method. The nodes between the two outer rings compose shell 1 ($k_s = 1$), while the nodes between the two inner rings compose shell 2 ($k_s = 2$). The nodes within the central ring constitute the core, in this case $k_s = 3$.

2.2 *K*-Shell Decomposition for Weighted Networks

This method applies the same pruning steps that is described above. Instead, this measure considers both the degree of a node and the weights of its links, and we assign for each node a weighted degree k'. The weighted degree of a node i is defined as

$$k' = [k_i^\alpha (\sum_j^{k_i} \omega_{ij})^\beta]^{\frac{1}{\alpha+\beta}} \tag{1}$$

where $\alpha = \beta = 1$, which treats the weight and the degree equally.

2.3 Degree Centrality

A weighted network can generally be represented as a set $G = (V, E, W)$ [22]. Here, V and E are the sets of all nodes and edges, respectively. Let k_i and ω_i

be the corresponding degree and weight of node i with its neighbors. W is the weight set of E, i.e., the link E_{ij} from nodes i to j has a weight $\omega_{ij} \in W$.

Definition 1 (Degree Centrality in a weighted network). *The DC of node i, denoted as $C_d^\omega(i)$, is defined as*

$$C_d^\omega(i) = \sum_j^N \omega_{ij} \tag{2}$$

where ω_{ij} is the weighted of the edges between node i and j, which is greater than 0 when the node i is connected to node j. Here, N is the number of neighbors of node i.

3   Formulation of the Proposed Method

In the proposed method, we integrate analytics data of users' profiles within a social network with targeted advertising campaigns. We collect analytics data of users' profiles and utilizes the data to filter through the users' profiles to select desired user profiles for delivery of advertisements targeted the interests and personality of the desired user profiles. Utilization of the analytics data includes assigning labels with ranking values compared with other users.

Step 1: put labels for each user and assign values, L_{ij}, which represents the value of the i-th users' j-th label. In this step, we construct a networks in which individuals are labeled with different attribute values. For example, if there is a company trying to advertise selling basketball shoes, all the users should be labeled closely related to sports, sports shoes, basketball etc. For those who have no interest in these labels or are never talking about these labels, we can consider drop these users to avoid sending advertisement to them. This could incur less size of networks.

Step 2: weights are assigned to each label. In this step, each type of labels should be assigned with possibly different weights since they are inequally contributed to the advertisement.

$$S_i = \sum_j^n \lambda_j L_{ij}, 0 < \lambda_j < 1 \tag{3}$$

where λ_j is the weight of j-th label and n is the number of labels. Until now, we sum up weighted label scores for each user profile to produce a profile score.

Step 3: identify a few of most influential users by weighted k-shell decomposition method. Before applying the weighted k-shell decomposition method, there is one question that need to be solved. Typically, the links of online social networks are directed. But both the unweighted k-shell decomposition and weighted k-shell decomposition methods avoid dealing with this problem. In this paper, we analyze and solve this problem as below. If there is a user located in the periphery of a network, it has only one link related to another user. No matter

Table 1. There are ten labels for each user (here, we only list 21 users' labels). For each label, it has been assigned integer value from 1 to 10. The high the value, the more the user are interested in the advertisement. Otherwise, the user has less interest in the advertising information.

	Label 1	Label 2	Label 3	Label 4	Label 5	Label 6	Label 7	Label 8	Label 9	Label 10
1	3	9	8	8	7	7	9	5	10	5
2	9	3	8	1	5	4	10	3	6	6
3	8	9	5	4	9	1	1	8	2	9
4	7	8	8	3	6	9	4	10	7	2
5	3	10	5	10	7	6	7	6	3	10
6	6	3	10	2	8	10	1	2	8	8
7	2	9	6	9	4	5	4	4	2	2
8	6	2	8	4	4	6	3	3	10	10
9	3	6	10	9	1	10	6	9	7	9
10	3	2	1	2	10	6	5	1	2	8
11	6	9	4	4	1	1	1	10	8	10
12	6	10	5	1	7	4	4	1	8	2
13	9	4	10	4	8	7	3	1	4	8
14	3	2	2	4	5	2	5	8	3	1
15	5	6	7	2	7	5	5	5	10	6
16	7	8	5	2	2	5	7	10	1	2
17	7	1	10	8	5	7	5	1	4	10
18	3	7	10	8	1	4	7	9	4	3
19	3	1	8	1	7	8	9	6	6	1
20	1	2	8	1	1	8	9	8	10	10
21	8	8	9	10	1	1	9	9	4	1

this link is in-degree link or out-degree one, it won't change the fact that this user is marginalized. Therefore, in this paper we regards the directed networks as undirected one during assigning weighted degree when applying the weighted k-shell decomposition method (Table 1).

Step 4: output the most influential users (core-positioned nodes).

4   Experiments and Results

In order to evaluate the performance of the proposed method and other centrality measures, we explore to employ SI model [32] to examine the spreading ability of the top ranked nodes. Although SI model cannot identify the influence of nodes, it reflects the spreading ability of nodes, which has been widely used for epidemic dynamics on networks. In the SI model, every node has two discrete states: (i) susceptible and (ii) infected, in which the infected nodes stay infected and spread the infection to the susceptible neighbors with rate I_θ. For epidemic spreading on networks, I_θ determines the range or scale over which a node can exert influence. Note that, in weighted networks, node i infects node j with probability $I_\theta = (\frac{w_{ij}}{w_M})^\theta$, where θ is a positive constant and w_M is the largest value of w_{ij} in the network [33]. In this paper, F(t) denotes the number of infected nodes after the time of t, which can be described as spreading ability.

4.1 Zachary Network

The data of Zachary's Karate Club Network are collected from the members of a university karate club by Wayne Zachary over two years [34]. Zachary constructed a weighted network by denoting each member in the club as a node. Each edge in the network represents the connected two members are friends outside the club activities and its weight indicates the relative strength of the associations (number of situations in and outside the club in which interactions occurred).

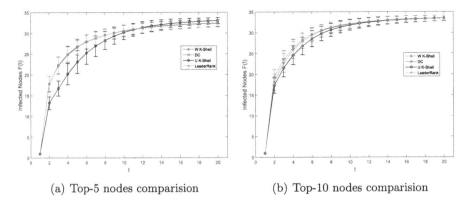

(a) Top-5 nodes comparision   (b) Top-10 nodes comparision

Fig. 2. Comparison of spreading ability of top-L nodes by three different methods. X-axis represents the time step and Y-axis represents the cumulative number of infected nodes with 15 steps in Zachary, which are obtained by averaging over 1000 implementations. Here $\theta = 6$.

In both Figs. 2(a) and (b), nodes identified by Weight k-shell decomposition (W K-Shell), LeaderRank and Degree Centrality (DC) have nearly equal influence, which are better than those identified by Unweighted k-shell (U K-Shell) decomposition method. Since the size of this network is quite small, we can just identify this for the purpose of illustration. In Fig. 3, it is obvious that the proposed method can assign relatively larger-scale values to than the other two methods, which discriminate the difference of influence between nodes more remarkably. Obviously, LeaderRank has the worst ability to discriminate its ranking value.

4.2 Artificial Networks

In this part, an Erdos and Renyi random network with 5000 nodes and 10000 edges is taken.

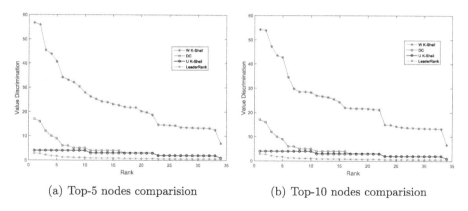

(a) Top-5 nodes comparision   (b) Top-10 nodes comparision

Fig. 3. Comparison of discrimination degree of ranking. X-axis represents the ranking of each node and Y-axis represents the values assigned by three different methods.

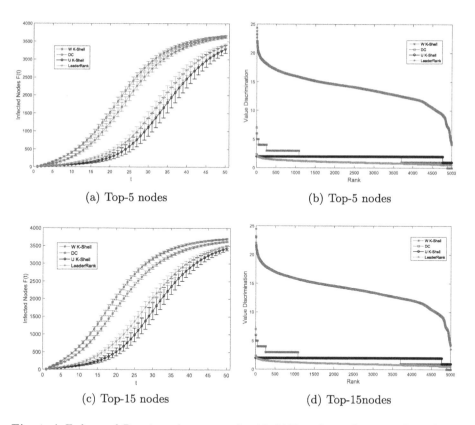

(a) Top-5 nodes   (b) Top-5 nodes

(c) Top-15 nodes   (d) Top-15nodes

Fig. 4. A Erdos and Renyi random network with 5000 nodes and 10000 edges. Comparison of spreading ability of three methods. X-axis represents the time step and For (a) and (c), Y-axis represents the cumulative number of infected nodes with 30 steps in Zachary, which are obtained by averaging over 1000 implementations. Here $\theta = 4$.

As we can see in Fig. 4, DC performs better than our method and unweighted *k*-shell decomposition method. However, the value discrimination shown by our method is the best. The assigned values for nodes decrease gradually, but the rest two methods have poor performance in discriminate the difference between the values of different ranks. Furthermore, the Unweighted K-Shell method has the worst ability to spread its influence, which means in large-scale cases, it becomes less useful than other methods.

5   Conclusions

In this paper, we propose a method to help advertise efficiently in online social networks. However, according to the present experimental results, the proposed method hasn't shown any outstanding performance, which is out of our expectation. We construct artificial networks in which there are only nodes and link relationship. Then assign attribute labels with random values to change the weight of the links in the network. If our method performs well based on the constructed networks assigned with different label values, which means from the identified most influential nodes the spreading of information is faster than other methods. However, the fact is not.

Even with not-so-good results, we can still get some significant insights. First, the ranking value discrimination shown by our method is significant when compared with other methods. Second, all the foregoing networks are not real online social networks. In fact, we need to recollect the data from real online networks, and if possible, advertising on real social networks is definitely the best way to test our method. In fact, all other test data is empirical and need to be substantiated by real data.

This paper gives rather than a good result, but need to be improved in other dimensions, such as collecting new data and revising weights of labels by training on real social network data.

References

1. Armstrong, G.: Marketing: An Introduction. Pearson Education, New York (2009)
2. Hays, L.: Technology: Using Computers to Divine Who Might Buy a Gas Grill. Wall Street Journal, New York (1994)
3. Sun, F., Shi, Y., Wang, W.: Content-based recommendation system based on vague sets, 294–297. IHMSC (2013)
4. Fang, P., Zheng, S.: A research on fuzzy formal concept analysis based collaborative filtering recommendation system. KAM, 352–355 (2009)
5. Domingos, P., Richardson, M.: Mining the network value of customers. ACM SIGKDD 57–66 (2001)
6. Richardson, M., Domingos, P.: Mining knowledge-sharing sites for viral marketing. ACM SIGKDD, 61–70 (2002)
7. Chen, D., Lv, L., Shang, M., Zhang, Y., Zhou, T.: Identifying influential nodes in complex networks. Phys. A **391**(4), 1777–1787 (2012)

8. Jordán, F., Benedek, Z., Podani, J.: Quantifying positional importance in food webs: a comparison of centrality indices. Ecol. Model. **205**(1–2), 270–275 (2007)
9. Kolaczyk, E.D., Chua, D.B., Barthlemy, M.: Group betweenness and co-betweenness: inter-related notions of coalition centrality. Soc. Net. **31**(3), 190–203 (2009)
10. Wang, F., Antipova, A., Porta, S.: Street centrality and land use intensity in Baton Rouge, Louisiana. J. Transp. Geogr. **19**(2), 285–293 (2011)
11. Zhang, H., Fiszman, M., Shin, D., et al.: Degree centrality for semantic abstraction summarization of therapeutic studies. J. Biomed. Inform. **44**(5), 830–838 (2011)
12. Linyuan, L., Jin, C.H., Zhou, T.: Similarity index based on local paths for link prediction of complex networks. Phys. Rev. E **80**(4), 046122 (2009)
13. Wang, J., Rong, L.: Similarity index based on the information of neighbor nodes for link prediction of complex network. Mod. Phys. Lett. B **27**(06), 1350039 (2013)
14. Lü, L., Zhou, T.: Link prediction in weighted networks: the role of weak ties. Europhys. Lett. **89**(1), 18001 (2010)
15. Pathak, P.H., Dutta, R.: Centrality-based power control for hot-spot mitigation in multi-hop wireless networks. Comput. Commun. **35**(9), 1074–1085 (2012)
16. Kitsak, M., Gallos, L.K., Havlin, S., et al.: Identification of influential spreaders in complex networks. Nat. Phys. **6**(11), 888–893 (2010)
17. Fan, W.L., Liu, Z.G., Hu, P.: A high robustness and low cost cascading failure model based on node importance in complex networks. Mod. Phys. Lett. B **28**(02), 1450011 (2014)
18. Zhang, Z.: Information entropy of diffusion processes on complex networks. Mod. Phys. Lett. **28**(17), 1450141 (2014)
19. Wei, D., Deng, X., Zhang, X., et al.: Identifying influential nodes in weighted networks based on evidence theory. Phys. A **392**(10), 2564–2575 (2013)
20. Nicosia, V., Criado, R., Romance, M., et al.: Controlling centrality in complex networks. Sci. Rep. **2**, 218–223 (2012)
21. Opsahl, T., Agneessens, F., Skvoretz, J.: Node centrality in weighted networks: generalizing degree and shortest paths. Soc. Net. **32**(3), 245–251 (2010)
22. Freeman, L.C.: Centrality in social networks conceptual clarification. Soc. Net. **1**(3), 215–239 (1978)
23. Bonacich, P., Lloyd, P.: Calculating status with negative relations. Soc. Net. **26**(4), 331–338 (2004)
24. Katz, L.: A new status index derived from sociometric analysis. Psychometrika **18**(1), 39–43 (1953)
25. Estrada, E., Rodrguez-Velzquez, J.A.: Subgraph centrality in complex networks. Phys. Rev. E **71**(5), 056103 (2005)
26. Brin, S., Page, L.: The pagerank citation ranking: bringing order to the web. Comput. Netw. 107–117 (1998)
27. Lv, L., Zhou, T.: Leaders in social networks, the delicious case. PloS ONE **6**(6), e21202 (2011)
28. Newman, M.: Scientific collaboration networks. II. Shortest paths, weighted networks, and centrality. Phys. Rev. E **64**(1), 016132 (2001)
29. Newman, M.: Analysis of weighted networks. Phys. Rev. E **70**(5), 056131 (2004)
30. Chu, X., Zhang, Z., Guan, J., et al.: Epidemic spreading with nonlinear infectivity in weighted scale-free networks. Phys. A **390**(3), 471–481 (2009)
31. Garas, A., Schweitzer, F., Havlin, S.: AK-shell decomposition method for weighted networks. New Phys. **14**(8), 083030 (2012)

32. Zhou, T., Liu, J.G., Bai, W.J., et al.: Behaviors of susceptible-infected epidemics on scale-free networks with identical infectivity. Phys. Rev. E **74**(5), 056109 (2006)
33. Yan, G., Zhou, T., Wang, J., et al.: Epidemic spread in weighted scale-free networks. Chin. Phys. Lett. **22**(2), 510–513 (2005)
34. Zachary, W.W.: An information flow model for conflict and fission in small groups. J. Anthropol. Res. **33**(4), 452–473 (1977)

Supervised/Unsupervised/Reinforcement Learning/Deep Learning

Semantic Modulation Based Residual Network for Temporal Language Queries Grounding in Video

Cheng Chen and Xiaodong Gu$^{(\boxtimes)}$ (iD)

Department of Electronic Engineering, Fudan University, Shanghai 200433, China
xdgu@fudan.edu.cn

Abstract. Temporal language queries grounding in video aims to retrieve one specific moment in a long, untrimmed video by a query sentence. This is a challenging issue as a video may contain multiple moments of interests which have complex temporal dependencies with other temporal moments. To preserve the original video moment information during the multiple layers convolution operations, this paper introduces residual learning into the issue and proposes a novel semantic modulation based residual network (SMRN) that incorporates dynamical semantic modulation and multiple prediction maps in a single-shot feed-forward framework. Semantic modulation mechanism dynamically modulates the residual learning by assigning where to pay the visual attention. And the mechanism is able to adjust the weights given to different video moments with the guide of query sentence. Additionally, the network combines multiple feature maps from different layers to naturally captures different temporal relationships for precisely matching video moment and sentence. We evaluate our model on three datasets, i.e., TACoS, Charades-STA, and ActivityNet Caption, with significant improvement over the current state-of-the-arts. Furthermore ablation experiments were performed to show the effectiveness of our model.

Keywords: Language queries grounding in video · Residual learning · Semantic modulation · Single-shot feed-forward network

1 Introduction

Detecting activities in videos is a fundamental issue of video understanding. Several researches such as video captioning, temporal action localization, video summarization and temporal language queries grounding in videos have been proposed for different scenarios. Temporal Language Queries Grounding in Videos (TLQG) aims to retrieve the start and end timestamps of one specific moment which best matches the language query in a long, untrimmed video. As untrimmed videos usually contain complex temporal activities and the language queries are flexible, it's challenging to model the fine-grained interactions between the two modalities. For example, The video shown in Fig. 1 mainly

© Springer Nature Switzerland AG 2020
M. Han et al. (Eds.): ISNN 2020, LNCS 12557, pp. 119–129, 2020.
https://doi.org/10.1007/978-3-030-64221-1_11

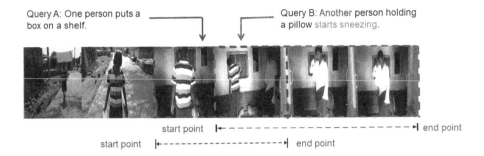

Fig. 1. Example of temporal language queries grounding in video.

contains two distinct activities with complex temporal dependencies, the contextual information is crucial for distinguishing the two activities and localizing the start point of the second activity. Recently, some methods [5,13] are proposed, they designed a single-shot Convolutional Neural Network (CNN) to aggregate video information. In these models, they only take into account the last CNN layer which highly integrate a large range of context information within the video. Although these models can capture temporal relationships, they are still suffering from inferior effectiveness. First, imagine that we are looking for a target segment in the video. The first thing we will do is to remember the information of one segment, then we will judge it with context segments information. As a CNN aggregate video information through a hierarchy of visual abstractions by stacking the layers, the last CNN layer almost 'forgets' the information of the original segments. Second, as showing in Fig. 1, a target segment only relates to partial regions of a video, without the guide of the query semantic meaning, it's not easy for a CNN to classify where to pay more attention. So it is crucial to make full use of the query semantics for finding the semantic-related regions.

In this paper, we propose a novel semantic modulation based residual network (SMRN). As the shortcut connection within the hierarchical residual network allows the information from low-level feature maps directly flows into high-level feature maps, the SMRN can effectively aggregates original video segment and contextual segment information for more accurate results. To better attend to the query-related video contents, semantic modulation is leveraged to modulate the temporal convolution processes by weighting different regions of video feature maps with referring to the query semantics. Additionally, our SMRN combines predictions from multiple feature maps with different ranges of temporal context information to naturally handle moments of complex temporal dependencies. Considering them as whole, SMRN is able to precisely summarize contextual video information, leading to an effective architecture for TLQG task. Our contributions are as follows:

(1) We introduce residual learning into the temporal language queries grounding in videos and propose a novel semantic modulation based residual network. The shortcut connection within the residual network can preserve the

information of the original video moment. On the other hand, the semantic modulation is able to help the model pay more attention to the sentence-related video contents.

(2) We combine predictions from different levels of feature maps to utilize the hierarchical structure of Convolution Neural Networks, gaining useful and effective information for better capture different ranges of temporal relationships.

(3) Extensive experiments were performed on three benchmarks: TACoS [3], Charades-STA [2], and ActivityNet Captions [1], the results show every component is effective for localizing precise temporal boundaries and the SMRN significantly outperform the state-of-the-art methods.

2 Proposed Method

2.1 The Framework of Semantic Attention Based Residual Networks

In the following, We first introduce the basic temporal residual convolution block and then incorporate semantic modulation to residual network. Afterward, the details of multiple prediction maps will be described.

Temporal Residual Convolution Block. As shown in Fig. 2, a video was first composed to a sequence of clips $V = \{v_i\}_{i=0}^{T}$, each clip consisting of N frames is represented as a tensor $f \in R^{d^v}$ (d^v is the feature dimension) which was extracted from a pre-trained CNN model (i.e., C3D model [6], VGG model), see Experiment section for details). By maxpooling the clips over a specific time span, corresponding moment candidates are obtained. Specifically, the moment candidate feature $f_{a,b}^M$ (a and b respectively represents the start and end time of the moment) is obtained by $maxpool(f_a^V, f_{a+1}^V, ..., f_b^V)$. And the sentence representation f^s can be obtained by sequentially feeding the word embedding (we use GloVe word2vec model to generate word embedding) into a three-layer LSTM. The moment candidates features and sentence feature serve as the basic elements for 2D video feature map.

Follow 2D-TAN [5] we restructure the whole moments sampled from video clips to a 2D-temporal feature map, denoted as $F^M \in \mathbb{R}^{T \times T \times d^V}$. The first two dimensions represent the coordinate indexes of the start and end clips of the moments, and d^V denoted the feature dimension. Different locations in the 2D-temporal feature map represent moments of different start and end timestamps. Given the 2D video feature map, we first fused it with sentence representation f^s as:

$$F^f = \left\| (w^s \cdot f^s) \odot (W^M \cdot F^M) \right\|_F, \tag{1}$$

where w^s and W^f are the learnable parameters, $\|\cdot\|_F$ is Frobenius normalization, and $\odot$ is Hadamard product. With such a cross modal fusion processing, the fused feature map $F^f \in \mathbb{R}^{T \times T \times d^f}$ is able to capture fine-grained interactions between query and video moments. As aforementioned, localizing one moment

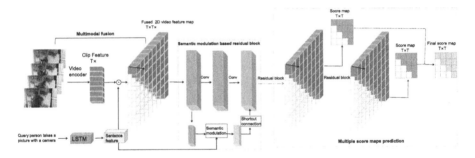

Fig. 2. The framework of our proposed Semantic Modulation based Residual Network (Color figure online).

needs to remember the information of the moment first, and then judge it with adjacent moment candidates. Inspired by the efficient residual learning, we established the hierarchical temporal residual network to perceive more context information though the large receptive filed of the network. Deeper in the network, moment candidates at each location of the feature map are deeper blended with other locations. Moreover, each location is able to remember the information of itself even in the deepest feature map with the shortcut connection in the residual network. Taking the fused feature map F, the standard residual temporal convolution block in this paper is defined as:

$$y = \mathcal{F}(x, \{W_i\}) + x, \tag{2}$$

where x and y denote the input 2D video feature map and the output feature map of the layer considered. $\mathcal{F}(x, \{W_i\})$ are the standard convolution operation, $\mathcal{F} + x$ is performed by a shortcut connection and element-wise addition. The convolution operation is denoted as $Conv(\Theta_k, \Theta_s, d_h)$, where Θ_k, Θ_s, d_h are the kernel size, stride size and channel numbers, respectively. By setting the Θ_k as 3 and Θ_s as 1, each residual temporal convolution block will preserve the temporal dimension of the 2D feature map and meanwhile introduce more context information of adjacent moment candidates to each location within the feature map. Moreover, the shortcut connection in the residual block will preserve lower layers information during convolution without introducing computation complexity.

Semantic Modulation Based Residual Network. As aforementioned, a target moment described by a language query only relates to partial regions of a video. Therefore, considering each location in the feature map equally may lead to sub-optimal result. To full exploit the semantic of the query sentence, semantic modulation attempts to adjust the weights given to the regions in the shortcut feature map.

Specifically, as shown in Fig. 2, given the feature map F^f extracted from one residual convolution block (without loss of generality, the block number is omitted here) and the sentence representation f^s, we first reshape $F^f = [f_1^f, f_2^f \dots f_m^f]$ (Orange cube in Fig. 2) by flattening the original F^f along the temporal dimension, where $f_i^f \in d^f$ and $m = T \cdot T$, here f_i^f is the fused video feature of the $i - th$ moment candidates, then we attentively weight the regions in the feature map adaptive to the sentence context:

$$\beta = softmax(\mathrm{W}^b \tanh(\mathrm{W}_f \mathrm{F}^f + \mathrm{W}_s f^s + b)), \tag{3}$$

where $\mathrm{W}^b, \mathrm{W}_f, \mathrm{W}_s$ and b are learnable parameters, β is the semantic modulation weights. Based on the weights, the video feature F^f is then modulated as:

$$\mathrm{F}_m^f = f(\mathrm{F}^f, \beta), \tag{4}$$

and $f(\cdot)$ denotes element-wise multiplication here. As Fig. 2 shows, the modulation feature maps F_m^M are then applied to the shortcut connection in the residual block. And the basic residual blocks are then updated to:

$$\mathrm{y} = \mathcal{F}(\mathrm{x}^M, \{W_i\}) + \mathrm{x}_m^M, \tag{5}$$

where x^M is the original feature map and x_m^M is the modulated feature map. As such, each video feature map will absorb the prior feature map information with the guide of sentence, and further activate the following semantic modulation based residual block to pay more attention to the sentence related regions.

Multiple Feature Maps for Prediction. As the temporal convolution operations preserve the temporal dimension of the 2D feature maps and meanwhile gradually expand the scope of context-aware information at each location within the maps, different feature maps can be used to predict activities that require different scopes of context information. Given a feature map F_m^f from the semantic modulation residual block, we use a fully connected layer followed by a sigmoid function to produce the confidence score map:

$$\mathrm{C}^M = sigmoid(\mathrm{W}_m \mathrm{F}_m^M + \mathrm{b}_m). \tag{6}$$

As such, N score maps are generated through once forward propagation, here N denotes the number of residual block within the SMRN. The value s_i of the ith valid location on the score map represents the similarity between the moment candidate and the language query.

In practice, we use weighting parameter λ_l, λ_s to combine the last two blocks score maps, which could result in better performance. The final score map is calculated by:

$$\mathrm{C}_f^M = \lambda_l \mathrm{C}_l^M + \lambda_s \mathrm{C}_s^M, \tag{7}$$

where C_l^M and C_s^M represents the last, the second last score map, respectively, and the C_f^M is final score map. λ_l and λ_s are set to 0.5 through cross-validation.

Training and Inference. During the training of SMRN, the training sample consists of three part: a language query, a input untrimmed video, and the ground truth moment. We first compute tIoU t_i (temporal Intersection-over-Union) of each moment candidates within different feature maps with the ground truth moment. Then scale the tIoU by two thresholds o_{min} and o_{max} as:

$$g_i = \begin{cases} 0, & t_i \leq o_{min}, \\ \frac{t_i - o_{min}}{o_{max} - o_{min}}, & o_{min} < t_i < o_{max}, \\ 1, & t_i \geq o_{max}. \end{cases} \tag{8}$$

The overlap prediction loss for each score map is realized as a binary cross entropy loss as:

$$Loss = \frac{1}{N} \sum_{i=1}^{N} g_i \log p_i + (1 - g_i) \log(1 - p_i), \tag{9}$$

where N is the total number of valid candidates within a score map and p_i is the predicted overlap score of a moment. As the network utilizes the last two score maps for prediction, the final training loss is defined as $L_l + L_S$, here L_l and L_S is the prediction loss calculated by the last and second last score map, respectively.

At inference time, the final confidence score map C_f^M can be generated in one forward pass. The moment candidates within the map are filtered by non-maximum suppression(NMS) according to the scores. Afterwards, the top n moments in each video are obtained.

3   Experiments

3.1   Datasets and Evaluation Metrics

We conduct experiments to evaluate the proposed SMRN model on three public large-scale datasets: TACoS [3], Charades-STA [2], and ActivityNet Captions [1]. As previous works [2–5], we adopt Rank@n, IoU@m to evaluate our model. The metric is defined as the percentage of the testing queries having at least one hitting retrieval (with IoU large than m) in the top-n retrieved results.

3.2   Implementation Details

For fair comparisons, we use the same video encoder as the previous methods. Specifically, C3D features for ActivityNet Captions and TACoS, VGG features for Charades-STA. The time dimensions of the 2D video feature map is set to 128 for TACoS, 64 for ActivityNet Captions and 16 for Charades-STA according to the video duration statistics. For SMRN network architecture, 12 layers convolution networks for TACoS and Charades-STA, and 16 layers convolution networks for ActivityNet Captions, each residual block contains two layers of convolution networks so 6 residual blocks for TACoS and Charades-STA, and 8 residual blocks for ActivityNet Captions. The semantic modulation mechanism is performed on the second residual block and the fourth residual block. The dimension of video, sentence and fused features (i.e. d^V, f^s) is set to 512. The whole framework is optimized in an end-to-end way with Adam.

3.3 Comparison to State-of-the-Art Methods

We compare our SMRN with recently proposed state-of-the-art baseline methods:

- Sliding window based methods: CTRL [2], ACRN [7], ACL [4] and MCN [3].
- Reinforcement learning based methods: SM-RL [8] and TripNet [9].
- RNN-based methods: CBP [10], TGN [11], and CMIN [12].
- Others: 2D-TAN [5], MAN [13] QSPN [14], ROLE [15], ABLR [16], and SAP [17]. Tables 1, 2 and 3 summarize the results on the aforementioned three benchmarks. Among all four evaluation metrics, our model achieves the highest accuracy on most datasets. Notably, for localizing moment in TACoS, Ours-SMRN outperform the recent state-of-art, i.e. 2D-TAN and CMIN by around 7 points and 9 points in the Rank@1, IoU@0.5 and Rank@5, IoU@0.3 metrics, respectively. As the videos is very long (around 7 min), and the activities take place in the same kitchen scenarios with some slightly varied cooking object in TACoS. The improvement in TACoS demonstrates that our model is able to learn the fine-grained differences between different moments. Moreover, SMRN ranks the first on Charades-STA which contains complex human activities from different scenarios. For large-scale ActivityNet Captions, SMRN also outperforms the state-of-art methods in the strict metric Rank@1, IoU@0.7 and Rank5 IoU@0.5.

Table 1. Performance comparisons on TACoS. The top-2 results are highlighted by **bold** and *italic* fonts, respectively

Methods	Rank@1			Rank@5		
	IoU@0.1	IoU@0.3	IoU@0.5	IoU@0.1	IoU@0.3	IoU@0.5
CTRL	24.32	18.32	13.30	48.73	36.69	25.42
MCN	14.42	-	5.58	37.35	-	10.33
TGN	41.87	21.77	18.9	53.40	39.06	31.02
ACRN	24.22	19.52	14.62	47.42	34.97	24.88
ROLE	20.37	15.38	9.94	45.45	31.17	20.13
ACL	31.64	24.17	20.01	57.85	42.15	30.66
CMIN	32.48	24.64	18.05	62.13	38.46	27.02
SM-RL	26.51	20.25	15.95	50.01	38.47	27.84
CBP	-	27.31	24.79	-	43.64	37.40
2D-TAN	*47.59*	*37.29*	*25.32*	*70.31*	*57.81*	*45.04*
SAP	31.15	-	18.24	53.51	-	28.11
Ours – SMRN	**50.44**	**42.49**	**32.07**	**77.28**	**66.63**	**52.84**

3.4 Ablation Studies

In this section, we perform ablation studies to evaluate the effects of each components in our proposed SMRN. Specifically, we re-train our model on

Table 2. Performance comparisons on Charades-STA.

Method	Rank@1		Rank@5	
	IoU@0.5	IoU@0.7	IoU@0.5	IoU@0.7
CTRL	23.63	8.89	58.92	29.52
MCN	17.46	8.01	48.22	26.73
MAN	*41.24*	20.54	83.21	*51.85*
ACRN	20.26	7.64	71.99	27.79
ROLE	21.74	7.82	70.37	30.06
ACL	30.48	12.20	64.84	35.13
2D-TAN	40.94	*22.85*	*83.84*	50.35
SM-RL	24.36	11.17	61.25	32.08
TripNet	36.61	14.50	-	-
CBP	36.80	18.87	70.94	50.19
ABLR	24.36	9.01	-	-
SAP	27.42	13.36	66.37	38.15
QSPN	35.60	15.80	79.40	45.40
SMRN	**43.58**	**25.22**	**86.45**	**53.39**

Table 3. Performance comparisons on Activity Captions.

Method	Rank@1		Rank@5	
	IoU@0.5	IoU@0.7	IoU@0.5	IoU@0.7
CTRL	29.01	10.34	59.17	37.54
MCN	21.36	6.43	53.23	29.70
CBP	35.76	17.80	65.89	46.20
2D-TAN	**44.14**	*26.54*	**77.13**	**61.96**
ACRN	31.67	11.25	60.34	38.57
CMIN	*43.40*	23.88	67.95	50.73
TripNet	32.19	13.93	-	-
QSPN	33.26	13.43	62.39	40.78
ABLR	36.79	-	-	-
SMRN	*42.97*	**26.79**	*76.46*	*60.51*

Charades-STA with the following four settings: **Base**: The shortcut connections and Semantic modulation mechanism are removed. Only the last score map is used to prediction. **Base + SC**: Two convolutional layers are used to construct a residual block, the short connections between the blocks are connected. The network only uses the last score map for prediction. **Base + SC + MP**: Based on the residual block, the last two score maps are combined for prediction. **Base + SC + SM**: Semantic modulation mechanism is integrated to modulate the residual learning, and only the last score map is used for prediction.

As shown in Table 4, the Base+SC significantly outperforms the baseline Base, it indicates the importance of the information from low convolutional layers. Comparing Base+SC with Base+SC+MP, we can find that combining multiple feature map is beneficial for improving performance. This comparison validates that different layers in the hierarchical residual network have different capabilities for capturing temporal dependencies. With semantic modulation introducing sentence information to the residual block, Base+SC+SM outperform the original network in all metrics. The full model SMRN gets highest results in all metrics.

Table 4. Ablation studies on Charades-STA dataset

Method	Rank1@		Rank5@	
	IoU@0.5	IoU@0.7	IoU@0.5	IoU@0.7
BASE	40.91	23.63	84.46	49.89
BASE+SC	43.09	24.78	85.73	51.61
BASE+SC+MP	43.31	25.03	86.05	*53.20*
BASE+SC+SM	43.49	25.19	85.81	52.69
SMRN	**43.58**	**25.22**	**86.45**	**53.39**

3.5   Qualitative Result

We illustrate some qualitative examples of our model in Fig. 3. Evidently, combining score maps from different residual block can produce more accurate result. For example, in the first video, the person first begins to play on his phone and then starts fixing the doorknob. The two activities are very similar. It is very difficult for the model to predict the temporal boundary without proper reference to the context adjacent moment. As our model exploits multiple context information from different residual block and modulates the residual learning by sentence feature. It performs better than the baseline.

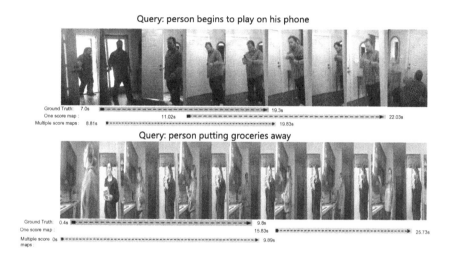

Fig. 3. Prediction examples from one score map and multiple score maps.

4   Conclusion

In this paper, we study temporal language queries grounding in video. A novel Semantic Modulation based Residual Network is proposed to tackle the issue. We introduce residual learning into the temporal language queries grounding in video for making full use of the original video moment information. To better select the target moment related moment candidates, sentence semantics are leveraged to modulate the short connections within the residual blocks. An important feature of SMRN is using multiple score maps attached to multiple feature maps at the top of the hierarchical residual network. This representation allows the model to precisely capture different temporal relationships. The experimental results obtained on three datasets show the superiority of our proposed model.

Acknowledgments. This work was supported in part by National Natural Science Foundation of China under grant 61771145 and 61371148.

References

1. Krishna, R., Hata, K., Ren, F., Li, F.F.: Dense-captioning events in videos. In: Proceedings of International Conference on Computer Vision, pp. 706–715. IEEE, Venice (2017)
2. Gao, J., Sun, C, Yang, Z., Nevatia, R.: TALL: temporal activity localization via language query. In: Proceedings of International Conference on Computer Vision, pp. 5267–5275. IEEE, Venice (2017)
3. Hendricks, L.A., Wang, O., Shechtman, E., Sivic, J., Darrell, T., Russell, B.: Localizing moments in video with temporal language. In: Proceedings of International Conference on Computer Vision, pp. 5803–5812. IEEE, Venice (2017)
4. Ge, R., Gao, J., Chen, K., Nevatia, R.: MAC: mining activity concepts for language-based temporal localization. In: Proceedings of the Winter Conference on Applications of Computer Vision, pp. 245–253. IEEE, Waikoloa (2019)
5. Zhang, S., Peng, H., Fu, J., Luo, J.: Learning 2D temporal adjacent networks for moment localization with natural language. In: proceedings of AAAI Conference on Artificial Intelligence. AAAI, New York (2020)
6. Tran, D., Bourdev, L., Fergus, R., Torresani, L., Manohar, P.: Learning spatiotemporal features with 3D convolutional networks. In: Proceedings of the IEEE International Conference on Computer Vision, pp. 4489–4497. IEEE, Boston (2015)
7. Liu, M., Wang, X., Nie, L., He, X., Chen, B., Chua, T.: Attentive moment retrieval in videos. In: Proceedings of the International ACM SIGIR Conference on Research and Development in Information Retrieval, pp. 15–24. ACM, Ann Arbor (2018)
8. Wang, W., Huang, Y., Wang, L.: Language driven temporal activity localization: a semantic matching reinforcement learning model. In: Proceedings of Computer Vision and Pattern Recognition, pp. 334–343. IEEE, California (2019)
9. Hahn, M., Kadav, A., Rehg, J.M., Graf, H.P.: Tripping through time: efficient localization of activities in videos. arXiv preprint arXiv:1904.09936
10. Wang, J.W., Ma, L., Jiang, W.H.: Temporally grounding language queries in videos by contextual boundary-aware prediction. In: Proceedings of AAAI Conference on Artificial Intelligence. AAAI, New York (2020)
11. Chen, J., Chen, X., Ma, L., Jie, Z., Chua, T.S.: Temporally grounding natural sentence in video. In: Proceedings of Conference on Empirical Methods in Natural Language Processing, pp. 162–171. ACL, Belgium (2018)
12. Zhang, Z., Lin, Z., Zhao, Z., Xiao, Z.: Cross-modal interaction networks for query-based moment retrieval in videos. In: Proceedings of the International ACM SIGIR Conference on Research and Development in Information Retrieval, pp. 655-664. ACM, Paris (2019)
13. Zhang, D., Dai, X., Wang, X., Wang, Y.F., Davis, L.S.: MAN: moment alignment network for natural language moment retrieval via iterative graph adjustment. In: Proceedings of Computer Vision and Pattern Recognition, pp. 1247–1257. IEEE, California (2019)
14. Xu, H., He, K., Plummer, B.A., Sigal, L., Sclaroff, S., Saenko, K.: Multilevel language and vision integration for text-to-clip retrieval. In: Proceedings of AAAI Conference on Artificial Intelligence, pp. 9062–9069 AAAI, Arlington (2019)

15. Liu, M., Wang, X., Nie, L., Tian, Q., Chen, B., Chua, T.S.: Cross-modal moment localization in videos. In: Proceedings of ACM International Conference on Multimedia, pp. 843–851. ACM, Seoul (2018)
16. Yuan, Y., Mei, T., Zhu, W.: To find where you talk: temporal sentence localization in video with attention based location regression. In: Proceedings of AAAI Conference on Artificial Intelligence, pp. 9159–9166. AAAI, Arlington (2019)
17. Chen, S., Jiang, Y.G.: Semantic proposal for activity localization in videos via sentence query. In: Proceedings of AAAI Conference on Artificial Intelligence, pp. 8199–8206. AAAI, Arlington (2019)

AlTwo: Vehicle Recognition in Foggy Weather Based on Two-Step Recognition Algorithm

Fengxin Li[1(✉)], Ziye Luo[1(✉)], Jingyu Huang[1], Lingzhan Wang[1], Jinpu Cai[1], and Yongping Huang[2]

[1] Software College, Jilin University, Changchun, China
lifengxin999@163.com, luoziye666@126.com, huangjy5517@163.com,
wlz799493967@163.com, caijp5517@gmail.com
[2] Computer Science and Technology College, Jilin University, Changchun, China
hyp@jlu.edu.cn

Abstract. This paper focuses on the application of vehicle recognition in foggy weather, and proposes a two-step recognition algorithm based on deep learning, hoping to still have a good recognition result under the influence of fog. In AlOne, we use the current popular defogging algorithms: adaptive histogram equalization, single-scale Retinex, dark channel priori for image defogging. In AlTwo, we build a convolutional neural network model based on AlexNet to recognize vehicle image and obtain prediction accuracy. By comparing different performance indicators, the best performing dark channel prior algorithm is selected as the defogging algorithm. In the end, we improved the accuracy of vehicle recognition under the influence of low, medium and high fog concentrations to more than 97% .

Keywords: Vehicle recognition · Image defogging · Adaptive histogram equalization · Single-scale Retinex · Dark channel prior · AlexNet

1 Introduction

According to data as of the end of 2019, the number of motor vehicles in China had reached 348 million [1]. The increasing number of cars year by year makes the construction and improvement of Intelligent Traffic Systems (ITS) more and more urgent. Vehicle recognition is an important part of ITS, and it has important applications in driverless technology, intelligent parking lot construction, and intelligent vehicle traffic statistics.

Since the operating environment of ITS is mainly located outdoors, it is inevitable to encounter some complicated weather conditions. In the process of vehicle recognition, it is necessary to rely on outdoor monitoring to collect images. Under rain, snow, fog and other weather, the collected images often appear degraded. In this paper, we take the foggy weather as an example to study how to improve the universality of vehicle recognition algorithms in complex weather.

In real life, the number of complex weather conditions is often much less than normal weather, so we cannot affect the recognition model during processing images to avoid affecting the accuracy of vehicle recognition under normal weather.

© Springer Nature Switzerland AG 2020
M. Han et al. (Eds.): ISNN 2020, LNCS 12557, pp. 130–141, 2020.
https://doi.org/10.1007/978-3-030-64221-1_12

At this stage, there is no systematic or perfect vehicle dataset in the foggy weather, so we choose to use the normal vehicle dataset and use the method of building a convolutional neural network to complete the task of vehicle recognition. Performance of image defogging algorithms based on image enhancement and image restoration is compared and studied.

2 Dataset

2.1 Vehicle Dataset Collection

We chose to use the GTI vehicle dataset as the dataset to complete vehicle recognition. The dataset was provided by Universidad Politécnica de Madrid (UPM) and widely used in computer vision research. Because the original data set is relatively large, we only used part of the original data set. Among them, there are 881 vehicle images, including images taken in multiple directions such as front, back, and near and far directions and there are 898 non-vehicle images, including driving-related non-vehicle objects including highways, greening, and traffic signs. The image size is 64×64. We randomly divide the dataset into training set and test set in a 3: 2 ratio. The training set consists of 1067 images and the test set has a total of 712 images (Figs. 1 and 2).

Fig. 1. Examples of vehicle images in the dataset

Fig. 2. Example of non-vehicle images in the dataset

2.2 Fog Image Synthesis

Due to the lack of relevant fog image data sets, we can only obtain foggy weather images through artificial synthesis. In the atmospheric scattering model [2], we observe that the images have the following relationships:

$$I(x, y) = J(x, y)t(x, y) + A(1 - t(x, y)) \tag{1}$$

Among them, $I(x, y)$ is the image after imaging, that is the fog image we observed. $J(x, y)$ is the object image, that is the image without fog. $t(x, y)$ is the refractive index of the corresponding pixel, that is the impact of fog on the imaging of the object. And A is the atmospheric illumination strength. According to Formula (1), as long as an appropriate refractive index is determined, a fog image can be obtained from the original image. Here we complete it according to the estimation formula:

$$t(x, y) = e^{-\beta d(x, y)} \tag{2}$$

Among them $d(x, y)$ is the pixel depth, $\beta \in (0, 1)$ is a parameter and the larger the value, the lower the visibility. According to the formulas (1) and (2), we can complete the fogging process of the image. By adjusting the parameter β, we can simulate the foggy weather of various visibility encountered in life. We use β to take 0.05, 0.10, 0.15 to simulate low fog concentration, medium fog concentration and high fog concentration respectively. The effect of fogging is shown in Figs. 3 and 4.

(a)Normal   (b)$\beta = 0.05$ (c)$\beta = 0.10$ (d)$\beta = 0.15$

Fig. 3. Synthesis of vehicle fog images with different concentrations

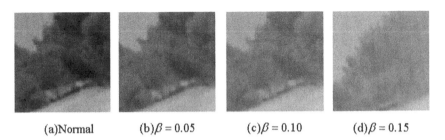

(a)Normal   (b)$\beta = 0.05$ (c)$\beta = 0.10$ (d)$\beta = 0.15$

Fig. 4. Synthesis of non-vehicle fog images with different concentrations

From the above images, we find that the edges of objects with fog images are blurred, the colors are dim, and the image features are reduced. This makes the difference between the vehicle image and the non-vehicle image smaller, which brings great challenges to the recognition process. As the fog concentration increases, the image degradation becomes more and more serious.

3   Method

3.1   AlOne: Image Defogging

There are two main types of image defogging algorithms at this stage, one is the traditional defogging method based on image processing, and the other is the artificial neural network defogging method based on deep learning. The defogging method based on deep learning mainly adopts the method of generative adversarial network for defogging. In reference 3, the author proposed the use of superposition conditional generative adversarial network for image defogging [3], but the process of training generative adversarial network is complex and unstable, and the success rate of training is not high. In terms of vehicle recognition, our goal is to make the trained network model better complete the task of vehicle recognition in foggy weather, and optimize the robustness of the model. Considering comprehensively, the feasibility of defogging algorithms based on deep learning is lower than the traditional defogging algorithms. Therefore, we choose three methods based on the traditional algorithms for comparative research.

Adaptive Histogram Equalization Algorithm
The boundary between the foreground and background of an image captured in foggy weather often becomes blurred, and the contrast is low. Adaptive histogram equalization defogging algorithm is a defogging algorithm based on image enhancement. As the name suggests, the core idea of the algorithm is to divide the image into many small subregions, and redistribute the brightness based on the calculation of the histogram of the subregion to improve the contrast of the image and achieve the purpose of image defogging [4]. The algorithm flow is shown below.

```
Algorithm: Adaptive Histogram Equalization Algorithm
Input: Image taken in foggy weather
Output: Defogging image
Step1: Convert the RGB image to the YUV space
Step2: Separate brightness channel Y
Step3: Redistribute the brightness
Step4: Merge the YUV channels and convert to RGB image
```

Single-Scale Retinex Algorithm
The single-scale Retinex algorithm is based on Retinex theory. Retinex theory is proposed by American physicist Edwin Land, who believes that the color of objects observed by humans is only related to the reflective property of the object itself, and has nothing to do with the external lighting situation [5].

For a color picture, it is determined by the original object image and the reflective property of the object. Our job is to estimate the reflective property of objects reasonably through the obtained images.

Jobson et al. proposed single-scale Retinex algorithm [6], which combines Gaussian Wrapping Function and Convolution Operation to estimate reflective property. The algorithm flow is shown below.

```
Algorithm: Single-scale Retinex Algorithm
Input: Image taken in foggy weather
Output: Defogging image
Step 1: Separate RGB channels
Step 2: Perform logarithmic operation
Step 3: Determine the parameter
Step 4: Process according to the formula
Step 5: Perform antilogarithmic operation
Step 6: Merge RGB channels
```

Dark Channel Prior Algorithm

The dark channel prior algorithm is a defogging algorithm based on image restoration. Based on the atmospheric scattering model, the defogging is performed based on the imaging principle on the foggy day and the dark channel prior theory.

Therefore, our task in the defogging algorithm is to reasonably estimate $t(x, y)$ and solve $J(x, y)$ through $I(x, y)$ and A. The dark channel prior theory [7] is a theoretical model proposed by He et al. Based on a large number of fog images and no fog images. The dark channel prior algorithm is also the most common defogging algorithm at this stage. The theoretical model states that for object image $J(x, y)$, the following relationships always exist:

$$J^{\text{dark}}(x, y) = \min_{(x_i, y_j) \in \Omega(x, y)} \left(\min_{c \in \{R, G, B\}} (J^c(x, y)) \right) \to 0 \qquad (3)$$

Where $J^{\text{dark}}(x, y)$ is the dark channel image, $\Omega(x, y)$ is the area centered at (x, y), and $J^c(x, y)$ is one of R, G, B channel images. According to the form of Eq. (3), we can transform Eq. (1) and get the $t(x, y)$ estimation formula:

$$\tilde{t}(x, y) = 1 - \frac{\min\limits_{(x_i, y_j) \in \Omega(x, y)} \left(\min\limits_{c \in \{R, G, B\}} (I^c(x, y)) \right)}{A} \qquad (4)$$

The algorithm flow is shown below.

```
Algorithm: Dark Channel Prior Algorithm
Input: Image taken in foggy weather
Output: Defogging image
Step 1: Separate RGB channels
Step 2: Solve for dark channel
Step 3: Calculate the top 0.1% pixel of the brightness of the dark
channel
Step 4: Take the highest brightness point as A
Step 5: Estimated refractive index t
Step 6: Process according to the formula
Step 7: Merge RGB channels
```

3.2 AlTwo: Alexnet-Based Convolutional Neural Network

There are two main types of algorithms currently used in the field of image recognition: one is traditional machine learning methods, and the other is deep learning methods. In reference 8, the author uses traditional machine learning methods, extracts the image feature by the HOG feature extraction algorithm, and completes the classification of vehicle signs by using the SVM classifier to classify the extracted feature vectors [8]. In reference 9, the author uses deep learning to complete the classification of CT images of medical cerebral hemorrhage by building a convolutional neural network and directly training the network [9]. Compared with traditional methods, the method of building a convolutional neural network eliminates the need for manual feature extraction, is simpler, and performs better in accuracy. Here we choose to build a convolutional neural network to complete the vehicle recognition.

Among the convolutional neural network models, the more classic network models include LeNet, AlexNet, VGG and other models [10–12]. These network models all perform well. In this paper, we choose to complete the task of vehicle recognition based on the AlexNet structure. The Alexnet structure is the winner of the 2012 ImageNet competition, and it is far ahead of the artificial feature extraction method in accuracy.

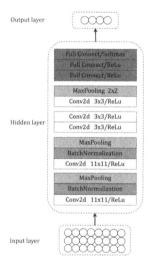

Fig. 5. AlexNet network structure

4 Experimental Design

4.1 Experiment Procedure

Combined with the technical route of the recognition process, we designed a simulation experiment flow is shown in the figure.

Use Python3.6 as the programming language, use tensorflow 1.15.0, keras 2.2.0 framework to build the neural network model structure shown in Fig. 5. the operating system CentOS, GPU model GeForce GTX 1080 (Fig. 6).

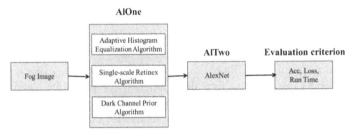

Fig. 6. Two-step recognition experiment flow

4.2 Evaluation Criterion

Recognition Accuracy
The recognition accuracy rate is the most intuitive indicator for evaluating the quality of the model. It measures the proportion of the model that is correctly classified on the test set and can intuitively reflect the generalization ability of the model.

Loss Function
The loss function is an important indicator of the neural network training process, and it is a criterion for measuring whether the network converges. When the loss of the neural network is in a state of dynamic equilibrium, the network model reaches a state of convergence. There are many ways to calculate the loss function. Here we choose the most commonly used cross entropy function.

Runtime
The processing speed of the defogging algorithm in practical applications also has a certain impact on vehicle recognition. If the time complexity of the algorithm running is too high, the collected images cannot be recognized in time, so we include the running time of AlOne as one of the evaluation criteria.

5   Results and Discussion

5.1   Recognition of Vehicle Images Under Normal Conditions

The neural network is trained 100 times, the loss function value of the training process is recorded, and a line chart of the loss function is drawn. The line chart is shown in the figure. From the image, it can be seen that during the early stages of training, the network loss function decreased steadily, and a small amplitude oscillation occurred at the 10th iteration, but did not affect the entire descent process. At the 45th training, the loss function began to be in a dynamic equilibrium state, and the network is close to convergence; at the end of 100 trainings, the network loss has dropped to very small, indicating that the network training was successful. According to the accuracy formula, the accuracy of the model in the test set is 99.27%, which indicates that the model can well complete the task of vehicle recognition under normal circumstances (Fig. 7).

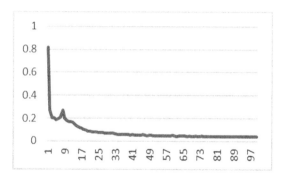

Fig. 7. Line chart of neural network training loss function

Experiments show that under normal conditions, the existing classic convolutional neural network structure can already complete the task of vehicle recognition well. The model has good generalization ability and can realize the recognition of various types of vehicles. Simulate the effects of smog on vehicle recognition. Next, we will simulate the impact of fog on vehicle recognition.

5.2 Impact of Fog on Vehicle Recognition

After fogging the test set images, we use the fog image as a test set to test the network model. The recognition accuracy of the three different fog concentrations is shown in the following Table 1:

Table 1. The recognition accuracy of different fog concentrations

Normal image	Low fog concentration	Medium fog concentration	High fog concentration
99.27%	79.63%	55.47%	49.43%

According to the experimental results, we find that in the foggy weather, the generalization ability of the model is reduced, the recognition accuracy of the vehicle is significantly reduced, and the lower the visibility, the lower the recognition accuracy. At low fog concentrations, the accuracy of the model recognition has been lower than 80%, indicating that the model can no longer be directly applied to the recognition of foggy weather vehicles recognition; under medium and high fog concentrations, the model recognition rate has decreased It is about 50%, indicating that the model has lost its recognition ability in this case. This further illustrates the impact of foggy weather on vehicle recognition.

5.3   Performance Comparison of Defogging Algorithm

Subjective Evaluation of Defogging Effect

We run different defogging algorithms on the fogged data set, and select different degrees of fog images of the same image for comparative analysis (Figs. 8, 9 and 10).

(a)   (b)   (c)   (d)

Fig. 8. Comparison of fogging effect at low fog concentration: (a) low fog concentration image (b) adaptive histogram equalization algorithm (c) single-scale Retinex algorithm (d) dark channel prior algorithm

(a)   (b)   (c)   (d)

Fig. 9. Comparison of fogging effect at medium fog concentration: (a) medium fog concentration image (b) adaptive histogram equalization algorithm (c) single-scale Retinex algorithm (d) dark channel prior algorithm

(a)   (b)   (c)   (d)

Fig. 10. Comparison of fogging effect at high fog concentration: (a) high fog concentration image (b) adaptive histogram equalization algorithm (c) single-scale Retinex algorithm (d) dark channel prior algorithm

The defogging effect of the adaptive histogram equalization algorithm has a good defogging effect at low fog concentrations, but when the fog concentration becomes larger, the defogging effect becomes inconspicuous, and a local detail loss phenomenon occurs; single-scale Retinex algorithm does not perform well in the intuitive effect of defogging, and even the phenomenon of blurred and halo appears, but the single-scale Retinex algorithm is relatively complete for the preservation of image specific details and will be helpful for the subsequent recognition process; dark The channel a priori algorithm has the most obvious defogging effect, but when processing images with high fog density, artifacts and dark colors appear. Through the comparison of the images, we can find that no matter the low fog density or the high fog density, the dark channel prior algorithm has better defogging effect.

Vehicle Recognition Accuracy

In order to better observe the generalization ability of the model, the impact on recognition accuracy is the performance index that we should pay most attention to. We use the trained vehicle model to test the recognition accuracy of the data set after running the AlOne defogging algorithm. The test results are as follows (Table 2 and Fig. 11):

Table 2. Comparison of the accuracy of defogging image recognition

	Fogging image	Normal image	Adaptive histogram equalization algorithm	Single scale Retinex algorithm	Dark channel prior algorithm
Low fog concentration	79.63%	99.27%	95.36%	93.67%	98.17%
Medium fog concentration	55.47%	99.27%	78.08%	92.69%	97.89%
High fog concentration	49.43%	99.27%	59.55%	90.58%	97.19%

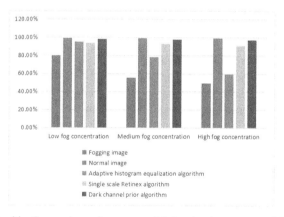

Fig. 11. Comparison of accuracy of defogging image recognition

In three foggy weathers with different concentrations, the three algorithms have certain improvement effects on the recognition accuracy. At low fog concentrations, the recognition accuracy of the three algorithms after defogging all exceeds 90%, which can meet the application requirements; but at medium fog concentrations and high fog concentrations, the adaptive histogram equalization algorithm is not obvious. The performance is improved, and the recognition accuracy is only 78% and 59%, but the recognition accuracy of the other two algorithms can still remain above 90%. By comparison, we find that the dark channel prior algorithm has the most significant improvement in accuracy, followed by the single-scale Retinex algorithm, and the improvement effect of the adaptive histogram equalization algorithm is poor.

Algorithm Runtime

We recorded the running time of the three defogging algorithms on the test set and calculated the average running time of a single picture. Table 3 is the time taken by the three algorithms to complete processing a 64×64 picture.

Table 3. Comparison of running time of defogging algorithm

	Adaptive histogram equalization algorithm	Single scale Retinex algorithm	Dark channel prior algorithm
RunTime	3.31 ms	47.61 ms	21.72 ms

By comparison, we find that the adaptive histogram equalization algorithm has the fastest processing speed and takes only 3.31 ms. The dark channel prior algorithm runs about 7 times as long as the adaptive histogram equalization algorithm. The single-scale Retinex algorithm reaches 15 times. It is obvious that the running time of the single-scale Retinex algorithm has not reached the real-time requirements when processing larger-scale images.

6   Conclusion

The AlexNet-based convolutional neural network has a vehicle recognition accuracy rate of 99.27% under normal conditions, which can be applied as a solution, but the generalization ability of the model in foggy weather is reduced, and the recognition effect is poor. Therefore, an appropriate defogging algorithm must be selected to preprocess the image, and then perform recognition of vehicle. Combining the intuitive defogging effect, recognition accuracy, and algorithm runtime, the dark channel prior algorithm performs the most prominently in the defogging process of vehicle recognition in foggy weather. The average recognition accuracy is 97.75%, which can be combined with a convolutional neural network model to complete recognition task.

References

1. Jiang, L.: The number of private car ownership in the country exceeded 200 million for the first time; 66 city car ownership exceeded one million; The number of private car ownership in China exceeded 200 million for the first time; 66 city car ownership exceeded 1 million. People's public security report (2020)

2. Fattal, R.: Single image dehazing. ACM Trans. Graph. (TOG) **27**(3), 1–9 (2008)
3. Suarez, P.L., et al.: Deep learning based single image dehazing. In: Proceedings of the IEEE Conference on Computer Vision and Pattern Recognition Workshops (2018)
4. Land, E.H.: Recent advances in Retinex theory and some implications for cortical computations: color vision and the natural image. Proc. Natl. Acad. Sci. U.S.A. **80**(16), 5163 (1983)
5. Land, E.H., McCann, J.J.: Lightness and retinex theory. Josa **61**(1), 1–11 (1971)
6. Jobson, D.J., et al.: Properties and performance of a center/surround Retinex. IEEE Trans. Image Process. **6**(3), 451–462 (1997)
7. He, K., et al.: Single image haze removal using dark channel prior. IEEE Trans. Pattern Anal. Mach. Intell. **33**(12), 2341–2353 (2010)
8. Llorca, D.F., et al.: Vehicle logo recognition in traffic images using HOG features and SVM. In: 16th International IEEE Conference on Intelligent Transportation Systems (ITSC 2013) (2013)
9. Jnawali, K., et al.: Deep 3D convolution neural network for CT brain hemorrhage classification. In: SPIE (2018)
10. Lecun, Y., et al.: Gradient-based learning applied to document recognition. Proc. IEEE **86**(11), 2278–2324 (1998)
11. Krizhevsky, A., et al.: Imagenet classification with deep convolutional neural networks. In: Advances in Neural Information Processing Systems (2012)
12. Simonyan, K., Zisserman, A.: Very deep convolutional networks for large-scale image recognition. arXiv preprint arXiv:1409.1556 (2014)

Development of a Drought Prediction System Based on Long Short-Term Memory Networks (LSTM)

Nian Zhang[1(✉)], Xiangguang Dai[2], M. A. Ehsan[1], and Tolessa Deksissa[3]

[1] Department of Electrical and Computer Engineering, University of the District of Columbia, Washington, D.C. 20008, USA
{nzhang,mdamimul.ehsan}@udc.edu
[2] Key Laboratory of Intelligent Information Processing and Control of Chongqing Municipal Institutions of Higher Education, Chongqing Three Gorges University, Chongqing 404000, China
daixiangguang@163.com
[3] Water Resources Research Institute (WRRI), University of the District of Columbia, Washington, D.C. 20008, USA
tdeksissa@udc.edu

Abstract. As streamflow quantity and drought problem become increasingly severe, it's imperative than ever to seek next generation machine learning models and learning algorithms which can provide accurate prediction. Reliable prediction of drought variables such as precipitation, soil moisture, and streamflow has been a significant challenge for water resources professionals and water management districts due to their random and nonlinear nature. This paper proposes a long short-term memory networks (LSTM) based deep learning method to predict the historical monthly soil moisture time series data based on the MERRA-Land from 1980 to 2012. The proposed LSTM model learns to predict the value of the next time step at each time step of the time sequence. We also compare the predication accuracy when the network state is updated with the observed values and when the network state is updated with the predicted values. We find that the predictions are more accurate when updating the network state with the observed values instead of the predicted values. In addition, it demonstrated that the proposed method has much lower MSE than the autoregressive integrated moving average model (ARIMA) model and autoregressive model (AR) model.

Keywords: Convolutional neural networks · Long short-term memory networks (LSTM) · Deep learning · Time series prediction · Drought prediction

1 Introduction

The streamflows of a river basin may be near or below normal, influenced by lower than normal precipitation and much below normal soil moisture contents. If below average rainfall continues then further degradation is expected to occur. Monthly monitoring of

© Springer Nature Switzerland AG 2020
M. Han et al. (Eds.): ISNN 2020, LNCS 12557, pp. 142–153, 2020.
https://doi.org/10.1007/978-3-030-64221-1_13

a river basin will prepare for the possibility that serious drought conditions may develop in the future [1]. As drought and streamflow quantity problem become increasingly severe, it's imperative to provide an effective drought early warning system which uses the historical data to make prediction of the probability of flows dropping below drought trigger levels [2]. Reliable estimation of streamflow has been a significant challenge for water resources professionals and water management districts. This is very much essential to manage water supply, floods, and droughts efficiently. Streamflow characteristics are primarily governed by climatic and watershed characteristics. Over the last decade it has been recognized that climate is changing and there can be significant impacts on the streamflow. The hydraulic consequences of a climate change can cause natural disaster, such as drought that occurs when there is a significant deficit in precipitation. It will also have serious impact to flooding, water quality, and ecosystems that are closely related to the society of human beings. Accurate estimation of streamflow quantity from a watershed will provide important information to determine urban watershed modeling, water quantity management, development of legislation, and strategies on water supply. In addition, drought prediction is one of most complicated and difficult hydrological problems because the nature of the drought variables is random and unpredictable and the physical processes underneath the phenomenon are too complex. It is also because of the insufficient knowledge on the driving factors and their impact on streamflow, as well as the lack of reliable prediction and design methodologies. Therefore, accurate drought prediction including streamflow quantity prediction, precipitation and soil moisture prediction are all critical to enhance the water resource management plan and operational performance assessment.

Recently some deep learning algorithms have been successfully applied to the water quantity prediction and drought prediction problems. A deep Belief Network layered by coupled Restricted Boltzmann Machines was proposed for long-term drought prediction across the Gunnison River Basin [3]. By using time lagged standardized streamflow index (SSI) sequence, it demonstrated lower error rate than multilayer perceptron neural network and support vector regression. A long short-term memory (LSTM) network was presented for streamflow prediction using previous streamflow data for a particular period [4]. It showed that the LSTM model can not only predict the relatively steady streamflow in the dry season, but can also capture data characteristics in the rapidly changing streamflow in the rainy season. However, the performance of LSTM hasn't been proved on the effect of drought variables, such as precipitation, soil moisture, streamflow for long-term drought prediction.

The novelty of this paper is the inclusion of a wider ranged hydrological variables to predict soil moisture content (%) for a higher elevation of interest using existing regression models, which differentiates this work from previously done research works as described in the literature. In addition, this paper presents how to design the architecture of the model and layer specifications to the time series prediction problem. Further it customizes the LSTM based time series model to solve the drought prediction problem. It describes the proposed long short-term memory networks (LSTM) based deep learning method to predict the historical monthly soil moisture time series data.

The rest of this paper is organized as follows. Section 2 describes the methodology including deep learning approach, deep neural network, and convolutional neural network. In Sect. 3, time series prediction using LSTM network is discussed. Performance evaluation metrics are presented in Sect. 4. In Sect. 5, the Modern-Era Retrospective analysis for Research and Applications (MERRA)-Land data set is described. The simulations and experimental results are demonstrated. In Sect. 6, the conclusions are given.

2 Methodology

Many state-of-the-art machine learning techniques, such as neural network, support vector machine, radial basis function, naive Bayes, decision tree, k-nearest neighbors, and deep learning have been applied to the time series prediction. However, few of them has been applied to the forecast of the probability of streamflows. These machine learning methods have been proven effective in predicting time series. Since streamflow prediction is a special case of time series prediction, therefore, they should be very promising in the streamflow prediction problems.

2.1 Deep Learning Approach

Deep learning algorithms are now applied to solve problems of a diverse nature, including prediction [5]. Therefore, we are considering deep learning algorithms for this research. Firstly, we would like to review a few basics of deep learning. The building blocks of deep learning or artificial neural networks are called perceptron, which mimics an equivalent functionality (in computation) as neuron (a biological cell of the nervous system that uniquely communicates with each other) [6].

Now, perceptron or artificial neurons receive input signals $(x_1, x_2, \ldots, x_m)$, multiply input by weight $(w_1, w_2, \ldots, w_m)$, add them together with a pre-determined bias, and pass through the activation function, $f(x)$. The signal goes to output as 0 or 1 based on the activation function threshold value. A perceptron with inputs, weights, summation and bias, activation function, and output all together forms a single layer perceptron. However, in common neural network diagrams, only input and output layers are shown. In a practical neural network, hidden layers are added between the input and output layers. The number of hidden layers is a hyperparameter and usually determined by evaluating the model performance. If the neural network has a single hidden layer, the model is called a shallow neural network, while a deep neural network consists of several hidden layers. In this research, we have considered DNN, convolutional neural network, and recurrent neural network in the form of long short-term memory, all of which will be discussed in the following sections.

2.2 Deep Neural Network (DNN)

DNN is composed of three neural network layers, namely an input layer, hidden layers, and an output layer. The number of hidden layers is tuned through trial and error [6]. Figure 1 illustrates such a model structure with two hidden layers consisting of three neurons each, five input neurons, and one output neuron. The number of neurons depends on the number of inputs and outputs. In Fig. 1,

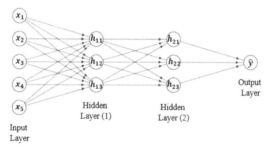

Fig. 1. Simplified architecture of a deep neural network

Inputs: $[x_1, x_2, x_3, x_4, x_5]$
Hidden layer weights: h
Output: $\hat{y}$

$$f(x; W, c, w, b) = w^T \max(0, W^T + c) + b \tag{1}$$

$$h = g(W^T x + c) \tag{2}$$

$$f(x) = \max(0, x) \tag{3}$$

A simplified DNN kernel is formulated in (1) that considers linear modeling. x, W, and c symbolize input, weights, and bias, respectively, while w and b are linear model parameters. The hidden layer parameter h is shown in (2), where g is the activation function. For DNN modeling, ReLu (3) is used as the hidden layer activation function.

3 Proposed Method

This section describes the proposed long short-term memory networks (LSTM) based deep learning method to predict the historical monthly soil moisture time series data. It presents how to design the architecture of the model and layer specifications to the time series prediction problem. Further it customizes the LSTM based time series model to solve the drought prediction problem.

3.1 Time Series Prediction Using LSTM Network

An LSTM network inherits the characteristic of memory from the recurrent neural network (RNN) [7]. This memory unit enables long-term feature retention between time steps of sequence data [8]. Figure 2 illustrates the flowchart of a time series X with C features of length S through an LSTM layer. The output layer will generate the predicted values, which contains D features of length S. In the diagram, for the tth LSTM block, h_t and c_t denote the output, i.e. the hidden state and the cell state at time step t, respectively.

Initially, the states of all the LSTM blocks will be initialized to all zeros. The first LSTM block to the left most uses the initial state of the network and the first time step

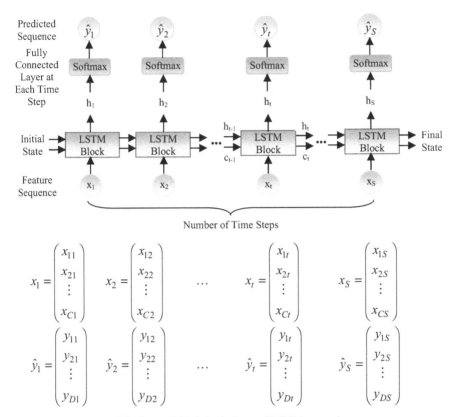

Fig. 2. Unfolded single layer of LSTM network

of the sequence to compute the first output, h_1 and the updated cell state, c_1. At time step t, the tth LSTM block uses the current state of the network (h_{t-1}, c_{t-1}) and the tth time step of the sequence to compute the output state, h_t, and the cell state, c_t.

An LSTM layer contains an array of LSTM blocks. For each LSTM block, it is represented by two states, including an output state, i.e. the hidden state and a cell state. The hidden state at time step t not only contains the output of the current LSTM block for the time step, but also serves as the input for the LSTM block at the next time step. The cell state contains time dependent information extracted from the previous time steps.

Different from the classic RNN, the LSTM is a recurrent neural network equipped with gates [9]. At each time step, the LSTM layer can choose either add information to or removes information from the cell state. The layer controls these updates using gates. The gated circuit of the LSTM is proposed to implement the flow of data at time step t, as illustrated in Fig. 3. LSTM introduces self-loops to produce paths where the gradient can flow for a long duration; thus, it is capable of learning long-term dependencies [6].

The equations describing the operations are listed below.

$$f(t) = \sigma_g (W_f x_t + U_f h_{t-1} + b_f) \tag{4}$$

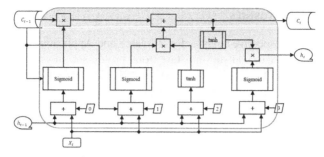

Fig. 3. Block diagram of LSTM operations on a time series sequence

$$i_t = \sigma_g(W_i x_t + U_i h_{t-1} + b_i) \tag{5}$$

$$o_t = \sigma_g(W_o x_t + U_o h_{t-1} + b_o) \tag{6}$$

$$c_t = f_t \circ c_{t-1} + i_t \circ \sigma_c(W_c x_t + U_c h_{t-1} + b_c) \tag{7}$$

$$h_t = o_t \circ \sigma_h(c_t) \tag{8}$$

where,

$x_t \in \Re^d$: Input vector to the LSTM unit
$f_t \in \Re^h$: Forget states activation vector
$i_t \in \Re^h$: Input/update gate's activation vector
$o_t \in \Re^h$: Output gate's activation vector
$h_t \in \Re^h$: Hidden state vector
$c_t \in \Re^h$: Cell state vector
$W \in \Re^{h \times d}$, $U \in \Re^{h \times h}$, $b \in \Re^h$: Weight matrices and bias vector parameters which will be adjusted during the training
σ_g: Sigmoid function
σ_c, σ_g: hyperbolic tangent function

In the performance evaluation, some commonly used accuracy parameters, such as root mean square error are employed to evaluate how well a model is performing to predict the intended parameter. Root mean square error (RMSE) is considered to investigate the model performances on the test set by comparing the differences between the predicted values by a model and the actual values. RMSE is the square root of the mean of the square of error terms (the difference between actual response (y_i) and predicted response ($\hat{y}_i$). n is the number of total input sets. The lower this value is, the better the model performance, while the desired is 0 or close value for this term. The formula for this measure is in (9).

$$RMSE = \sqrt{\frac{\sum_{i=1}^{n}(y_i - \hat{y}_i)^2}{n}} \tag{9}$$

4 Experimental Results

The Modern-Era Retrospective analysis for Research and Applications (MERRA) data set is used to use the historical soil moisture (total profile soil moisture content) from 1980 to 2012 to predict the future soil moisture [10]. The data set is plotted in Fig. 4. We train on the first 90% of the time series sequence and test on the last 10%. In order to obtain the identical data scale for different features, it is necessary to pre-process the raw data by standardizing the data to a normalized distribution. Within the scope of zero mean and unit variance, we prevent the training data, test data, and predicted responses from diverging.

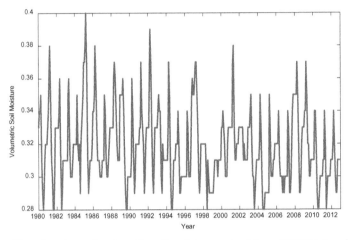

Fig. 4. Monthly soil moisture (total profile soil moisture content) from MERRA-Land from 1980 to 2012

To forecast the values of a sequence at future time steps, we use the responses with values lagged by one time step to be the training sequences. The nonlinear autoregressive (NAR) model can be represented mathematically by predicting the values of a sequence at future time steps, $\hat{y}$ from the historical values of that time series, as shown in Fig. 5. Time series without the final time step are used as the training sequences. The form of the prediction can be expressed as follows:

$$\hat{y}(t) = f(y(t-1), \ldots, y(t-d)) \tag{10}$$

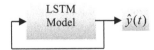

Fig. 5. Nonlinear autoregressive LSTM prediction model

We set up a LSTM network in the sequence-to-one regression mode. The output of the LSTM layer is the last element of the sequence and will be fed into the fully connected layer. For example, if the input sequence is $\{x_1, x_2, x_3, x_4\}$, the output of the LSTM layer will be the hidden state,h_4. In this LSTM network, it consists of a sequence input layer, an LSTM layer, a fully connected layer, and a regression output layer. In the LSTM layer array, a sequence input layer inputs one sequence data to a network at a time. This LSTM layer contains 200 hidden units. We use the adam optimization algorithm featured with adjustable learning rate to train the dynamic neural networks for 600 epochs. To ensure a steady gradient change, we limit the threshold of the gradient to 1.

After several trials, we decide to set the initial learning rate to 0.005 to gain better performance. We slow down the learning rate to 20% of its original value when it has elapsed 150 epochs. Then we train the LSTM network with these parameter selections. The training progress is plotted in Fig. 6. The top subplot reveals the root-mean-square error (RMSE) calculated from the standardized data. The bottom subplot displays the error between the actual values and the predicted values.

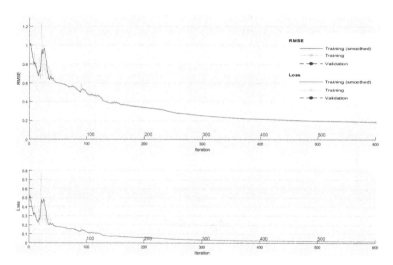

Fig. 6. Training progress on the monthly soil moisture using LSTM network

Once the LSTM network has been trained, we will predict time steps one at a time and update the network state at each prediction. Therefore, we can forecast the values of multiple time steps in the future. Like what we did for the training data, we standardize the test data using the same mean of the population, μ and the standard deviation of the population, σ. In order to initialize the network state, h, we first predict on the training data. Then *we use the value at the last time step of the training response to make the very first prediction.* We then use Eq. (10) to *use the previous prediction* to predict value at the next time step one at a time for the remaining predictions. We un-standardize the predictions in order to observe the real world values of the soil moisture. The combination of training time series (in blue) with the forecasted values (in red) is shown in Fig. 7.

In order to visually compare the forecasted values with the actual data, we plot the first 40 predicted values at the time steps over the actual values, as shown in Fig. 8. We

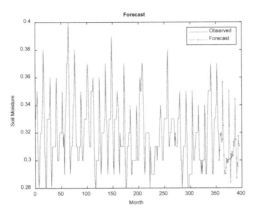

Fig. 7. Training time series with the forecasted values (Color figure online)

also display the difference between them at each time step and the RMSE, i.e. 0.019717 from the unstandardized predictions in the lower subplot in Fig. 8.

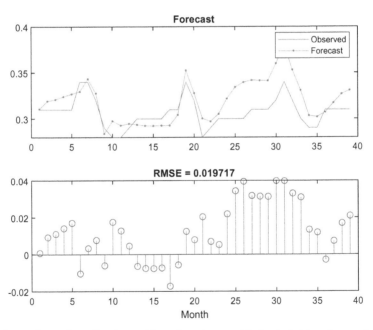

Fig. 8. Comparison of predicted monthly soil moisture with the test data when updating the network state with previous predictions

We further explore the prediction performance by updating network state with observed values. Unlike the previous study where we used the previous prediction to predict value, we update the network state with the actual (observed) values instead of the predicted values. We first initialize the network state by *resetting the network state to an initial state of zeros.* Therefore, previous predictions will not affect the predictions on the new time sequence. We then initialize the network state by start predicting on the training data. At each time step, *we predict the value on the next time step using the observed value of the previous time step.* In order to retrieve the soil moisture information, we un-standardize the predictions using the same mean and the standard deviation of the population as before.

Similarly, we compare the forecasted values with the actual test data for the first 40 time steps, as shown in Fig. 9. We also demonstrate the difference between them at each time step, as well as the RMSE, i.e. 0.0087584 from the unstandardized predictions in the bottom subplot. After comparing Fig. 8 and Fig. 9, we find that the prediction accuracy is much higher when we update the network state with the observed values instead of using the predicted values.

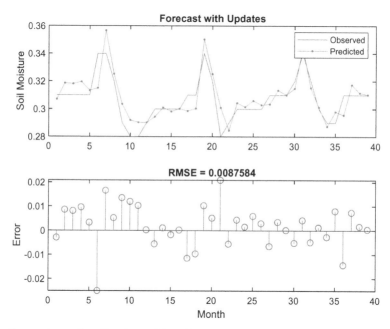

Fig. 9. Comparison of predicted monthly soil moisture with the test data when updating the network state with the observed values

We also compare the performance of the proposed LSTM deep learning model with other popular predictive models, such as autoregressive integrated moving average model (ARIMA) and autoregressive model (AR). Table 1 depicts the root mean squared error (RMSE) for each algorithm on the test data. We found that our proposed algorithm has the lowest error rate.

Table 1. Comparative model performances

Algorithm	Root Mean Squared Error (MSE)
Autoregressive Integrated Moving Average model (ARIMA)	0.0950
Autoregressive model (AR)	0.0246
Proposed LSTM model	**0.0088**

5 Conclusions

This paper proposes a long short-term memory networks (LSTM) based deep learning method to predict the historical monthly soil moisture time series data based on the MERRA-Land from 1980 to 2012. The proposed LSTM model learns to predict the value of the next time step at each time step of the time sequence. We customize the dynamic LSTM model to solve the soil moisture prediction problem. We also compare the predication accuracy when the network state is updated with the observed values and when the network state is updated with the predicted values. We find that the predictions are more accurate when updating the network state with the observed values instead of the predicted values. Furthermore, we also compare the proposed method with other time series prediction methods. We find that it has much lower MSE than the autoregressive integrated moving average model (ARIMA) model and autoregressive model (AR) model. The future study will to obtain the soil moisture index, and use it to predict the Drought index. The drought prediction system will have profound impact to the water resources management, agriculture, and urban construction.

Acknowledgment. This work was supported in part by the National Science Foundation (NSF) grants #1505509 and #2011927, DoD grants #W911NF1810475 and #W911NF2010274, NIH grant #1R25AG067896-01, and USGS grant #2020DC142B, and in part by the Foundation of Chongqing Municipal Key Laboratory of Institutions of Higher Education ([2017]3), Foundation of Chongqing Development and Reform Commission (2017[1007]), Scientific and Technological Research Program of Chongqing Municipal Education Commission (Grant Nos. KJQN201901218 and KJQN201901203), Natural Science Foundation of Chongqing (Grant No. cstc2019jcyj-bshX0101), Foundation of Chongqing Three Gorges University.

References

1. Water Supply Outlook and Status, Interstate Commission on the Potomac River Basin. https://www.potomacriver.org
2. Hao, Z., AghaKouchak, A., Nakhjiri, N., Farahmand, A.: Global integrated drought monitoring and prediction system. Sci. Data 1(140001), 1–10 (2014)
3. Agana, N.A., Homaifar, A.: A deep learning based approach for long-term drought prediction. In: SoutheastCon 2017, Charlotte, USA, 30 March–2 April, pp. 1–8 (2017)
4. Fu, M., Fan, T., Ding, Z., Salih, S.Q., Al-Ansari, N., Yaseen, Z.M.: Deep learning data-intelligence model based on adjusted forecasting window scale: application in daily streamflow simulation. IEEE Access 8, 32632–32651 (2020)

5. Filik, Ü.B., Filik, T.: Wind speed prediction using artificial neural networks based on multiple local measurements in eskisehir. Energy Procedia **107**, 264–269 (2017)
6. Goodfellow, I.: Deep Learning. MIT Press, Cambridge (2016)
7. Rochac, J.F.R., Zhang, N., Xiong, J.: A spectral feature based CNN long short-term memory approach for classification. In: The Tenth International Conference on Intelligent Control and Information Processing (ICICIP), Marrakesh, Morocco, 14–19 December (2019)
8. Ehsan, M.A., Shahirinia, A., Zhang, N., Oladunni, T.: Interpretation of deep learning on wind speed prediction and visualization. In: The 10th International Conference on Information Science and Technology (ICIST), Bath, London, and Plymouth, United Kingdom, 9–15 September (2020)
9. Rochac, J.F.R., Zhang, N., Xiong, J, Zhong, J., Oladunni, T.: Data augmentation for mixed spectral signatures coupled with convolutional neural networks. In: The 9th International Conference on Information Science and Technology (ICIST 2019), Hulunbuir, Inner Mongolia, China, 2–5 August (2019)
10. Koster, R.D., De Lannoy, G.J.M., Forman, B.A., Liu, Q., Mahanama, S.P.P., Touré, A.: Assessment and enhancement of MERRA land surface hydrology estimates. J. Clim. **24**, 6322–6338 (2011)

Deep Point Cloud Odometry: A Deep Learning Based Odometry with 3D Laser Point Clouds

Chi Li[1], Yisha Liu[2], Fei Yan[1], and Yan Zhuang[1(✉)]

[1] The School of Control Science and Engineering, Dalian University of Technology, Dalian 116024, China
lichiduter@mail.dlut.edu.cn, {fyan,zhuang}@dlut.edu.cn
[2] Information Science and Technology College, Dalian Maritime University, Dalian 116026, Liaoning, China
liuyisha@dlmu.edu.cn

Abstract. Deep learning-based methods have attracted more attention to the pose estimation research that plays a crucial role in location and navigation. How to directly predict the pose from the point cloud in a data-driven way remains an open question. In this paper, we present a deep learning-based laser odometry system that consists of a network pose estimation and a local map pose optimization. The network consumes the original 3D point clouds directly and predicts the relative pose from consecutive laser scans. A scan-to-map optimization is utilized to enhance the robustness and accuracy of the poses predicted by the network. We evaluated our system on the KITTI odometry dataset and verified the effectiveness of the proposed system.

Keywords: Pose estimation · 3D point clouds · Deep learning

1 Introduction

Laser odometry is widely used for autonomous driving and robot localization, which has been achieved great success. Classic laser odometry systems estimate poses by the laser registration methods, such as Iterative Closest Point (ICP) [1], Normal Distribution Transform (NDT) [11], and their variants [16,17,19]. Registration methods tend to be unreliable in some challenging scenarios, e.g., featureless places and motion with significant angular changes. Because of the sparsity of the point clouds caused by the low resolution of the laser scanner, the matching algorithm may not find the corresponding points or features, which may bring the drifts or even errors to the pose estimation.

Y. Zhuang—This work was supported in part by the National Natural Science Foundation of China under grant 61973049 and U1913201.

ⓒ Springer Nature Switzerland AG 2020
M. Han et al. (Eds.): ISNN 2020, LNCS 12557, pp. 154–163, 2020.
https://doi.org/10.1007/978-3-030-64221-1_14

In recent years, the deep learning-based methods have attracted much attention in the research of geometry problems such as localization, relative pose estimation and odometry system. Many learning-based works are achieving state-of-the-art results in the field of visual odometry. Zhou et al. [24] presented a unsupervised training method to estimate the ego-motion from video. A novel Recurrent Convolutional Neural Network based VO system is proposed by Wang et al. [21] for dealing with the sequences data. [10] developed a unsupervised visual odometry which can estimate absolute scale and dense depth map simultaneously. Moreover, there are also a few laser odometry systems achieved in a data-driven fashion. [12] utilized the vanilla CNN (Convolutional Neural Network) for a laser odometry. Deep learning based 2D scan matching method is proposed by Li et al. [8] and [22] integrated deep semantic segmentation for the pose estimation.

Unlike regular data formats like images, the point cloud is unordered and sparse, which makes it difficult for the laser odometry to use the verified pipeline of the data-driven visual odometry. Some methods convert the point clouds into a structured representation for using the 2D or 3D convolution to extract the feature to estimate the ego-motion. [20] transformed the spare point clouds into the multi-channel dense matrix and employed the CNN to achieved the IMU assisted laser odometry. Qing Li et al. encoded the point clouds into the image-like formats by cylindrical projection and constructed a learning-based laser odometry. [9] DeepLO [2] proposed a deep LiDAR odometry via supervised and unsupervised frameworks using the regular point cloud representation. The projection lost the information of the original point cloud, so it is worth exploring to use point clouds to directly estimate odometry. Some works, like PointNet [14,15], have made deep learning based on point cloud directly become a research hotspot.

In this paper, we propose a deep learning-based laser odometry using the point clouds as the input. Our main contributions are as follows: 1) We propose a scan-to-scan laser pose estimation network that directly consumes the irregular point clouds. 2) We use local map optimization to improve the robustness of network estimation, which makes up the laser odometry.

The rest of this paper is organized as follows. Section 2 shows an overview of the system. In Sect. 3, the proposed the system is presented. Experimental results are given in Sect. 4. The conclusions are drawn in Sect. 5.

2   System Overview

In this section, we briefly show our system, which is composed of a relative pose estimator and a local map pose optimizer, as shown in Fig. 1.

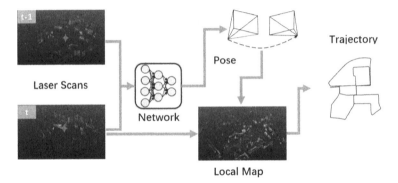

Fig. 1. System overview of the proposed deep point cloud odometry. The laser point cloud input into a network directly, and relative pose prediction of the network is further optimized by a local map matching.

The pose estimator is a PointNet-based CNN architecture, which is used to process the point cloud directly. It takes two consecutive point clouds as input and predicts the relative 6-DoF pose between them.

The pose optimizer is based on the ICP algorithm, which is used for point registration. The inputs of it are the relative pose predicted by pose estimator, the current point cloud, and the local map, and then it fine-tunes the pose by matching the point cloud to the local map.

Pose estimation only accumulating the scan-to-scan estimation tends to bring the errors over time, so the local map optimization is utilized to reduce the impact of cumulative errors.

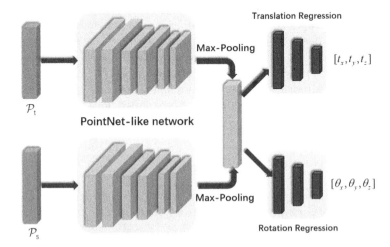

Fig. 2. Architecture of the network in proposed laser odometry system.

3   Pose Estimation with the Point Clouds

This section presents the proposed point clouds odometry composed of the deep pose estimation and local map pose optimization in detail.

3.1   Relative Pose Regression Through Convolutional Neural Networks

To estimate the relative pose of two consecutive laser scans, we train a network consisted of CNN-based feature extraction and a pose regression. The original points are used as the input of the network because they contain all the information which is needed to match.

The PointNet-like CNN architecture is employed to extract the feature of the point cloud, and then the features from different scans are combined and sent to the regressor to estimate the relative pose. As the Fig. 2 shows, the network takes two point clouds from consecutive laser scans: target point cloud $\mathcal{P}_t$ and source point cloud $\mathcal{P}_s$ as inputs and produce the 6-DoF relative pose: translation $t = [t_x, t_y, t_z]^T$ and rotation in the form of Euler angle $\boldsymbol{\theta} = [\theta_{roll}, \theta_{pitch}, \theta_{yaw}]^T$ as output

$$t, \boldsymbol{\theta} = \mathcal{F}(\mathcal{P}_t, \mathcal{P}_s). \tag{1}$$

We use $\mathcal{L}_t$ and $\mathcal{L}_r$ as the loss function to train the network.

$$\begin{aligned} \mathcal{L}_t &= \|\hat{t} - t^\star\|_2^2 \\ \mathcal{L}_r &= \|\hat{\boldsymbol{\theta}} - \boldsymbol{\theta}^\star\|_2^2 \end{aligned} \tag{2}$$

where $\hat{t}$ and $\hat{\boldsymbol{\theta}}$ are the output of the network, $t^\star$ and $\boldsymbol{\theta}^\star$ are the ground truth. We use the ℓ_2-norm in this work.

For training the network to learn the translation and rotation simultaneously, it is necessary to use a weight regularizer λ to balance the rotational loss with translational loss, because the scale and units between the translational and rotational pose components are different. To learn translation and rotation without including any hyperparameters, [6] presented a loss function that can learn the weight regularizer.

$$\mathcal{L}_{pose} = \mathcal{L}_t \exp(-s_t) + s_t + \mathcal{L}_r \exp(-s_r) + s_r \tag{3}$$

where s_t and s_r are the learnable parameters to regularize the scale between the translational and rotational losses.

3.2   Pose Optimization with Local Map

The pose optimization employs a scan-to-map matching with the geometry method to fine-tune the poses predicted by the network.

If the scan-to-scan matching creats errors, the rest of the trajectory will be affected by the errors. We propose maintaining a local map that can be used to match the current scan for geometric constraints to modify the errors. The local map can improve the robustness of the odometry when some scan-to-scan matching creates errors.

An ICP is designed to register the current scan to the local map in the pose optimization, which takes the current scan, local map, and relative pose as input and computes the refined pose as output.

$$\Delta \hat{T} = \arg \min_{\Delta T} \frac{1}{2} \sum_{j=1}^{N} \| \Delta T p_j - p_{m(j)} \|_2^2 \tag{4}$$

where $p_j \in \mathcal{P}_s$ is the point in the source point cloud, $p_{m(j)} \in \mathcal{P}_m$ is p_i's corresponding point in the local map, and $\Delta \hat{T}$ is the refined relative pose in the form of the special Euclidean group $SE(3)$ of transformations. The pose predicted by the network is used as the initial pose of the ICP. The ICP uses Eq. (4) as the cost function to match the scan to the local map iteratively and estimates the refined pose.

The local map contains historical point clouds over time, which needs to be maintained and updated. The local map updating comprises two steps: one step is removing the points that are outside the field of view from the local map which keeps the number of points in the local map not large, thereby map points culling can improve computational efficiency by reducing the computational complexity of searching for corresponding points, the other one is to add the points of the current scan to the local map, so that makes the local map has more extra feature points.

Table 1. Absolute translation errors (RMSE) of the test data from KITTI

	03	04	05	07	10
Ours	4.873021	1.258067	5.221578	1.186617	14.592466
LeGO-LOAM	2.965074	0.511566	9.223725	0.921545	9.844019
FGR+ICP	3.173614	1.608129	29.039456	4.574337	14.550793

4   Experimental Results

In this section, we evaluate the performance of the proposed point cloud odometry. The network model is trained and tested by using publicly available datasets, KITTI odometry dataset [4]. The experimental results of local map optimization are also given in this section.

4.1   Implementation

We implemented the proposed system using PyTorch [13] and PyTorch Geometric [3], and trained the network with an NVIDIA RTX 2080ti. The optimizer employed the Adam Optimizer [7] to train the network with parameter $\beta_1 = 0.9$ and $\beta_2 = 0.99$. The learning rate was initialized with 0.001 and decreased by 0.1 every 10 epochs until $1 * 10^{-6}$. The parameters s_t and s_r in Eq. (3) were set 0.0 and -2.0 respectively.

4.2   Dataset

The KITTI odometry dataset is a well-known public dataset of odometry benchmark. The dataset provides camera images, point clouds, Inertial Measurement Unit and other sensor data. We mainly use the point clouds which are captured by a Velodyne HDL-64E laser sensor. The dataset includes many driving scenarios, such as urban, streets, and highways. Sequence 00–10 of all 22 sequences of the dataset provide the ground-truth pose collected by the GPS/IMU sensor.

Our network was trained on sequences 00, 01, 02, 06, 08, and 09 and tested on sequences 03, 04, 05, 07, and 10. The point clouds inputted to the network were removed the grounds that may bias the evaluation results.

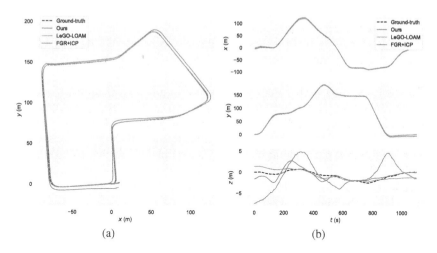

Fig. 3. Trajectories of KITTI sequence 07. The left figure (a) plots in XY plane. The right one (b) shows trajectories on the X, Y and Z axis respectively.

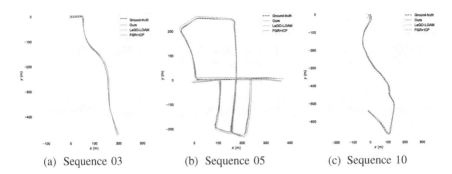

| (a) Sequence 03 | (b) Sequence 05 | (c) Sequence 10 |

Fig. 4. Trajectories on test datasets.

4.3 Odometry Evaluation

We use averaged Root Mean Square Errors (RMSEs) of the pose errors to evaluate our system's performance. The results of the evaluation of test datasets are
shown in Table 1. The algorithms used to compare are LeGO-LOAM [18] and
Fast global registration [23] with ICP fine-tuning, and all of the algorithms do
not implement the loop closure detection. LeGo-LOAM is a state-of-the-art laser
odometry system, which is the variant of LOAM, the top laser-based method in
the KITTI odometry dataset. Fast global registration (FGR) is a global matching
algorithm that is insensitive to an initial value and combines the local matching
algorithm, ICP, to improve the pose estimation accuracy. We use the Evo tool
[5], a python package for the evaluation of odometry and SLAM, to evaluate the
experimental results of odometry.

Figures 3 and 4 show the predicted trajectories of the test datasets, in which
the black dashed line is ground truth, the blue line is the proposed method, the
green line is the LeGO-LOAM, and the red one is the FGR + ICP. It can be
seen that the proposed system can provide nice results on the test datasets. This
proves the proposed point cloud odometry is capable of learning to estimate the
poses. From the details of the results in Table 1, we can see our system is not the
best of all methods, so our algorithm also needs to improve performance, which
can be achieved by training with more data.

4.4 Pose Optimization Evaluation

Figure 5 shows the comparisons of the pose predicted by the network with and
without the local map optimization on the test dataset. The trajectories are on
the top row, where the black dashed line is ground truth, the blue line is the
result of after pose optimization, and the purple line is the output of the network.
We utilized the box plot to show the error statistics on the bottom row. The top
and bottom of the box are the 25th and 75th percentiles; the centerline is the
median, and whiskers show the minimum and maximum errors.

From Fig. 5, it can be seen that the trajectories after optimization are more
accurate than before optimization. Meanwhile, pose optimization also improves

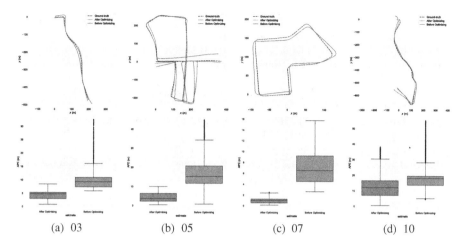

(a) 03   (b) 05   (c) 07   (d) 10

Fig. 5. Top row: trajectories of the pose estimation with and without local map optimization. Bottom row: box plot of the pose estimation without and with local map optimization which show the error statistics.

the system's robustness. This proves that pose optimization is useful for the whole system and can help the network improve performance.

5   Conclusions

In this paper, we have presented deep point cloud odometry, a deep learning-based odometry with the point clouds. It estimated the poeses by using the irregular point clouds directly and employed the local map optimization to improve the accuracy and robustness of odometry estimation. The results of the experiment showed that the proposed system could estimate the trajectories on the public dataset. In our future work, we plan to improve the generalization ability of the network to adapt to different resolution of the laser sensors and implement the deep learning-based method to map the point clouds.

References

1. Besl, P.J., McKay, N.D.: Method for registration of 3-D shapes. In: Sensor Fusion IV: Control Paradigms and Data Structures, vol. 1611, pp. 586–606. International Society for Optics and Photonics (1992)
2. Cho, Y., Kim, G., Kim, A.: DeepLO: geometry-aware deep lidar odometry. arXiv preprint arXiv:1902.10562 (2019)
3. Fey, M., Lenssen, J.E.: Fast graph representation learning with PyTorch geometric. In: ICLR Workshop on Representation Learning on Graphs and Manifolds (2019)
4. Geiger, A., Lenz, P., Urtasun, R.: Are we ready for autonomous driving? The KITTI vision benchmark suite. In: 2012 IEEE Conference on Computer Vision and Pattern Recognition (CVPR), pp. 3354–3361. IEEE (2012)

5. Grupp, M.: EVO: Python package for the evaluation of odometry and SLAM (2017). https://github.com/MichaelGrupp/evo
6. Kendall, A., Cipolla, R.: Geometric loss functions for camera pose regression with deep learning. In: 2017 IEEE Conference on Computer Vision and Pattern Recognition (CVPR), pp. 6555–6564 (2017)
7. Kingma, D.P., Ba, J.: Adam: A method for stochastic optimization. arXiv preprint arXiv:1412.6980 (2014)
8. Li, J., Zhan, H., Chen, B.M., Reid, I., Lee, G.H.: Deep learning for 2D scan matching and loop closure. In: 2017 IEEE/RSJ International Conference on Intelligent Robots and Systems (IROS), pp. 763–768. IEEE (2017)
9. Li, Q., Chen, S., Wang, C., Li, X., Wen, C., Cheng, M., Li, J.: LO-Net: deep real-time lidar odometry. In: Proceedings of the IEEE Conference on Computer Vision and Pattern Recognition (CVPR), pp. 8473–8482 (2019)
10. Li, R., Wang, S., Long, Z., Gu, D.: UnDeepVO: monocular visual odometry through unsupervised deep learning. In: 2018 IEEE International Conference on Robotics and Automation (ICRA), pp. 7286–7291. IEEE (2018)
11. Magnusson, M.: The three-dimensional normal-distributions transform: an efficient representation for registration, surface analysis, and loop detection. Ph.D. thesis, Örebro universitet (2009)
12. Nicolai, A., Skeele, R., Eriksen, C., Hollinger, G.A.: Deep learning for laser based odometry estimation. In: RSS workshop Limits and Potentials of Deep Learning in Robotics, vol. 184 (2016)
13. Paszke, A., et al.: PyTorch: an imperative style, high-performance deep learning library. In: Advances in Neural Information Processing Systems, pp. 8026–8037 (2019)
14. Qi, C.R., Su, H., Mo, K., Guibas, L.J.: PointNet: deep learning on point sets for 3D classification and segmentation. In: Proceedings of the IEEE Conference on Computer Vision and Pattern Recognition (CVPR), pp. 652–660 (2017)
15. Qi, C.R., Yi, L., Su, H., Guibas, L.J.: PointNet++: deep hierarchical feature learning on point sets in a metric space. In: Advances in Neural Information Processing Systems, pp. 5099–5108 (2017)
16. Segal, A., Haehnel, D., Thrun, S.: Generalized-ICP. In: Robotics: Science and Systems, Seattle, WA , vol. 2, p. 435 (2009)
17. Serafin, J., Grisetti, G.: NICP: dense normal based point cloud registration. In: 2015 IEEE/RSJ International Conference on Intelligent Robots and Systems (IROS), pp. 742–749. IEEE (2015)
18. Shan, T., Englot, B.: LeGO-LOAM: lightweight and ground-optimized lidar odometry and mapping on variable terrain. In: 2018 IEEE/RSJ International Conference on Intelligent Robots and Systems (IROS), pp. 4758–4765. IEEE (2018)
19. Stoyanov, T., Magnusson, M., Andreasson, H., Lilienthal, A.J.: Fast and accurate scan registration through minimization of the distance between compact 3D NDT representations. Int. J. Robot. Res. **31**(12), 1377–1393 (2012)
20. Velas, M., Spanel, M., Hradis, M., Herout, A.: CNN for IMU assisted odometry estimation using velodyne LiDAR. In: 2018 IEEE International Conference on Autonomous Robot Systems and Competitions (ICARSC), pp. 71–77. IEEE (2018)
21. Wang, S., Clark, R., Wen, H., Trigoni, N.: End-to-end, sequence-to-sequence probabilistic visual odometry through deep neural networks. Int. J. Robot. Res. **37**(4–5), 513–542 (2018)
22. Wong, J.M., et al.: SegICP: integrated deep semantic segmentation and pose estimation. In: 2017 IEEE/RSJ International Conference on Intelligent Robots and Systems (IROS), pp. 5784–5789. IEEE (2017)

23. Zhou, Q.-Y., Park, J., Koltun, V.: Fast global registration. In: Leibe, B., Matas, J., Sebe, N., Welling, M. (eds.) ECCV 2016. LNCS, vol. 9906, pp. 766–782. Springer, Cham (2016). https://doi.org/10.1007/978-3-319-46475-6_47
24. Zhou, T., Brown, M., Snavely, N., Lowe, D.G.: Unsupervised learning of depth and ego-motion from video. In: Proceedings of the IEEE Conference on Computer Vision and Pattern Recognition (CVPR), pp. 1851–1858 (2017)

Models, Methods and Algorithms

Imputation of Incomplete Data Based on Attribute Cross Fitting Model and Iterative Missing Value Variables

Jinchong Zhu[1], Liyong Zhang[2(✉)], Xiaochen Lai[1,4], and Genglin Zhang[3]

[1] School of Software, Dalian University of Technology, Dalian 116620, China
`zjc20181225@gmail.com, laixiaochen@dlut.edu.cn`
[2] School of Control Science and Engineering,
Dalian University of Technology, Dalian 116624, China
`zhly@dlut.edu.cn`
[3] Dalian Chinacreative Technology Co., Ltd., Dalian 116021, China
`zhanggl@china-creative.net`
[4] Key Laboratory for Ubiquitous Network and Service Software of Liaoning Province,
Dalian 116620, China

Abstract. The problem of missing values is often encountered in tasks such as machine learning, and imputation of missing values has become an important research content in incomplete data analysis. In this paper, we propose an attribute cross fitting model (ACFM) based on auto-associative neural network (AANN), which enhances the fitting of regression relations among attributes of incomplete data and reduces the dependence of imputation values on pre-filling values. Besides, we propose a model training scheme that takes missing values as variables and dynamically updates missing value variables based on optimization algorithm. The imputation accuracy is expected to be gradually improved through the dynamic adjustment of missing values. The experimental results verified the effectiveness of proposed method.

Keywords: Incomplete data · Auto-associative neural network · Missing value · Imputation

1 Introduction

There are various factors in the process of data collection, transmission and storage, etc., that may cause data loss in different degrees. The incompleteness of the data leads to the result that most of the computational intelligence technologies cannot be applied directly [1]. In cases where incomplete records cannot be deleted directly, an effective method is needed to fill the missing values.

The neural network is flexible in construction and can mine complex association relationships within data attributes efficiently. Sharpe and Solly propose to construct a multi-layer perceptron (MLP) for each missing pattern, which is used to fit the regression relation between missing attributes and existing attributes [2]. Ankaiah and Ravi

© Springer Nature Switzerland AG 2020
M. Han et al. (Eds.): ISNN 2020, LNCS 12557, pp. 167–175, 2020.
https://doi.org/10.1007/978-3-030-64221-1_15

propose an improved MLP imputation method that takes each missing attribute as output and the rest as input respectively [3]. The MLP imputation model can fit regression relations among data attributes, but model training is time-consuming when there are many missing patterns.

AANN is a type of network structure with the same number of nodes in output layer and input layer. One model can impute the data in all missing patterns [4]. Marwala et al. propose an imputation method combining AANN and genetic algorithm, then apply to real dataset [5]. Nelwamondo et al. use principal component analysis to select a reasonable number of nodes in hidden layer based on Marwala's framework [6]. Ravi and Krishna put forward four improved models based on AANN [7], in which the general regression auto-associative neural network not only improves the efficiency of imputation, but also outperforms MLP and three other models in most datasets. Gautam et al. propose a counter propagation auto-associative neural network [8] and find that the method of local learning can get better imputation results.

The number of complete records is small when the missing rate of dataset is high, we will lose a lot of information about the existing data in incomplete records if we only use complete records to train network. Silva-Ramírez et al. impute a fixed value into each missing value so that incomplete records can participate in the process of model training [9]. García-Laencina et al. initialize missing values by zero then apply all data to train a multi-tasking network [10].

As mentioned above, the training efficiency of AANN-based method is higher than that of MLP-based method in the multi-missing patterns. Consequently, this paper models incomplete dataset based on the AANN architecture, and fits regression relations among attributes of incomplete data. There will be an initial estimation error if the pre-filling process is used when incomplete records are input into the model. Hence, this paper proposes a model training scheme that takes missing values as variables and updates missing values iteratively during model training process (UMVDT). The improved model and training scheme make optimal use of the existing data in incomplete records, and gradually reduce the estimation error of missing value variables. The accuracy of imputation is improved by local learning and global approximation.

The rest of this paper is organized as following. Section 2 introduces MLP and AANN imputation models. Section 3 proposes ACFM based on AANN and a model training scheme named UMVDT. Section 4 analyzes the imputation performance of ACFM and UMVDT. And the full text is summarized in Sect. 5.

2   MLP and AANN Imputation Models

The imputation method based on MLP needs to build a subnet for each missing pattern. If the indices of missing attributes are P_t at the t missing pattern, the cost function of the subnet will be

$$E_k = \frac{1}{2} \sum_{x_i \in X_C} \sum_{j \in P_t} \left(f_{ij} \left(\sum_{k \notin P_t} w_{jk} \cdot x_{ik} \right) - x_{ij} \right)^2 , \tag{1}$$

where $x_i = [x_{i1}, x_{i2}, \cdots, x_{is}]^T$ represents the i - th record, s represents the dimension of attributes, X_C represents the subset of complete records, $f_{ij}(\cdot)$ represents the nonlinear mapping of the model, and w_{jk} represents the weight of the model. Each model fits the regression relation between missing attributes and existing attributes in each missing pattern.

The AANN model requires that the number of nodes in output layer are equal to that in input layer. The imputation method based on AANN makes one structure impute incomplete records at all missing patterns. The cost function can be expressed as

$$E = \frac{1}{2} \sum_{x_i \in X_C} \sum_{j=1}^{s} \left(f_{ij} \left(\sum_{k=1}^{s} w_{jk} \cdot x_{ik} \right) - x_{ij} \right)^2, \tag{2}$$

Each output neuron of the AANN model is calculated by all input neurons. The output values are easier to learn the input values at the same position with the model training. As a result, the quality of imputation values depends on the quality of pre-filling ones. The MLP model takes missing attributes as output and takes existing attributes as input, and fills missing values through this regression network. Hence, compared with the MLP model, AANN lacks an explicit regression relation to guide the training of model and the imputation of missing values.

3 Proposed Architecture

3.1 AANN-Based ACFM

The AANN imputation model implements the imputation of multiple missing patterns through one network architecture, but it does not establish a clear regression relation among data attributes. Inspired by AANN, this paper fits all regression relations between one of attributes and the rest of attributes in incomplete dataset by one network architecture, so that the output value of the model no longer depends on the input value at the same position. The cost function of the proposed model is

$$E = \frac{1}{2} \sum_{x_i \in X} \sum_{j \notin M_i} \left(f_{ij} \left(\sum_{k=1, k \neq j}^{s} w_{jk} \cdot x_{ik} \right) - x_{ij} \right)^2, \tag{3}$$

where X represents an incomplete dataset and M_i represents indices of missing values in record x_i. Pre-filling processing is required when incomplete records are input into the model. Since pre-filling value has estimation error compared with original data, the model should limit the training error between pre-filling data and its predicted data to optimize the model parameters. We define this error as the missing value error. In the formula (3), $j \notin M_i$ means that missing value error is no more used to optimize model parameters. The model based on this cost function can fit regression relations among data attributes at a network architecture, thus the model is called attribute cross fitting model.

The data transmission process for output neurons in ACFM is shown in Fig. 1. An incomplete record with two missing values is input into ACFM. ACFM does not use the missing value error to optimize model parameters, hence the output values y_{i1} and y_{i2} will not be calculated. And the output value y_{i3} of ACFM is calculated by input values except for x_{i3}. Meanwhile, the output values from y_{i4} to y_{is} have a similar calculation rule as y_{i3}.

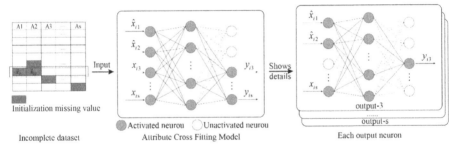

Fig. 1. Schematic diagram of data transmission for output neurons in ACFM

3.2 A Novel Training Scheme

The missing value error has been limited to optimize the ACFM model, but there is still pre-filling data with estimated errors when incomplete records are input into the model. We propose a model training scheme that takes missing values as variables and dynamically adjusts missing value variables during model training. The missing value variables can gradually match the fitting relationship determined by existing data. The principle of UMVDT is shown in Fig. 2. Firstly, the missing values in incomplete records are initialized as variables, and then incomplete records are input into ACFM for calculating the errors between network output and input. Finally, the backpropagation algorithm is used to update missing value variables and network parameters.

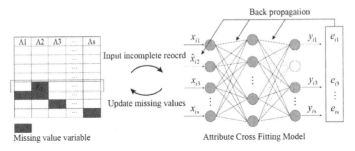

Fig. 2. Schematic diagram of UMVDT training scheme

Suppose that the input layer of ACFM is layer 1 and the output layer is layer $n + 1$, w^l and b^l represent weights and thresholds from layer l to layer $l + 1$ of ACFM respectively ($1 \leq l \leq n$). For the model output j', the output of the j - th neuron in the hidden layer of ACFM is

$$\begin{cases} a_{ij}^l = g\left(z_{ij}^l\right) = g\left(b_j^{l-1} + \sum_{k=1}^{s_{l-1}} w_{jk}^{l-1} \cdot a_k^{l-1}\right) & 2 < l \leq n \\ a_{ij}^l = g\left(z_{ij}^l\right) = g\left(b_j^{l-1} + \sum_{k=1,k\neq j'}^{s_{l-1}} w_{jk}^{l-1} \cdot x_{ik}\right) & l = 2 \end{cases}, \tag{4}$$

where $g(\cdot)$ is the activation function, z^l_{ij} is the linear summation of the j - th neuron in layer l, and s_l is the number of neurons in layer l. The output of ACFM is linear, so the output value $y_{ij} = z^{n+1}_{ij}$. We define the intermediate variable δ^{n+1}_{ij} as

$$\begin{cases} \delta^{n+1}_{ij} = \frac{\partial e_i}{\partial z^{n+1}_{ij}} = \frac{\partial e_{ij}}{\partial z^{n+1}_{ij}} = \left(y_{ij} - x_{ij}\right) & j \notin M_i \\ \delta^{n+1}_{ij} = 0, & j \in M_i \end{cases}, \tag{5}$$

where e_i is the error between the i - th sample x_i and the network output y_i. When $2 \le l \le n$, δ^l_{ij} is

$$\delta^l_{ij} = \frac{\partial e_{ij}}{\partial z^l_{ij}} = \sum_{k=1}^{s_{l+1}} \delta^{l+1}_{ik} \cdot w^l_{kj} \cdot g'\left(z^l_{ij}\right), \tag{6}$$

Assuming the learning rate is η, when the gradient descent method is used to optimize the model, the update rule of model parameters is

$$\begin{cases} w^l_{jk} = w^l_{jk} - \eta \cdot \frac{\partial e_i}{\partial w^l_{jk}} = w^l_{jk} - \eta \cdot \delta^{l+1}_{ij} \cdot a^l_k \\ b^l_j = b^l_j - \eta \cdot \frac{\partial e_i}{\partial b^l_j} = b^l_j - \eta \cdot \delta^{l+1}_{ij} \end{cases}, \tag{7}$$

Missing value variables are updated with model parameters during model training. When $k \in M_i$, the update rule of missing value variable $\hat{x}_{ik}$ is

$$\hat{x}_{ik} = \hat{x}_{ik} - \eta \cdot \frac{\partial e_i}{\partial \hat{x}_{ik}} = \hat{x}_{ik} - \eta \cdot \sum_{j=1}^{s_2} \delta^2_{ij} \cdot w^1_{jk}, \tag{8}$$

4 Experiment

4.1 Experimental Design

In order to verify the imputation performance of proposed method, our experiment uses four complete datasets obtained from the UCI database [11], and the description of datasets is shown in Table 1. We select the continuous features in the dataset, which are described in detail on the official website. For the sake of forming incomplete datasets, partial data are deleted randomly according to specified deletion rates which are set as 5%, 10%, 15%, 20%, 25%, and 30%. It is generally believed that data sets with higher missing rate are not of great value.

We implement five kinds of imputation methods based on MLP, AANN and ACFM models, which are MLP-I, AANN-I, ACFM-I, AANN-UMVDT and ACFM-UMVDT. The imputation method based on MLP model adopts traditional training scheme (MLP-I), i.e. the model is trained by complete data, then the model output is used to fill missing values. Based on AANN and ACFM models, we implement the traditional training scheme (AANN-I, ACFM-I) and the UMVDT training scheme (AANN-UMVDT, ACFM-UMVDT). All models are optimized by gradient descent with momentum. The

Table 1. Description of datasets

Datasets	Records	Attributes	Datasets	Records	Attributes
Blood	748	4	Iris	150	4
Seeds	210	7	Abalone	4177	7

learning rate is set to 0.2 and the momentum is set to 0.9. For each missing rate, all imputation methods are repeated 10 times, and the average values of errors are used as experimental results. The mean absolute percentage error (MAPE) is used to evaluate imputation error:

$$\text{MAPE} = \frac{1}{|M|} \sum_{x_i \in X} \sum_{j \in M_i} \left| \frac{y_{ij} - x_{ij}}{x_{ij}} \right|, \tag{9}$$

where $|M|$ represents the number of missing values.

4.2 Experimental Results

Experimental results are shown in Table 2, 3, 4 and 5. The suboptimal results are indicated in bold font, and the optimal results are indicated in underline and bold font.

Table 2. The MAPE values of Blood dataset

Missing rates	MLP-I	AANN-I	AANN-UMVDT	ACFM-I	ACFM-UMVDT
5%	1.113	0.983	0.941	**0.488**	**0.449**
10%	1.122	1.087	0.968	**0.537**	**0.510**
15%	0.872	1.274	1.005	**0.620**	**0.599**
20%	0.680	1.114	0.974	**0.671**	**0.633**
25%	0.861	1.188	1.026	**0.728**	0.754
30%	0.985	1.212	1.124	**0.764**	0.800

4.3 Experimental Discussion

Imputation Performance of ACFM. It can be seen from experimental that the imputation results of MLP-I are better than those of AANN-I. The imputation results of ACFM-I are better than those of MLP-I and AANN-I. And ACFM-UMVDT are better than AANN-UMVDT. Because MLP establishes an exclusive regression network for each missing attribute, it can describe the regression relation in the data more accurately. ACFM enhances the ability to fit the regression relation among attributes of incomplete data compared with AANN. At the same time, ACFM fits regression relations on a network architecture compared with MLP, which increases the generalization of the model.

Table 3. The MAPE values of Iris dataset

Missing rates	MLP-I	AANN-I	AANN-UMVDT	ACFM-I	ACFM-UMVDT
5%	0.157	0.298	**0.150**	0.153	**0.139**
10%	0.150	0.358	0.158	**0.139**	**0.128**
15%	0.237	0.376	**0.190**	0.219	**0.167**
20%	0.234	0.386	**0.188**	0.217	**0.173**
25%	0.272	0.401	**0.189**	0.237	**0.186**
30%	0.335	0.455	**<u>0.234</u>**	**0.272**	**0.234**

Table 4. The MAPE values of Seeds dataset

Missing rates	MLP-I	AANN-I	AANN-UMVDT	ACFM-I	ACFM-UMVDT
5%	0.071	0.083	**0.067**	**0.067**	**0.062**
10%	0.093	0.096	**0.077**	**0.077**	**0.068**
15%	0.104	0.095	**0.072**	0.076	**0.067**
20%	0.114	0.109	**0.084**	0.090	**0.081**
25%	0.096	0.097	**<u>0.076</u>**	0.083	**0.076**
30%	0.151	0.122	**<u>0.085</u>**	0.090	**0.088**

Table 5. The MAPE values of Abalone dataset

Missing rates	MLP-I	AANN-I	AANN-UMVDT	ACFM-I	ACFM-UMVDT
5%	0.155	0.567	**0.133**	0.145	**0.119**
10%	0.219	0.547	**<u>0.114</u>**	0.163	**0.137**
15%	0.352	0.605	**0.154**	0.196	**0.125**
20%	0.499	0.633	**0.189**	0.223	**0.168**
25%	0.451	0.632	**0.191**	0.243	**0.171**
30%	0.631	0.535	0.337	**0.246**	**<u>0.193</u>**

Comparison Between UMVDT and Traditional Training Scheme. The   UMVDT training method is superior to the traditional training scheme except for the imputation results of Iris dataset at the missing rates of 25% and 30%. The UMVDT training scheme makes full use of the whole existing data in incomplete records and takes missing values as variables to make missing values gradually match the fitting relationship. The missing value variables and model parameters are updated alternately, so the imputation effect can be improved significantly. The imputation of ACFM-I, ACFM-UMVDT and the variation of the missing value variables (MVV) of ACFM-UMVDT in each round of training are shown in Fig. 3. It can be found that not only the MAPE values calculated by missing value variables are more accurate than those of original model, but also the imputation accuracy can be further improved by the model which is trained by the data updated iteratively.

Comparing the MAPE values of ACFM-I, ACFM-UMVDT and Missing value variables

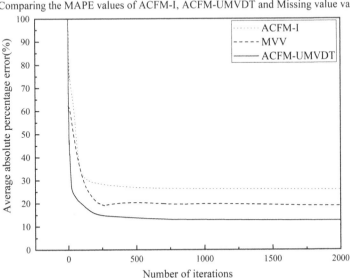

Fig. 3. The imputation and updating results of ACFM-I, ACFM-UMVDT, and missing value variables on Iris dataset

5   Conclusion

In this paper, we propose an imputation method of incomplete data based on ACFM and UMVDT. ACFM enhances the fitting of regression relation of AANN and reduces the dependence of output value on input value in the corresponding position. When incomplete records are input into the model, the missing values are set as variables, which can not only increase the imputation accuracy, but also reduce the deviation of the model. Experimental results show that, ACFM model can obtain more accurate imputation results compared with MLP and AANN models. UMVDT improves the accuracy of imputation on AANN and ACFM models compared with traditional training scheme. Moreover, the combination of ACFM and UMVDT achieve the most optimal imputation result. Despite these findings, the portability of UMVDT on other models and the practicability of proposed method need to be further proved.

Acknowledgement. This work was supported by National Key R&D Program of China under Grant 2018YFB1700200.

References

1. Nelwamondo, F.V., Golding, D., Marwala, T.: A dynamic programming approach to missing data estimation using neural networks. Inf. Sci. **237**, 49–58 (2013)
2. Sharpe, P.K., Solly, R.J.: Dealing with missing values in neural network-based diagnostic systems. Neural Comput. Appl. **3**(2), 73–77 (1995)
3. Ankaiah, N., Ravi, V.: A novel soft computing hybrid for data imputation. In: Proceedings of the International Conference on Data Mining, pp. 65–70 (2011)
4. Gondara, L., Wang, K.: MIDA: multiple imputation using denoising autoencoders. In: Phung, D., Tseng, Vincent S., Webb, Geoffrey I., Ho, B., Ganji, M., Rashidi, L. (eds.) PAKDD 2018. LNCS (LNAI), vol. 10939, pp. 260–272. Springer, Cham (2018). https://doi.org/10.1007/978-3-319-93040-4_21
5. Marwala, T., Chakraverty, S.: Fault classification in structures with incomplete measured data using auto associative neural networks and genetic algorithm. Curr. Sci., 542–548 (2006)
6. Mistry, F.J., Nelwamondo, F.V., Marwala, T.: Missing data estimation using principle component analysis and auto associative neural networks. J. Syst. Cybernatics Inform. **7**(3), 72–79 (2009)
7. Ravi, V., Krishna, M.: A new online data imputation method based on general regression auto associative neural network. Neurocomputing **138**, 106–113 (2014)
8. Gautam, C., Ravi, V.: Counter propagation auto-associative neural network based data imputation. Inf. Sci. **325**, 288–299 (2015)
9. Silva-Ramírez, E.L., Pino-Mejías, R., López-Coello, M., Cubiles-de-la-Vega, M.: Missing value imputation on missing completely at random data using multilayer perceptrons. Neural Netw. **24**(1), 121–129 (2011)
10. García-Laencina, P.J., Sancho-Gómez, J.L., Figueiras-Vidal, A.R.: Classifying patterns with missing values using multi-task learning perceptrons. Expert Syst. Appl. **40**(4), 1333–1341 (2013)
11. UC Irvine Machine Learning Repository. http://archive.ics.uci.edu/ml/index.php

Adaptive Gaussian Noise Injection Regularization for Neural Networks

Yinan Li and Fang Liu[✉]

Department of Applied and Computational Mathematics and Statistics,
University of Notre Dame, Notre Dame, IN 46556, USA
`yli28@alumni.nd.edu, fliu2@nd.edu`

Abstract. We propose whiteout, a family of Noise injection (NI) regularization techniques through injecting adaptive Gaussian noises. Through theoretical analysis, we 1) establish the regularization effect of whiteout in the framework of generalized linear models with a broad range of closed-form penalty terms, including l_γ (for $\gamma \in (0, 2]$), the adaptive lasso, the group lasso, among others; 2) show that whiteout stabilizes the training of NNs with robustness or decreased sensitivity to small perturbations in the input; 3) prove that the noise-perturbed loss function with whiteout converges almost surely to the ideal loss function, and the minimizer of the former is consistent for the minimizer of the latter; 4) derive the tail bound on the noise-perturbed loss function to establish the practical feasibility for optimization. The superiority of whiteout over the Bernoulli NI techniques (dropout and shakeout) in prediction accuracy (up by 2~3%) in relatively small-sized training data and comparability in large-sized training data are demonstrated thorough experiments. This work is the first in-depth theoretical and methodological examination of the regularization effects of Gaussian NI in NNs in general.

Keywords: Regularization · Sparsity · Stability · Adversarial robustness · Consistency · Convergence

1 Introduction

1.1 Background

Neural networks (NNs) are prone to over-fitting given their complex structures and the large amount of parameters. Earlier techniques for regularizing NN estimation include weight decay, early stopping, l_1 and l_2 regularizations, etc. Examples of recent developments include the input gradient regularization [1], $l_{0.5}$ regularization for smoothing interval NNs [2], regularization that lowers the parameter magnitude with low sensitivity [3], regularization using the Frobenius

Electronic supplementary material The online version of this chapter (https:// doi.org/10.1007/978-3-030-64221-1_16) contains supplementary material, which is available to authorized users.

© Springer Nature Switzerland AG 2020
M. Han et al. (Eds.): ISNN 2020, LNCS 12557, pp. 176–189, 2020.
https://doi.org/10.1007/978-3-030-64221-1_16

norm of the Jacobian of the NN [4], information-theory-based regularization [5], among others.

Most of the above mentioned work has explicit regularization terms for NN parameters. In this paper, we focus on the implicit noise injection (NI) regularization that has also been proven effective against overfitting. Early work of NI in NNs appeared in the 1980's and 1990's [6–10]. A recent popular NI regularization approach is dropout or the Bernoulli noise injection [11,12], where nodes in the input and hidden layers are randomly "dropped" from a NN with some fixed probabilities during its training. Various extensions to dropout have been proposed. Maxout facilitates optimization and improves the accuracy of dropout with a new activation function [13]. Fast dropout speeds up the computation of dropout via a Gaussian approximation [14]. Dropconnect applies Bernoulli noises to weights instead of nodes [15]. Partial dropout regularizes restricted Boltzmann machines (RBMs) by combining weight decay, model averaging, and network pruning [16]. Standout trains NNs jointly with a binary belief network that selectively sets nodes to zero [17]. Shakeout applies multiplicative adaptive Bernoulli noises to input and hidden nodes during training [18].

1.2   Our Contributions

We propose *whiteout* that injects adaptive Gaussian noises to input and hidden nodes during the training of NNs. "Adaptive" here refers to that the variance of the injected noises in whiteout differs by NN weight parameters and gets continuously updated during training. To the best of our knowledge, this is the first in-depth work that explores the regularization effects and robustness of Gaussian NI techniques. Our contributions are summarized as follows.

1. Whiteout offers a wide range of regularizers. We establish theoretically that with properly designed variance for the injected noise, whiteout is equivalent to minimizing penalized loss functions with closed-from penalty terms for model complexity such as the l_γ regularization ($\gamma \in (0,2]$) including l_1 and l_2 as special cases, $l_1 + l_2$, adaptive lasso, group lasso, among others. By offering more sparsity regularization types, whiteout can effectively discount noisy and irrelevant input and hidden features in prediction, especially for small-sized training data.
2. We show whiteout improves adversarial robustness and decreases the sensitivity of a learned NN to external perturbation in the input features.
3. We provide a thorough investigation of the asymptotic properties of the whiteout noise perturbed loss function and its minimizer when the size of training data n and the number of epochs k go to infinity, and derive the tail bound for the perturbed loss function with finite n and k. Some of the conclusions are also applicable to other NI techniques under the same regularity conditions.
4. We show empirically that whiteout outperforms Bernoulli NI (dropout and shakeout) when the training size is small, given its effectiveness in imposing flexible sparsity regularization on NN parameters.

5. We develop a back propagation procedure that can easily accommodate the whiteout NI during training. We also show that whiteout can be applied to unsupervised learning, such as RBMs and auto-encoders.

1.3   Related Work

[6] and [7] demonstrate experimentally that adding noise to the input during the training of a NN has a remarkable effect on the generalization capability of the network. [9] formulates NI to input layers as a way of decreasing the sensitivity of a learned NN to small perturbations (adversarial attacks). [8] examines the theoretical properties and offers an explanation on the generalizability of constant-variance Gaussian NI to input layers by connecting NI with heat kernels. [10] examines feedforward NNs and suggests that injecting noises to input layers can be regarded as drawing samples from the kernel density estimator of the true density; but does not establish that NI decreases the generalization error of a trained NN.

Dropout is shown to yield the l_2 regularization on the NN parameters in the framework of generalized linear models (GLMs) [11]. [12] empirically compares the performance of multiplicative Gaussian NI with mean 1 and a constant variance with Bernoulli NI and shows that the former is comparable or slightly better; but provides minimal theoretical exploration. [18] demonstrates empirically that multiplicative Gaussian NI with a constant variance yields similar performance as dropout in NN regularization. Shakeout achieves l_1 and l_2 regularization [18]. The l_1 and l_2 regularization can also be realized by imposing the Laplace and Gaussian priors on NN parameters, respectively, in a variational Bayesian framework [19]. Gaussian NI in NNs is briefly discussed in [19] from a Bayesian perspective.

In summary, existing research on Gaussian NI focus on Gaussian noises with a constant variance and is limited in scope and depth. There also lacks a theoretical explanation on how Gaussian NI or NI in general regularizes NN estimation. We examine a much broader range of Gaussian noises the variance of which is adaptive to trained parameters in Sect. 2. Our theoretical analysis provides insights on why Gaussian NI or NI in general is an effective overfitting mitigation technique (Sect. 3) and is robust to adversarial attacks (Sect. 4). In addition, we present in Sect. 5 the asymptotic properties of whiteout noise-perturbed loss functions, which should hold for other NI techniques under the same regularity conditions, and investigate the tail bound of the loss function distribution to understand its practical trainability.

2   Whiteout: Adaptive Gaussian Noise Injection

Let l be the index for layers ($l = 1, \ldots, L - 1$; L is the output layer), and j be the index for the nodes in layer l ($j = 1, \ldots, m^{(l)}$). The weight connecting the j-th node $X_j^{(l)}$ in layer l and the k-th node $X_k^{(l+1)}$ in layer $l + 1$ is denoted by $w_{jk}^{(l)}$, where $X_k^{(l+1)} = f^{(l)}\left(u_k^{(l+1)}\right)$ with $u_k^{(l+1)} = b_k^{(l)} + \sum_{j=1}^{m^{(l)}} w_{jk}^{(l)} X_j^{(l)}$ and $f^{(l)}$ is the activation function connecting layers l and $l+1$. During the updating of $w_{jk}^{(l)}$ in training, whiteout replaces $X_j^{(l)}$ with noise-perturbed $\tilde{X}_{j(k)}^{(l)}$; that is,

$$u_k^{(l+1)} = b_k^{(l)} + \sum_{j=1}^{m^{(l)}} w_{jk}^{(l)} \tilde{X}_{j(k)}^{(l)}, \text{ where} \tag{1}$$

$$\text{additive: } \tilde{X}_{j(k)}^{(l)} = X_j^{(l)} + e_{jk}, \text{ where } e_{jk} \overset{ind}{\sim} N\left(0, V\left(|w_{jk}^{(l)}|, \boldsymbol{\lambda}\right)\right), \tag{2}$$

$$\text{multiplicative: } \tilde{X}_{j(k)}^{(l)} = X_j^{(l)} e_{jk}, \text{ where } e_{jk} \overset{ind}{\sim} N\left(1, V\left(|w_{jk}^{(l)}|, \boldsymbol{\lambda}\right)\right), \tag{3}$$

where $V(|w_{jk}^{(l)}|, \boldsymbol{\lambda})$ is the variance of the injected noise to X_j to obtain $\tilde{X}_{j(k)}^{(l)}$ in Eq. (1), and $\boldsymbol{\lambda}$ contains the tuning parameters. In other words, the variance is not a constant term but varies by $w_{jk}^{(l)}$. Since $w_{jk}^{(l)}$ is unknown, its estimate will be used, after an initialization, which is continuously updated during training until convergence. With properly designed $V(|w_{jk}^{(l)}|, \boldsymbol{\lambda})$, we get different regularization effects on $\mathbf{w}$. Some examples are

$$V(|w_{jk}^{(l)}|, \boldsymbol{\lambda}) = \sigma^2 |w_{jk}^{(l)}|^{-\gamma} + \lambda, \tag{4}$$

$$V(|w_{jk}^{(l)}|, \boldsymbol{\lambda}) = \sigma^2 (|w_{jk}^{(l)}|)^{-1} |\hat{w}_{jk}^{(l)}|^{-\gamma}, \tag{5}$$

$$V(|w_{jk}^{(l)}|, \boldsymbol{\lambda}) = \sigma^2 (\mathbf{w}_g^{(l)T} \mathbf{K}_g \mathbf{w}_g^{(l)})^{1/2} \left(p_g w_{jk}^{(l)2}\right)^{-1}, \tag{6}$$

where $\sigma^2 \geq 0$, $\lambda \geq 0$, and $\gamma \in (0,2)$ are hyperparameters; $\hat{w}_{jk}^{(l)}$ is a reasonable estimate on w_{jk} in Eq. (5) obtained, through, such as weight decay or early stopping or dropout; and g is the index of groups that nodes in layer l belong to, p_g is the size of group g, and K_g is a positive definite matrix in Eq. (6). We will establish the closed-form regularizers associated with different noise types in Eqs. (4) to (6) in Sect. 3.

3   Whiteout Leads to Closed-Form Model Regularization

A common framework where NI is established as a regularization technique is through GLMs and the exponential family distribution [18, 20–22]. In a GLM, the conditional distribution of output Y given inputs $\mathbf{X} \in \mathcal{R}^p$ is

$$f(Y|\mathbf{X}, \mathbf{w}) = h(Y, \tau) \exp\left((\boldsymbol{\eta} \mathbf{T}(Y) - A(\boldsymbol{\eta}))/d(\tau)\right), \tag{7}$$

where $\boldsymbol{\eta} = \mathbf{Xw}$ is the natural parameter, $\mathbf{w}$ captures the relation between $\mathbf{X}$ and Y, and τ is the dispersion parameter. The loss function is often defined as the negative log-likelihood of the GLM given independent training cases $(\mathbf{x}_i, y_i)$ for $i = 1, \cdots, n$

$$l(\mathbf{w}|\mathbf{x}, \mathbf{y}) = \sum_{i=1}^{n} (A(\boldsymbol{\eta}_i) - \boldsymbol{\eta} \mathbf{T}(y_i))/d(\tau) - \log(h(y_i, \tau)). \tag{8}$$

Whiteout injects noises defined in Eqs. (2) or (3) to $\mathbf{x}_i$ to obtain $\tilde{\mathbf{x}}_i$ and works with the noise-perturbed loss function

$$l_p(\mathbf{w}|\tilde{\mathbf{x}}, \mathbf{y}) = \sum_{i=1}^{n} l(\mathbf{w}|\tilde{\mathbf{x}}_i, y_i). \tag{9}$$

Lemma 1 establishes that the expected $l_p(\mathbf{w}|\tilde{\mathbf{x}}, \mathbf{y})$ over the distribution of injected noises $\mathbf{e}$ is a penalized loss function given $(\mathbf{x}, \mathbf{y})$ with a closed-form regularization term.

Lemma 1 (regularized loss function with whiteout for GLMs). The expectation of Eq. (9) over the distribution of injected noise is

$$E_{\mathbf{e}}\left(\sum_{i=1}^{n} l_p(\mathbf{w}|\tilde{\mathbf{x}}_i, y_i)\right) = \sum_{i=1}^{n} l(\mathbf{w}|\mathbf{x}_i, y_i) + \frac{R(\mathbf{w})}{d(\tau)}, \tag{10}$$

where $R(\mathbf{w}) \triangleq \sum_{i=1}^{n} E_{\mathbf{e}}(A(\tilde{\mathbf{x}}_i \mathbf{w})) - A(\boldsymbol{\eta}_i) \approx \frac{1}{2} \sum_{i=1}^{n} A''(\boldsymbol{\eta}_i) \mathrm{Var}(\tilde{\mathbf{x}}_i \mathbf{w}).$ $\tag{11}$

The proof is provided in the supplementary materials to this paper. The approximation in Eq. (11) is obtained via the second-order Taylor-expansion of the expected perturbed loss function around $\boldsymbol{\eta}_i = 0$ (and is exactly "=" for linear models). $A(\boldsymbol{\eta}_i) = A(\mathbf{x}_i \mathbf{w})$ is convex and smooth in $\mathbf{w}$ in GLMs [23], and $R(\mathbf{w})$ is always positive per the Jensen's inequality [22]. Based on Lemma 1, we explore further the regularization term $R(\mathbf{w})/d(\tau)$ in Eq. (11) for some specific types of the noise variance, and the results are given in Corollary 2, which are easy to obtain based on Lemma 1.

Corollary 2 (regularization on w with whiteout NI). Define $\boldsymbol{\Lambda}(\mathbf{w}) \triangleq \mathrm{diag}(A''(\mathbf{x}_1 \mathbf{w}), \cdots, A''(\mathbf{x}_n \mathbf{w}))$ and $\Gamma(\mathbf{w}) \triangleq \mathrm{diag}(\mathbf{x}^{\mathrm{T}} \boldsymbol{\Lambda}(\mathbf{w}) \mathbf{x})$ in the GLM framework. Whiteout NI can realize the following regularizers for the noise variance terms in Eqs. (4) to (6).

a). $l_{2-\gamma}$ *bridge regularizer for* $\gamma \in [0, 2)$: Let $\lambda = 0$ in Eq. (4), then $V(e_{jk}) = \sigma^2 |w_{jk}|^{-\gamma}$ and

$$\text{additive: } R(\mathbf{w}) \approx (\sigma^2/2) \mathbf{1}^T \boldsymbol{\Lambda}(\mathbf{w}) \mathbf{1} \big| \big| |\mathbf{w}|^{2-\gamma} \big| \big|_1, \tag{12}$$

$$\text{multiplicative: } R(\mathbf{w}) \approx (\sigma^2/2) \big| \big| \Gamma(\mathbf{w}) |\mathbf{w}|^{2-\gamma} \big| \big|_1, \tag{13}$$

where $\mathbf{1}_{n \times 1}$ is a column vector of 1. The penalty term $R(\mathbf{w})$ is similar to the bridge penalization [24], which reduces to the l_1 (lasso) penalty [25] when $\gamma = 1$, and to the l_2 (ridge) penalty when $\gamma = 0$.

b). $l_1 + l_2$ *elastic net regularizer*: Let $\gamma = 1$ in Eq. (4), then $V(e_{jk}) = \sigma^2 |w_{jk}|^{-1} + \lambda$,

$$\text{additive: } R(\mathbf{w}) \approx \frac{1}{2} \mathbf{1}^T \boldsymbol{\Lambda}(\mathbf{w}) \mathbf{1} \big(\sigma^2 ||\mathbf{w}||_1 + \lambda ||\mathbf{w}||_2^2 \big), \tag{14}$$

$$\text{multiplicative: } R(\mathbf{w}) \approx \frac{\sigma^2}{2} \big| \big| \Gamma(\mathbf{w}) |\mathbf{w}| \big| \big|_1 + \frac{\lambda}{2} \big| \big| \Gamma(\mathbf{w}) |\mathbf{w}|^2 \big| \big|_1. \tag{15}$$

The penalty term $R(\mathbf{w})$ contains similar norm on $\mathbf{w}$ as in the elastic net $(l_1 + l_2)$ regularization [26].

c). *adaptive lasso regularizer*: Given Eq. (5), then

$$\text{additive: } R(\mathbf{w}) \approx (\sigma^2/2) \mathbf{1}^T \boldsymbol{\Lambda}(\mathbf{w}) \mathbf{1} \big| \big| |\mathbf{w}| |\hat{\mathbf{w}}|^{-\gamma} \big| \big|_1, \tag{16}$$

$$\text{multiplicative: } R(\mathbf{w}) \approx (\sigma^2/2) \big| \big| \Gamma(\mathbf{w}) |\mathbf{w}| |\hat{\mathbf{w}}|^{-\gamma} \big| \big|_1, \tag{17}$$

which contain a similar norm on $\mathbf{w}$ as the adaptive lasso regularization [27].

d). *group lasso regularizer*: Given Eq. (6), then

$$\text{additive: } R(\mathbf{w}) \approx \frac{\sigma^2}{2} \mathbf{1}^T \mathbf{\Lambda}(\mathbf{w}) \mathbf{1} \left(\sum_{g=1}^{G} ||(\mathbf{w}_g' \mathbf{K}_g \mathbf{w}_g)^{\frac{1}{2}} p_g^{-1}|| \right), \qquad (18)$$

$$\text{multiplicative: } R(\mathbf{w}) \approx \frac{\sigma^2}{2} \sum_{g=1}^{G} ||\Gamma_g(\mathbf{w})|| ||(\mathbf{w}_g' \mathbf{K}_g \mathbf{w}_g)^{\frac{1}{2}}||p_g^{-1}||_1. \qquad (19)$$

The penalty terms contain a similar norm on $\mathbf{w}$ as the group lasso penalization [28]. If every node in layer l is its own group, then it reduces to the lasso regularizer l_1 on $\mathbf{w}$; if every node belongs to one group, then it reduces to ridge regularizer l_2 on $\mathbf{w}$. Since there often lacks motivation or instruction to group hidden nodes, it make sense that the group lasso NI is applied to the predefined groups of input nodes only.

For additive noises, besides the various norms on $\mathbf{w}$, $R(\mathbf{w})$ also involves $\mathbf{\Lambda}(\mathbf{w})$. If $A''(\mathbf{x}_i \mathbf{w})$ does not depend on $\mathbf{w}$ (e.g., in linear models), $R(\mathbf{w})$ leads to the nominal regularization as defined by the norm on $\mathbf{w}$. Otherwise, the regularization effects on $\mathbf{w}$ through $R(\mathbf{w})$ are not exact as defined by the norms on $\mathbf{w}$ due to the scaling of $\mathbf{\Lambda}(\mathbf{w})$. For example, in the logistic regression with binary outcomes, $A(\eta_i) = \ln(1 + e^{\eta_i})$, $A''(\eta_i) = p_i(\mathbf{w})(1 - p_i(\mathbf{w}))$ with $p_i(\mathbf{w}) = \Pr(y_i = 1|\mathbf{x}_i) = (1 + \exp(-\mathbf{x}_i \mathbf{w}))^{-1}$, and $R(\mathbf{w}) \approx \frac{\sigma^2}{2} \sum_{i=1}^{n} (p_i(\mathbf{w})(1 - p_i(\mathbf{w}))) ||\mathbf{w}|^{2-\gamma}||_1$; that is, the norm on $\mathbf{w}$ is scaled by the total variance of the binary outcome. For multiplicative noise, the norms are on $\Gamma(\mathbf{w})\mathbf{w}$, a scaled version of $\mathbf{w}$, rather than on $\mathbf{w}$. Plugging in the optimizer of the loss function in Eq. (8) $\hat{\mathbf{w}}$, which is the MLE in the case of GLMs, $n^{-1}\mathbf{x}^T \Lambda(\hat{\mathbf{w}})\mathbf{x} = n^{-1} \sum_{i=1}^{n} \nabla^2 l(\hat{\mathbf{w}}|\mathbf{x}_i, y_i)$ is an estimator of the Fisher information matrix. Since $\Gamma(\mathbf{w}) = \text{diag}(\mathbf{x}^T \Lambda(\mathbf{w})\mathbf{x})$ per its definition, the multiplicative whiteout noise can then be regarded as regularizing $\mathbf{w}$ after it is scaled by the diagonal Fisher information matrix, similar to the interpretation on dropout offered by [22].

The hyperparameters in Eqs. (2) and (3) determine the type and degree of the regularization. For example, in Eq. (4), when $\lambda = 0$, sparsity regularization $l_{2-\gamma}$ will be imposed for $\gamma \in (0, 2)$, and which type specifically depends the value of γ; when $\sigma^2 \neq 0$ and $\lambda \neq 0$, the ratio of σ^2 and λ determines the relative regularization between sparsity $l_{2-\gamma}$ and l_2 on $\mathbf{w}$. Our empirical studies suggest that a good γ can often be found in the neighborhood of $(0.5, 1.5)$. For practical implementation, the hyperparameters can be chosen by the cross validation (CV), or be integrated into the Bayesian optimization process for specifying NN hyperparameters.

4   Whiteout Improves Robustness of Learned NNs

Denote the training data by $\mathbf{z}_i = (\mathbf{x}_i, \mathbf{y}_i)$ for $i = 1, \dots, n$, where $\mathbf{x}_i = (x_{i1}, \cdots, x_{ip})$ refers to p input nodes. Let $\mathbf{d}$ denote the external perturbation (adversarial attacks) on $\mathbf{x}_i$ with $E(d_{ij}) = 0$ and $V(d_{ij}) = \varpi^2$ for $j = 1, \dots, p$. If a trained NN is sensitive to external perturbation, then the predicted $\bar{\mathbf{y}}_i$ given $\mathbf{x}_i$ and $\bar{\mathbf{y}}_i^*$ given $\mathbf{x}_i + \mathbf{d}_i$ via the trained NN can be very different. We show that the predictions from the NN trained with whiteout NI are more robust to external perturbations, as the loss function minimized by whiteout takes into account the sensitivity of the NN.

Let $\mathbf{f}$ denote the NN with L layers. Suppose there are q output nodes, let $\mathbf{f} = \mathbf{f}^{(L-1)} \circ \mathbf{f}^{(L-2)} \circ \cdots \circ \mathbf{f}^{(1)} = \mathbf{f}^{(L-1):1}$ be a vector of q compound functions over the sequence of active functions connecting the layers 1 to $L-1$. Denote by $\hat{\mathbf{y}}_i$ the prediction given x_i via the NN trained with whiteout and $\hat{\mathbf{y}}_i^*$ the prediction given $\mathbf{x}_i + \mathbf{d}_i$ from the NN. Define

$$\boldsymbol{\Delta}_i = \hat{\mathbf{y}}_i^* - \hat{\mathbf{y}}_i. \tag{20}$$

Let $m^{(l)}$ denote the number of nodes in layer $l \geq 2$, $\frac{\partial f_j^{(1)}}{\partial \mathbf{x}_i} = \left(\frac{\partial f_j^{(1)}}{\partial x_{i1}}, \cdots, \frac{\partial f_j^{(1)}}{\partial x_{ip}} \right)^T$

for $j = 1, \ldots, m^{(2)}$, $\mathbf{h}_i^{(l)}$ denote the hidden nodes in layer l, $\frac{\partial f_j^{(l)}}{\partial \mathbf{h}_i^{(l)}} = \left(\frac{\partial f_j^{(l)}}{\partial h_{i1}^{(l)}}, \cdots, \right.$

$\left. \frac{\partial f_j^{(l)}}{\partial h_{i,m^{(l)}}^{(l)}} \right)^T$ for $j = 1, \ldots, m^{(l+1)}$ for $l = 2, \ldots, L-1$, and $\Psi_{i,q'}(\mathbf{w}) = \left(\frac{\partial \mathbf{f}^{(L-1):1}[q']}{\partial f_1^{(1)}} \right.$

$\frac{\partial f_1^{(1)}}{\partial \mathbf{x}_i}, \ldots, \frac{\partial \mathbf{f}^{(L-1):1}[q']}{\partial f_{m^{(2)}}^{(1)}} \frac{\partial f_{m^{(2)}}^{(1)}}{\partial \mathbf{x}_i}, \frac{\partial \mathbf{f}^{(L-1):2}[q']}{\partial f_1^{(2)}} \frac{\partial f_1^{(2)}}{\partial \mathbf{h}_i^{(1)}}, \ldots, \frac{\partial \mathbf{f}^{(L-1)}[q']}{\partial \mathbf{h}_i^{(L-1)}}, \ldots, \frac{\partial \mathbf{f}^{(L-1)}[q']}{\partial \mathbf{h}_i^{(L-1)}} \right)$ for

$q' = 1, \ldots, q$, and $\mathbf{n}_{q',i}^T = (\mathbf{d}_i + \mathbf{e}_{i1}^{(1)}, \ldots, \mathbf{d}_i + \mathbf{e}_{ip}^{(1)}, \mathbf{e}_i^{(2)}, \ldots, \mathbf{e}_i^{(L-2)}, \mathbf{e}_{q',i}^{(L-1)})$. Each element in $\boldsymbol{\Delta}_i$ in Eq. (20) can be approximated through the first-order Taylor expansion around $\mathbf{n}_{q',i} = \mathbf{0}$; that is,

$$\boldsymbol{\Delta}_i[q'] \approx \Psi_{i,q'}(\mathbf{w}) \cdot \mathbf{n}_{q',i} \text{ for } q' = 1, \ldots, q. \tag{21}$$

Note that $\Psi_{i,q'}(\mathbf{w})$ and $\mathbf{n}_{q',i}$ are of length $pm^{(2)} + \sum_{l=2}^{L-2} m^{(l)} m^{(l+1)} + m^{(L-1)}q$, so $\boldsymbol{\Delta}_i[q']$ is a scalar. We can now define sensitivity of a learned NN, on which the results on the stability and robustness of the trained NN, is based on.

Definition 3. The sensitivity of a NN is defined as the summed ratio over all cases $i = 1, \ldots, n$ between the variance of Δ_i and the variance of d_i,

$$S(\mathbf{w}) = \sum_{i=1}^{n} \frac{|\mathbf{V}(\Delta_i)|_1}{|\mathbf{V}(\mathbf{d}_i)|_1} \approx p^{-1} \sum_{i=1}^{n} \sum_{q'=1}^{q} \Psi_{i,q'}(\mathbf{w}) \begin{pmatrix} R+A & 0 \\ 0 & B_{q'} \end{pmatrix} \Psi_{i,q'}^T(\mathbf{w}), \tag{22}$$

where R is a symmetric matrix with $R_{i,i+1} = \cdots = R_{i,i+p-1} = R_{i+1,i} = \cdots = R_{i+p-1,i} = 1$ for $i = 1, \ldots, p(m^{(2)}-1)+1$, and 0 otherwise; $A = \text{diag}(\frac{1}{\varpi^2}+1)$ for $j = 1, \ldots, p$ and $k = 1, \ldots, m^{(2)}$, $B_{q'} = \text{diag}\left(D^{(2)}, \ldots, D^{(L-2)}, D_{q'}^{(L-1)} \right)$ with $D^{(l)} = \text{diag}\left(\frac{1}{\varpi^2} \mathbf{V}(|w_{jk}^{(l)}|, \boldsymbol{\lambda}) \right)$ for $l = 2, \ldots, L-2$ and $D_{q'}^{(L-1)} = \text{diag}\left(\frac{1}{\varpi^2} \mathbf{V}(|w_{jq'}^{(L-1)}|, \boldsymbol{\lambda}) \right)$.

The sensitivity quantifies the expected fluctuation in outcome prediction via the NN when there is external perturbation (adversarial attack), relative to the degree of the perturbation. It is desirable to have a robust system with low sensitivity to adversarial attacks when it comes to prediction. The smaller the sensitivity, the more robust the NN model and its outcome prediction are to adversarial attacks. Theorem 4 states whiteout helps to achieve that goal, as it minimizes the sum of the original loss function and the sensitivity of the NN.

Theorem 4 (Adversarial robustness with whiteout NI). Let $l(\mathbf{w}, \mathbf{b}| \mathbf{x}, \mathbf{y}) = \sum_{i=1}^{n} |\mathbf{y}_i - \bar{\mathbf{y}}_i|_2^2$ and $l_p(\mathbf{w}, \mathbf{b}|\mathbf{x}, \mathbf{e}, \mathbf{y}) = \sum_{i=1}^{n} |\mathbf{y}_i - \hat{\mathbf{y}}_i|_2^2$. The expectation of l_p over the distribution of $\mathbf{e}^*$ is

$$E_{\mathbf{e}}(l(\mathbf{w}, \mathbf{b}|\mathbf{e}, \mathbf{x}, \mathbf{y})) \approx l(\mathbf{w}, \mathbf{b}|\mathbf{x}, \mathbf{y}) + aS(\mathbf{w}), \tag{23}$$

where $S(\mathbf{w})$ is the sensitivity defined in Eq. (22) in Definition 3.

The proof of Theorem 4 is given in the supplementary materials to this paper. Equation (23) suggests minimizing the perturbed loss function with whiteout noise is approximately equivalent (via the first-order Taylor expansion) to minimizing the original loss function with a penalty term for the instability (sensitivity) of the NN to adversarial attacks. The tuning parameter a quantifies how much weight assigned to the sensitivity of the NN. When a is close to 0, there is the sensitivity of the NN does not carry a lot of weight during the optimization and can be arbitrarily large, which is unwelcome from a learning perspective. With non-zero a, whiteout automatically minimizes the sensitivity-adjusted loss function in Eq. (23) and delivers a trained NN of greater stability and robustness.

5   Theoretical Properties for Noise Perturbed Loss Function and Its Minimizer

We present in this section the asymptotic properties of noise-perturbed loss functions with whiteout and the estimates of NN parameters from minimizing the perturbed loss function. The main results are presented in Theorems 7 and 8, which eventually establishes that the minimizer of the perturbed loss function is consistent for the minimizer of the loss function if the distribution of $\mathbf{X}$ and $\mathbf{Y}$ were known (as $n \to \infty$) and the epoch number $k \to \infty$ for a given optimization algorithm. Note the proofs do not require the injected noises to be Gaussian. Therefore, the asymptotic conclusions and results presented in this section should hold for other NI techniques under the same regularity conditions. To show the perturbed loss function is trainable for practical implementation, we also investigate the tail bound with finite n and k. The proofs of all theoretical results in this section are provided in the supplementary materials to this paper.

We begin with defining and differentiating several types of loss functions to facilitate the investigation of theoretical properties in NI techniques in general (Table 1). WLOG, we present the definitions in terms of the l_2 loss, but the definitions are general to be applicable to any type of loss function (e.g., the l_1 loss and negative log-likelihood). Let $p(\mathbf{X}, \mathbf{Y})$ denote the unknown underlying distribution from which training data $(\mathbf{x}, \mathbf{y})$ are sampled. Let $f(\mathbf{Y}|\mathbf{X}, \mathbf{w}, \mathbf{b})$ be the NN that models the relation between $\mathbf{X}$ and $\mathbf{Y}$.

Table 1. Loss functions

Loss function	Definition
Ideal loss function (ilf)	$l(\mathbf{w},\mathbf{b})=\mathrm{E}_{\mathbf{x},\mathbf{y}}\lvert f(\mathbf{x}\lvert\mathbf{w},\mathbf{b})-\mathbf{y}\rvert_2^2$
Empirical loss function (elf)	$l(\mathbf{w},\mathbf{b}\lvert\mathbf{x},\mathbf{y})=n^{-1}\sum_{i=1}^{n}\lvert f(\mathbf{x}_i\lvert\mathbf{w},\mathbf{b})-\mathbf{y}_i\rvert_2^2$
Noise perturbed empirical loss function (pelf)	$l_p(\mathbf{w},\mathbf{b}\lvert\mathbf{x},\mathbf{y},\mathbf{e})$ $=(kn)^{-1}\sum_{k'=1}^{k}\sum_{i=1}^{n}\lvert f(\mathbf{x}_i,\mathbf{e}_{ik'}\lvert\mathbf{w},\mathbf{b})-\mathbf{y}_i\rvert_2^2$
Noise-marginalized pelf (nm-pelf)	$l_p(\mathbf{w},\mathbf{b}\lvert\mathbf{x},\mathbf{y})=\mathrm{E}_{\mathbf{e}}(l_p(\mathbf{w},\mathbf{b}\lvert\mathbf{x},\mathbf{y},\mathbf{e}))$
Fully marginalized pelf (fm-pelf)	$l_p(\mathbf{w},\mathbf{b})=\mathrm{E}_{\mathbf{x},\mathbf{y}}(l_p(\mathbf{w},\mathbf{b}\lvert\mathbf{x},\mathbf{y}))=\mathrm{E}_{\mathbf{x},\mathbf{y},\mathbf{e}}(l_p(\mathbf{w},\mathbf{b}\lvert\mathbf{x},\mathbf{y},\mathbf{e}))$

In an ideal world, one would minimize the ilf to obtain the estimation on $\mathbf{w}$ and $\mathbf{b}$; however, it is not computable in real life since $p(\mathbf{X},\mathbf{Y})$ is unknown. The empirical version of the ilf is the elf, the original loss function without any regularization, and elf$\to$ ilf as $n\to\infty$. NI regularization (e.g., dropout, shakeout, whiteout) minimizes the pelf ($\mathbf{e}_{ik'}$ in the pelf definition in Table 1 represents the collective noise injected into case i in the k'-th epoch during training). The nm-pelf can be interpreted as training a NN by minimizing the pelf with a finite n and infinite epochs ($k\to\infty$) and is approximately equal to the elf plus a penalty term to mitigate over-fitting. All the asymptotic properties below are established in NNs with one hidden layer. Extending to multiple layers is technically challenging; but it is a subject we will continue to explore.

Lemma 5 (almost sure convergence of pelf to nm-pelf, and of nm-pelf to fm-pelf). In a NN with one hidden layer and the hidden nodes are uniformly bounded, for any $\delta>0$ **(1)**. $\lvert\inf_{\mathbf{w},\mathbf{b}} l_p(\mathbf{w},\mathbf{b}\lvert\mathbf{x},\mathbf{y},\mathbf{e})-\inf_{\mathbf{w},\mathbf{b}} l_p(\mathbf{w},\mathbf{b}\lvert\mathbf{x},\mathbf{y})\rvert<\delta$ as $k\to\infty$ with probability 1 and convergence rate $\propto O(e^{-kn})$; **(2)**. $\lvert\inf_{\mathbf{w},\mathbf{b}} l_p(\mathbf{w},\mathbf{b}\lvert\mathbf{x},\mathbf{y})-\inf_{\mathbf{w},\mathbf{b}} l_p(\mathbf{w},\mathbf{b})\rvert<\delta$ as $n\to\infty$ with probability 1 and convergence rate $\propto O(e^{-n})$.

Lemma 5 suggests the almost sure (a.s.) convergence of the pelf to the nm-pelf, and the nm-pelf to the fm-pelf, which, taken together with the triangle inequality $\lvert\inf_{\mathbf{w},\mathbf{b}} l_p(\mathbf{w},\mathbf{b}\lvert\mathbf{x},\mathbf{y},\mathbf{e})-\inf_{\mathbf{w},\mathbf{b}} l_p(\mathbf{w},\mathbf{b})\rvert\le\lvert\inf_{\mathbf{w},\mathbf{b}} l_p(\mathbf{w},\mathbf{b}\lvert\mathbf{x},\mathbf{y},\mathbf{e})-\inf_{\mathbf{w},\mathbf{b}} l_p(\mathbf{w},\mathbf{b}\lvert\mathbf{x},\mathbf{y})\rvert$ $+\lvert\inf_{\mathbf{w},\mathbf{b}} l_p(\mathbf{w},\mathbf{b}\lvert\mathbf{x},\mathbf{y})-\inf_{\mathbf{w},\mathbf{b}} l_p(\mathbf{w},\mathbf{b})\rvert$, lead to the a.s. convergence of the pelf to the fm-pelf (Corollary 6).

Corollary 6 (almost sure convergence of pelf to fm-pelf). $\lvert\inf_{\mathbf{w},\mathbf{b}} l_p(\mathbf{w},\mathbf{b}\lvert\mathbf{x},\mathbf{y},\mathbf{e})-\inf_{\mathbf{w},\mathbf{b}} l_p(\mathbf{w},\mathbf{b})\rvert<\delta$ as $k\to\infty, n\to\infty$ for any $\delta>0$ with probability 1 and convergence rate $\propto O(e^{-kn})$.

The main results presented in Theorems 7 and 8 are based on Lemma 5 and Corollary 6; that is, the a.s. convergence of the pelf to the ilf and that the minimizer of the former is consistent for the minimizer of the later as $n\to\infty$ and $k\to\infty$. The consistency of the minimizer is a desirable asymptotic property and justifies NI theoretically as an approach for learning NNs.

Theorem 7 (almost sure convergence of pelf to ilf). Let $\sigma_{\max}(n)$ be the maximum noise variance among all injected noises. If $\sigma_{\max}(n) \to 0$ as $n \to \infty$, then $|\inf\limits_{\mathbf{w},\mathbf{b}} l_p(\mathbf{w},\mathbf{b}|\mathbf{x},\mathbf{y},\mathbf{e}) - \inf\limits_{\mathbf{w},\mathbf{b}} l(\mathbf{w},\mathbf{b})| < \delta$ as $k \to \infty, n \to \infty$ for any $\delta > 0$ with probability 1.

With finite n, the pelf always deviates from the ilf given the former has a regularization term. On the other hand, if the injected noise diminishes as $n \to \infty$ (e.g., if the variance of the noise goes to 0), so do the injected noise and the regularization, then the pelf will eventually get arbitrarily close to the ilf per Theorem 7. Regarding the condition $\sigma_{\max}(n) \to 0$ as $n \to \infty$, [29] shows that additive NI can be interpreted as generating kernel distribution estimate for the training data; and for the infima of two loss functions (pelf and ilf in our case) to get arbitrarily close as $n \to \infty$, it is sufficient (though not necessary) that the l_1 distance between the true density g from which $\mathbf{x}$ (of dimension p) are sampled and its kernel density estimate $\hat{g}$ with bandwidth $h = O(n^{-(p+4)^{-1}})$ has the minimum expected upper bound. By recasting whiteout as a kernel estimate problem, then $\sigma_{\max}(n) = O(n^{-(p+4)^{-1}})$.

Theorem 8 (consistency of minimizer of pelf to minimizer of ilf). Let $\hat{\mathbf{w}}_p^{r,n}$ and $\mathbf{w}^0$ denote the optimal weights from minimizing the pelf and the ilf respectively. Let $\mathbf{W}$ be the weight space, assumed to be compact. Define $\hat{\mathbf{W}}^0 = \{\mathbf{w}^0 \in \mathbf{W} | l_p(\mathbf{w}^0,\mathbf{b}) \le l(\mathbf{w},\mathbf{b})$ for all $\mathbf{w} \in \mathbf{W}\}$ that consists the minimizers of the ilf and is a non-empty subset of $\mathbf{W}$. Define the distance of $\mathbf{w}$ from $\hat{\mathbf{W}}^0$ as $d(\mathbf{w},\hat{\mathbf{W}}^0) = \min\limits_{\mathbf{w}^0 \in \hat{\mathbf{W}}^0} ||\mathbf{w} - \mathbf{w}^0||$ for any $\mathbf{w} \in \mathbf{W}$. Let r be the reciprocal of step length in an iterative weight updating algorithm (e.g., the learning rate in the BP algorithm) and $r \to \infty$ (i.e., infinite noises are generated and injected during the weight training). If $\Pr\left(\sup\limits_{f \in \mathbf{F}_n} |l(\mathbf{w},\mathbf{b}) - l_p(\mathbf{w},\mathbf{b}|\mathbf{x},\mathbf{y},\mathbf{e})| > t\right) \to 0$ as $r \to \infty$, $n \to \infty$, then $\Pr\left(\lim\limits_{n \to \infty}\left(\limsup\limits_{r \to \infty} d(\hat{\mathbf{w}}_p^{r,n}, \hat{\mathbf{W}}^0)\right) = 0\right) = 1$.

The optimization in NNs is a non-convex problem. Theorem 8 shows that the minimizer of the pelf converges to a parameter set that minimizes the ilf rather than a single parameter value. The condition $r \to \infty$ is a stronger requirement than $k \to \infty$ since $r \to \infty$ indicates $k \to \infty$, but not vice versa.

When implementing whiteout in practice, one minimizes the pelf with a finite number of epochs k given a training set with finite n. It is thus important to examine the fluctuation of the pelf around its expectation and the tail bound of its distribution to ensure that it is trainable in practice.

Theorem 9 (tail bound on pelf). Assume $\mathbf{Y}$ is bounded, the loss function is uniformly bounded, and the activation functions employed by a NN are Lipschitz continuous. Then $\exists B > 0$, such that $l_p(\mathbf{w},\mathbf{b}|\mathbf{x},\mathbf{y},\mathbf{e}) : \mathcal{R}^k \to \mathcal{R}$, which is a function of injected Gaussian whiteout noise $\mathbf{e}_{k \times 1}$, is $B/\sqrt{kn}$-Lipschitz with respect to the Euclidean norm, for any $\delta > 0$,

$$\Pr(\left|l_p(\mathbf{w},\mathbf{b}|\mathbf{x},\mathbf{y},\mathbf{e}) - \mathbf{E_e}\left(l_p(\mathbf{w},\mathbf{b}|\mathbf{x},\mathbf{y},\mathbf{e})\right)\right| > \delta) \le 2\exp\left(-kn\delta^2/(2B^2)\right). \tag{24}$$

Equation (24) suggests that the fluctuation of the pelf around its expectation is bounded in that the distribution on the difference between the two has a tail that decay to 0 exponentially fast in k, providing assurance on the plausibility of minimizing the pelf for practical applications.

6   Experiments

We compare the prediction performance of whiteout with dropout, shakeout, and no regularization (referred to as "no-reg" hereafter) in 4 experiments. In the Lee Silverman Voice Treatment (LSTV) experiment, we predict a binary phonation outcome post LSTV from 309 dysphonia attributes in Parkinson's patients. The employed NN has two fully connected hidden-layers (9 and 6 nodes) with the sigmoid activation function. We used a 10-fold CV to select hyper-parameters σ, and set $\lambda = \sigma^2$ and $\gamma = 1$ to yield the $l_1 + l_2$ regularization for the multiplicative whiteout. We set τ at $\tau = 2\sigma^2/(1 + 2\sigma^2)$ in dropout, and $c = 0.5$ and $\tau = 2\sigma^2/(1 + 2\sigma^2)$ in shakeout to make the l_2 regularization in dropout and the regularization effect in shakeout comparable to whiteout. We calculated the mean prediction accuracy on the validation set over 100 repetitions. In the LIBRAS experiment, we predict 15 hand movements in the Brazilian sign language from 90 attributes. We randomly chosen 240 samples as the training set and used the rest 120 samples as the testing set. The applied NN has two fully connected hidden layers NN (with 20 and 10 nodes) with the sigmoid and softmax activation functions. We set $\gamma = 1$ in whiteout, $c = 0.5$ in shakeout, and applied a 4-fold CV to select $\sigma^2 = \lambda$ in τ in shakeout, and τ in dropout. The mean prediction accuracy was summarized over 50 repetitions. For the MNIST and CIFAR-10 experiments, we employed the same NNs as used in [18]. The three NI techniques were applied to hidden nodes in the fully connected layers, and a 4-fold CV was applied to select the tuning parameter in each NI technique. The prediction accuracy rates in the testing sets were averaged over 50 repetitions.

The results are given in Table 2. Overall, whiteout has the best performance with the highest prediction accuracy in the LSTV and LIBRAS experiments (small training sets) and similar accuracy in the MNIST and CIFAR-10 experiment (large training sets). Due to space limitation, the empirical distributions of the final weight estimates in the LSRV and LIBRAS experiments are not presented. More weight estimates fall within the neighborhood of 0 with whiteout due the sparsity regularization compared to dropout and shakeout.

Table 2. Mean (SD) prediction accuracy (%) on testing or validation data

Experiment (n^*)	No-reg	Dropout	Shakeout	Whiteout
LSTV (126)	73.93 (12.13)	84.17 (10.37)	84.66 (8.98)	**87.27** (9.84)
LIBRAS (240)	62.15 (5.73)	66.35 (3.60)	66.85 (3.56)	**69.04** (2.08)
MNIST (50,000)	99.19 (0.03)	**99.23** (0.03)	**99.24** (0.05)	99.22 (0.08)
CIFAR-10 (40,000)	75.22 (0.16)	77.97 (0.22)	77.85 (0.58)	**78.57** (0.39)

$^*n$ is the number of samples in the training data.

7   Discussion

We have proposed whiteout for regularizing the training of NNs. Whiteout has connections with a wider range of regularizers due to the flexible and adaptive nature of the whiteout noise variance terms.

We have also worked out the back-propagation algorithm to incorporate the whiteout NI that is provided in the supplementary materials to this paper. Whiteout can also be applied to unsupervised learning (e.g., dimension reduction and pre-training of NNs). The supplementary materials to this paper illustrates the regularization effects of whiteout in RBMs and auto-encoders, where the expected perturbed loss functions with regard to the distribution of injected noises can be written as the sum of the regular loss function and a penalty term on model parameters.

Supplementary materials

The supplementary materials can be found at https://arxiv.org/abs/1612. 01490v4.

Acknowledgments. We thank Mr. Ruoyi Xu for running the MNIST and CIFAR-10 experiments and the three reviewers for there comments and feedback.

References

1. Finlay, C., Calder, J., Abbasi, B., Oberman, A.: Lipschitz regularized deep neural networks generalize and are adversarially robust. arXiv:1808.09540v4 (2019)
2. Yang, D.: YanLiu: L1/2 regularization learning for smoothing interval neural networks: algorithms and convergence analysis. Neurocomputing **272**, 122–129 (2018)
3. Tartaglione, E., Lepsø y, S., Fiandrotti, A., Francini, G.: Learning sparse neural networks via sensitivity-driven regularization. In: Bengio, S., Wallach, H., Larochelle, H., Grauman, K., Cesa-Bianchi, N., Garnett, R. (eds.) Advances in Neural Information Processing Systems, vol. 31, pp. 3878–3888 (2018)
4. Jakubovitz, D., Giryes, R.: Improving DNN robustness to adversarial attacks using Jacobian regularization. In: Ferrari, V., Hebert, M., Sminchisescu, C., Weiss, Y. (eds.) ECCV 2018, Part XII. LNCS, vol. 11216, pp. 525–541. Springer, Cham (2018). https://doi.org/10.1007/978-3-030-01258-8_32

5. Blot, M., Robert, T., Thome, N., Cord, M.: Shade: information-based regularization for deep learning. In: 2018 25th IEEE International Conference on Image Processing (ICIP), pp. 813–817 (2018)
6. Plaut, D.C., Nowlan, S.J., Hinton, G.E.: Experiments on learning by back-propagation. Technical report, CMU-CS-86-126 (1986)
7. Sietsma, J., Dow, R.J.F.: Neural network pruning - why and how. In: Proceedings of the IEEE International Conferences on Neural Networks, vol. 1, pp. 325–333 (1988)
8. Grandvalet, Y., Canu, S., Boucheron, S.: Noise injection: theoretical prospects. Neural Comput. **9**, 1093–1108 (1997)
9. Matsuoka, K.: Noise injection into inputs in back-propagation learning. IEEE Trans. Syst. Man Cybern. **22**, 436–440 (1992)
10. Holmstrom, L., Koistinen, P.: Using additive noise in back-propagation training. IEEE Trans. Neural Networks **3**, 24–38 (1992)
11. Hinton, G.E., Srivastava, N., Krizhevsky, A., Sutskever, I., Salakhutdinov, R.R.: Improving neural networks by preventing co-adaptation of feature detectors. arXiv:1207.0580 (2012)
12. Srivastava, N., Hinton, G., Krizhevsky, A., Sutskever, I., Salakhutdinov, R.: Dropout: a simple way to prevent neural networks from overfitting. J. Mach. Learn. Res. **15**, 1929–1958 (2014)
13. Goodfellow, I.J., Warde-Farley, D., Mirza, M., Courville, A., Bengio, Y.: Maxout networks. In: Proceedings of the 30th International Conference on Machine Learning (ICML), vol. 28, pp. 1319–1327 (2013)
14. Wang, S., Manning, C.D.: Fast dropout training. In: Proceedings of the 30th International Conference on Machine Learning, vol. 28, pp. 118–126 (2013)
15. Wan, L., Zeiler, M., Zhang, S., LeCun, Y., Fergus, R.: Regularization of neural networks using DropConnect. PLMR **28**, 1058–1066 (2013)
16. Wang, B., Klabjan, D.: Regularization for unsupervised deep neural nets. In: Proceedings of the Thirty-First AAAI Conference on Artificial Intelligence (2017)
17. Ba, J., Frey, B.: Adaptive dropout for training deep neural networks. In: Advances in Neural Information Processing Systems, pp. 1–9 (2013)
18. Kang, G., Li, J., Tao, D.: Shakeout: a new approach to regularized deep neural network training. IEEE Trans. Pattern Anal. Mach. Intell. **40**, 1245–1258 (2018)
19. Graves, A.: Practical variational inference for neural networks. In: Proceedings of the 24th International Conference on Neural Information Processing Systems, NIPS 2011, pp. 2348–2356 (2011)
20. An, G.: The effects of adding noise during backpropagation training on a generalization performance. Neural Comput. **8**, 643–674 (1996)
21. Bishop, C.M.: Training with noise is equivalent to Tikhonov regularization. Neural Comput. **7**, 108–116 (1995)
22. Wager, S., Wang, S., Liang, P.: Dropout training as adaptive regularization. In: Advances in Neural Information Processing Systems (NIPS), vol. 26, pp. 351–359 (2013)
23. Wainwright, M.J., Jordan, M.I.: Graphical Models, Exponential Families, and Variational Inference. Now Publishers Inc., Delft (2008)
24. Frank, I., Friedman, J.: A statistical view of some chemometrics regression tools. Technometrics **35**, 109–135 (1993)
25. Tibshirani, R.: Regression shrinkage and selection via the lasso. J. Roy. Stat. Soc. B **58**, 267–288 (2006)
26. Zou, H., Hastie, T.: Regularization and variable selection via the elastic net. J. Roy. Stat. Soc. B **67**, 301–320 (2005)

27. Zou, H.: The adaptive lasso and its oracle properties. J. Am. Stat. Assoc. Theory Methods **101**, 1418–1429 (2006)
28. Yuan, M., Lin, Y.: Model selection and estimation in regression with grouped variables. J. Roy. Stat. Soc. B **68**, 49–67 (2006)
29. Holmström, L., Klemelä, J.: Asymptotic bounds for the expected L1 error of a multivariate kernel density estimator. J. Multivar. Anal. **42**, 245–266 (1992)

Pattern Recognition Based on an Improved Szmidt and Kacprzyk's Correlation Coefficient in Pythagorean Fuzzy Environment

Paul Augustine Ejegwa[1,2], Yuming Feng[1(✉)], and Wei Zhang[3]

[1] Key Laboratory of Intelligent Information Processing and Control,
Chongqing Three Gorges University, Wanzhou, Chongqing 404100, China
`ejegwa.augustine@uam.edu.ng, yumingfeng25928@163.com`
[2] Department of Mathematics/Statistics/Computer Science,
University of Agriculture, P.M.B. 2373, Makurdi, Nigeria
[3] Chongqing Engineering Research Center of Internet of Things
and Intelligent Control Technology, Chongqing Three Gorges University,
Wanzhou, Chongqing 404100, China
`cqec126@163.com`

Abstract. Correlation measure is an applicable tool in Pythagorean fuzzy domain for resolving problems of multi-criteria decision-making (MCDM). Szmidt and Kacprzyk proposed a correlation coefficient in intuitionistic fuzzy domain (IFSs) by considering the orthodox parameters of IFSs. Nonetheless, the approach contradicts the axiomatic description of correlation coefficient between IFSs in literature. In this paper we modify the Szmidt and Kacprzyk's approach for measuring correlation coefficient between IFSs to satisfy the axiomatic description of correlation coefficient, and extend the modified version to Pythagorean fuzzy environment. Some numerical illustrations are considered to ascertain the merit of the modified version over Szmidt and Kacprzyk's approach. Finally, the proposed correlation coefficient measure is applied to resolve some pattern recognition problems. In recap, the goal of this paper is to modify Szmidt and Kacprzyk's correlation coefficient for IFSs, extend it to Pythagorean fuzzy context with pattern recognition applications.

Keywords: Correlation coefficient · Intuitionistic fuzzy set · Pattern recognition · Pythagorean fuzzy set

1 Introduction

Pattern recognition is the act of detecting arrangements of characteristics or data that produce information about a given system or data set. In a technological setting, a pattern is a recurrent of sequences of data over time that can be useful for forecast tendencies, alignments of features in images that ascertain objects, among others. The process of identifying patterns by using machine learning

© Springer Nature Switzerland AG 2020
M. Han et al. (Eds.): ISNN 2020, LNCS 12557, pp. 190–206, 2020.
https://doi.org/10.1007/978-3-030-64221-1_17

procedure is referred to pattern recognition. In fact, pattern recognition has a lots to do with artificial intelligence and machine learning. The idea of pattern recognition is important because of its application potential in neural networks, software engineering, computer vision, etc. However, the potential areas of application of pattern recognition are greeted with uncertainties and imprecisions. Thus fuzzy sets [52] is key in resolving/curbing the embedded uncertainties in pattern recognition. The theory of fuzzy sets has been applied to resolve some problems involving uncertainties [4, 6, 24].

In real life, some decision-making problems could not be handled with fuzzy set approach because fuzzy set only considered membership degree (MD) whereas, many real life problems have the component of both MD and non-membership degree (NMD) with the possibility of hesitation. This scenario can best be captured by a concept called intuitionistic fuzzy sets (IFSs) [1, 2]. This concept is described with MD μ, NMD ν and hesitation margin (HM) π with the property that their aggregate is one and $\mu + \nu$ is less than or equal to one. Due to the resourcefulness of IFS, it has been applied to tackle pattern recognition problems [30, 44] and other multi-criteria decision-making (MCDM) cases [3, 5, 22, 23, 33, 38, 39].

The idea of IFS is not applicable in a situation whenever a decision-maker wants to take decision in a multi-criteria problem in which $\mu + \nu$ is greater than one. In fact assume $\mu = 0.5$ and $\nu = 0.6$, clearly the notion of IFS cannot model such a case. This prompted [45, 48] to generalize IFS as Pythagorean fuzzy set (PFS) such that $\mu + \nu$ is also greater than one and $\mu^2 + \nu^2 + \pi^2 = 1$. In a nutshell, PFS is a special case of IFS with additional conditions and thus has more capacity to curb uncertainties more appropriately with higher degree of accuracy. The concept of PFSs have been sufficiently explored by different authors so far [12, 17, 47]. Many applications of PFSs have been discussed in pattern recognitions and other multi-criteria problems [11, 13–16, 18, 21, 26–28, 45, 46, 48, 53–55].

Correlation coefficient was first studied in statistics by Karl Pearson in 1895 to measure the interrelation between two variables or data. Since then, it has become widely used by statisticians, engineers, scientists, etc. As correlation coefficient is very proficient, it has been since strengthened to curb imprecisions/uncertainties in predictions, pattern recognitions, and medical diagnosis among others. And as such, correlation coefficient has been studied in fuzzy environment [7, 9, 10, 36, 51]. To equip correlation coefficient to better handle fuzzy data, the idea was encapsulated into intuitionistic fuzzy context and applied to many MCDM problems [29, 31, 32, 34, 35, 40, 41, 43, 50, 53]. By extension, authors have started to study correlation coefficient in Pythagorean fuzzy setting with applications in several multi-criteria problems [19, 25, 37, 42].

This work is motivated to strengthen the correlation coefficient in [40] to enhance its satisfaction of the axiomatic definition of correlation coefficient in intuitionistic fuzzy setting as presented in literature, and subsequently extend the improved version to Pythagorean fuzzy context. The objectives of this paper are to

- revisit Szmidt and Kacprzyk's correlation coefficient of IFSs,
- modify Szmidt and Kacprzyk's correlation coefficient of IFSs to satisfy the axiomatic description of correlation coefficient,
- extend the modified version of the correlation coefficient to Pythagorean fuzzy environment for better output,
- numerically show the superiority of the novel correlation coefficient in both intuitionistic/Pythagorean fuzzy contexts over Szmidt and Kacprzyk's approach, and
- demonstrate the applicability of the modified method in Pythagorean fuzzy environment to cases of pattern recognition.

The paper is outlined thus; Sect. 2 briefly revises the fundamentals of PFS and Sect. 3 discusses Szmidt and Kacprzyk's correlation coefficient for IFSs, its modification and numerical verifications. Section 4 extends the modified version of Szmidt and Kacprzyk's correlation coefficient to PFSs and numerically verifies its authenticity. Section 5 demonstrates the application of the novel correlation coefficient in pattern recognition cases captured in Pythagorean fuzzy environment. Section 6 gives conclusion and area for further research.

2   Basic Notions of Pythagorean Fuzzy Sets

Assume X is a fixed non-empty set and take *PFS(X)* to be the set of all *PFSs* defined in X throughout the paper.

Definition 1 [1]. An IFS $\hat{A}$ of X is characterized by

$$\hat{A} = \{\langle \frac{\mu_{\hat{A}}(x), \nu_{\hat{A}}(x)}{x} \rangle \mid x \in X\}, \tag{1}$$

where

$$\mu_{\hat{A}}(x) : X \to [0,1] \text{ and } \nu_{\hat{A}}(x) : X \to [0,1]$$

define MD and NMD of $x \in X$ to $\hat{A}$ in which,

$$0 \leq \mu_{\hat{A}}(x) + \nu_{\hat{A}}(x) \leq 1. \tag{2}$$

For every $\hat{A}$ of X,

$$\pi_{\hat{A}}(x) = 1 - \mu_{\hat{A}}(x) - \nu_{\hat{A}}(x)$$

is the IFS index or HM of $x \in X$. Then, $\pi_{\hat{A}}(x)$ is the grade of non-determinacy of $x \in X$, to $\hat{A}$ and $\pi_{\hat{A}}(x) \in [0,1]$. $\pi_{\hat{A}}(x)$ explains the indeterminacy of whether $x \in X$ or $x \notin X$. It follows that

$$\mu_{\hat{A}}(x) + \nu_{\hat{A}}(x) + \pi_{\hat{A}}(x) = 1. \tag{3}$$

Definition 2 [48]. A Pythagorean fuzzy set $\hat{A}$ of X is defined by

$$\hat{A} = \{\langle \frac{\mu_{\hat{A}}(x), \nu_{\hat{A}}(x)}{x} \rangle \mid x \in X\}, \tag{4}$$

where

$$\mu_{\hat{A}}(x) : X \to [0, 1] \text{ and } \nu_{\hat{A}}(x) : X \to [0, 1]$$

define MD and NMD of $x \in X$ to $\hat{A}$ where

$$0 \le (\mu_{\hat{A}}(x))^2 + (\nu_{\hat{A}}(x))^2 \le 1. \tag{5}$$

Taking $\mu_{\hat{A}}^2(x) + \nu_{\hat{A}}^2(x) \le 1$, we have an indeterminacy grade of $x \in X$ to $\hat{A}$ given as

$$\pi_{\hat{A}}(x) \in [0, 1] = \sqrt{1 - [(\mu_{\hat{A}}(x))^2 + (\nu_{\hat{A}}(x))^2]}. \tag{6}$$

Clearly,

$$(\mu_{\hat{A}}(x))^2 + (\nu_{\hat{A}}(x))^2 + (\pi_{\hat{A}}(x))^2 = 1. \tag{7}$$

Suppose $\pi_{\hat{A}}(x) = 0$ then $\mu_{\hat{A}}^2(x) + \nu_{\hat{A}}^2(x) = 1$.

Example 1 Assume $\hat{A} \in PFS(X)$. If $\mu_{\hat{A}}(x) = 0.7$ and $\nu_{\hat{A}}(x) = 0.5$ for $X = \{x\}$. Then, $0.7^2 + 0.5^2 \le 1$. Thus $\pi_{\hat{A}}(x) = 0.5099$, and so $\mu_{\hat{A}}^2(x) + \nu_{\hat{A}}^2(x) + \pi_{\hat{A}}^2(x) = 1$.

Definition 3 [48]. Suppose $\hat{A}, \hat{B} \in PFS(X)$, then

(i) $\overline{\hat{A}} = \{\langle \frac{\nu_{\hat{A}}, \mu_{\hat{A}}}{x} \rangle \mid x \in X\}$.

(ii) $\hat{A} \cup \hat{B} = \{\langle \max(\frac{\mu_{\hat{A}}(x), \mu_{\hat{B}}(x)}{x}), \min(\frac{\nu_{\hat{A}}(x), \nu_{\hat{B}}(x)}{x}) \rangle \mid x \in X\}$.

(iii) $\hat{A} \cap \hat{B} = \{\langle \min(\frac{\mu_{\hat{A}}(x), \mu_{\hat{B}}(x)}{x}), \max(\frac{\nu_{\hat{A}}(x), \nu_{\hat{B}}(x)}{x}) \rangle \mid x \in X\}$.

It follows that, $\hat{A} = \hat{B}$ iff $\mu_{\hat{A}}(x) = \mu_{\hat{B}}(x)$, $\nu_{\hat{A}}(x) = \nu_{\hat{B}}(x) \forall x \in X$, and $\hat{A} \subseteq \hat{B}$ iff $\mu_{\hat{A}}(x) \le \mu_{\hat{B}}(x)$, $\nu_{\hat{A}}(x) \ge \nu_{\hat{B}}(x) \forall x \in X$. We say $\hat{A} \subset \hat{B}$ iff $\hat{A} \subseteq \hat{B}$ and $\hat{A} \ne \hat{B}$.

3   Correlation Coefficient for Pythagorean Fuzzy Sets

To start with, we reiterate the axiomatic description of correlation coefficient for IFSs. Assume $\hat{A}$ and $\hat{B}$ are IFSs of $X = \{x_1, x_2, ..., x_n\}$.

Definition 4 [29]. The correlation coefficient for $\hat{A}$ and $\hat{B}$ denoted by $\rho(\hat{A}, \hat{B})$ is a measuring function $\rho : IFS \times IFS \to [0, 1]$ which satisfies the following conditions;

(i) $\rho(\hat{A}, \hat{B}) \in [0, 1]$,
(ii) $\rho(\hat{A}, \hat{B}) = \rho(\hat{B}, \hat{A})$,
(iii) $\rho(\hat{A}, \hat{B}) = 1$ if and only if $\hat{A} = \hat{B}$.

On the strength of this description, we present Szmidt and Kacprzyk's correlation coefficient for IFSs and check whether it satisfies Definition 4.

3.1 Szmidt and Kacprzyk's Correlation Coefficient for IFSs

Now, we present the correlation coefficient for IFSs in [40] as follows. This correlation coefficient completely described IFSs.

Definition 5 [40]. The correlation coefficient $\rho(\hat{A}, \hat{B})$ between IFSs $\hat{A}$ and $\hat{B}$ of X is

$$\rho(\hat{A}, \hat{B}) = \frac{\Delta_1(\hat{A}, \hat{B}) + \Delta_2(\hat{A}, \hat{B}) + \Delta_3(\hat{A}, \hat{B})}{3}, \tag{8}$$

where

$$\left. \begin{array}{l} \Delta_1(\hat{A}, \hat{B}) = \dfrac{\sum_{i=1}^{n}(\mu_{\hat{A}}(x_i) - \overline{\mu_{\hat{A}}})(\mu_{\hat{B}}(x_i) - \overline{\mu_{\hat{B}}})}{\sqrt{\sum_{i=1}^{n}(\mu_{\hat{A}}(x_i) - \overline{\mu_{\hat{A}}})^2}\sqrt{\sum_{i=1}^{n}(\mu_{\hat{B}}(x_i) - \overline{\mu_{\hat{B}}})^2}} \\[3ex] \Delta_2(\hat{A}, \hat{B}) = \dfrac{\sum_{i=1}^{n}(\nu_{\hat{A}}(x_i) - \overline{\nu_{\hat{A}}})(\nu_{\hat{B}}(x_i) - \overline{\nu_{\hat{B}}})}{\sqrt{\sum_{i=1}^{n}(\nu_{\hat{A}}(x_i) - \overline{\nu_{\hat{A}}})^2}\sqrt{\sum_{i=1}^{n}(\nu_{\hat{B}}(x_i) - \overline{\nu_{\hat{B}}})^2}} \\[3ex] \Delta_3(\hat{A}, \hat{B}) = \dfrac{\sum_{i=1}^{n}(\pi_{\hat{A}}(x_i) - \overline{\pi_{\hat{A}}})(\pi_{\hat{B}}(x_i) - \overline{\pi_{\hat{B}}})}{\sqrt{\sum_{i=1}^{n}(\pi_{\hat{A}}(x_i) - \overline{\pi_{\hat{A}}})^2}\sqrt{\sum_{i=1}^{n}(\pi_{\hat{B}}(x_i) - \overline{\pi_{\hat{B}}})^2}} \end{array} \right\}, \tag{9}$$

and

$$\left. \begin{array}{l} \overline{\mu_{\hat{A}}} = \dfrac{\sum_{i=1}^{n}\mu_{\hat{A}}(x_i)}{n}, \quad \overline{\mu_{\hat{B}}} = \dfrac{\sum_{i=1}^{n}\mu_{\hat{B}}(x_i)}{n} \\[3ex] \overline{\nu_{\hat{A}}} = \dfrac{\sum_{i=1}^{n}\nu_{\hat{A}}(x_i)}{n}, \quad \overline{\nu_{\hat{B}}} = \dfrac{\sum_{i=1}^{n}\nu_{\hat{B}}(x_i)}{n} \\[3ex] \overline{\pi_{\hat{A}}} = \dfrac{\sum_{i=1}^{n}\pi_{\hat{A}}(x_i)}{n}, \quad \overline{\pi_{\hat{B}}} = \dfrac{\sum_{i=1}^{n}\pi_{\hat{B}}(x_i)}{n} \end{array} \right\}. \tag{10}$$

3.2 Numerical Verifications

Now we consider some examples to illustrate the validity of the Szmidt and Kacprzyk's correlation coefficient for IFSs. The examples are as in [40]. For the sake of simplicity in computations, let

$$(\mu_{\hat{A}}(x_i) - \overline{\mu_{\hat{A}}}) = \alpha_1, \quad (\mu_{\hat{B}}(x_i) - \overline{\mu_{\hat{B}}}) = \beta_1,$$

$$(\nu_{\hat{A}}(x_i) - \overline{\nu_{\hat{A}}}) = \alpha_2, \quad (\nu_{\hat{B}}(x_i) - \overline{\nu_{\hat{B}}}) = \beta_2,$$

$$(\pi_{\hat{A}}(x_i) - \overline{\pi_{\hat{A}}}) = \alpha_3, \quad (\pi_{\hat{B}}(x_i) - \overline{\pi_{\hat{B}}}) = \beta_3,$$

$$(\mu_{\hat{A}}(x_i) - \overline{\mu_{\hat{A}}})(\mu_{\hat{B}}(x_i) - \overline{\mu_{\hat{B}}}) = \gamma_1,$$
$$(\nu_{\hat{A}}(x_i) - \overline{\nu_{\hat{A}}})(\nu_{\hat{B}}(x_i) - \overline{\nu_{\hat{B}}}) = \gamma_2,$$
$$(\pi_{\hat{A}}(x_i) - \overline{\pi_{\hat{A}}})(\pi_{\hat{B}}(x_i) - \overline{\pi_{\hat{B}}}) = \gamma_3.$$

Example I. Assume $\hat{A}$ and $\hat{B}$ are IFSs of $X = \{x_1, x_2, x_3\}$, let

$$\hat{A} = \{\langle \frac{0.1, 0.2, 0.7}{x_1} \rangle, \langle \frac{0.2, 0.1, 0.7}{x_2} \rangle, \langle \frac{0.3, 0.0, 0.7}{x_3} \rangle\},$$

and

$$\hat{B} = \{\langle\frac{0.3, 0.0, 0.7}{x_1}\rangle, \langle\frac{0.2, 0.2, 0.6}{x_2}\rangle, \langle\frac{0.1, 0.6, 0.3}{x_3}\rangle\}.$$

It follows that (Tables 1, 2, 3, 4, 5 and 6)

$$\overline{\mu_{\hat{A}}} = \overline{\mu_{\hat{B}}} = 0.2,\ \overline{\nu_{\hat{A}}} = 0.1,\ \overline{\nu_{\hat{B}}} = 0.2667,\ \overline{\pi_{\hat{A}}} = 0.7,\ \overline{\pi_{\hat{B}}} = 0.5333.$$

Table 1. Computation for membership grades

X	α_1	β_1	α_1^2	β_1^2	γ_1
x_1	−0.1000	0.1000	0.0100	0.0100	−0.0100
x_2	0.0000	0.0000	0.0000	0.0000	0.0000
x_3	0.1000	−0.1000	0.0100	0.0100	−0.0100

Thus $\Delta_1(\hat{A}, \hat{B}) = \dfrac{-0.0200}{\sqrt{0.0200 \times 0.0200}} = -1.$

Table 2. Computation for non-membership grades

X	α_2	β_2	α_2^2	β_2^2	γ_2
x_1	−0.1000	−0.2667	0.0100	0.0711	0.0267
x_2	0.0000	−0.0667	0.0000	0.0004	0.0000
x_3	−0.1000	0.3333	0.0100	0.1111	−0.0333

Thus $\Delta_2(\hat{A}, \hat{B}) = \dfrac{-0.0600}{\sqrt{0.0200 \times 0.1822}} = -9939.$

Table 3. Computation for hesitation grades

X	α_3	β_3	α_3^2	β_3^2	γ_3
x_1	0.0000	0.1667	0.0000	0.0278	0.0000
x_2	0.0000	0.0667	0.0000	0.0044	0.0000
x_3	0.0000	−0.2333	0.0000	0.0544	0.0000

Thus $\Delta_3(\hat{A}, \hat{B}) = \dfrac{0.0000}{\sqrt{0.0000 \times 0.0866}} = 0.0000.$ Hence $\rho(\hat{A}, \hat{B}) = -0.6646.$

Example II. Assume $\hat{A}$ and $\hat{B}$ are IFSs of $X = \{x_1, x_2, x_3\}$, let

$$\hat{A} = \{\langle\frac{0.1, 0.2, 0.7}{x_1}\rangle, \langle\frac{0.2, 0.1, 0.7}{x_2}\rangle, \langle\frac{0.29, 0.0, 0.71}{x_3}\rangle\},$$

and

$$\hat{B} = \{\langle \frac{0.1, 0.3, 0.6}{x_1} \rangle, \langle \frac{0.2, 0.2, 0.6}{x_2} \rangle, \langle \frac{0.29, 0.1, 0.61}{x_3} \rangle \}.$$

It follows that

$$\overline{\mu_{\hat{A}}} = \overline{\mu_{\hat{B}}} = 0.1967, \ \overline{\nu_{\hat{A}}} = 0.1, \ \overline{\nu_{\hat{B}}} = 0.2, \ \overline{\pi_{\hat{A}}} = 0.7033, \ \overline{\pi_{\hat{B}}} = 0.6033.$$

Table 4. Computation for membership grades

X	α_1	β_1	α_1^2	β_1^2	γ_1
x_1	-0.0967	-0.0967	0.0094	0.0094	0.0094
x_2	0.0033	0.0033	0.0000	0.0000	0.0000
x_3	0.0933	0.0933	0.0087	0.0087	0.0087

Thus $\Delta_1(\hat{A}, \hat{B}) = \dfrac{0.0181}{\sqrt{0.0181 \times 0.0181}} = 1.$

Table 5. Computation for non-membership grades

X	α_2	β_2	α_2^2	β_2^2	γ_2
x_1	0.1000	0.1000	0.0100	0.0100	0.0100
x_2	0.0000	0.0000	0.0000	0.0000	0.0000
x_3	-0.1000	-0.1000	0.0100	0.0100	0.0100

Thus $\Delta_2(\hat{A}, \hat{B}) = \dfrac{0.0200}{\sqrt{0.0200 \times 0.0200}} = 1.$

Table 6. Computation for hesitation grades

X	α_3	β_3	α_3^2	β_3^2	γ_3
x_1	-0.0033	-0.0033	0.00001	0.00001	0.00001
x_2	-0.0033	-0.0033	0.00001	0.00001	0.00001
x_3	-0.0067	0.0067	0.00004	0.00004	-0.00004

Thus $\Delta_3(\hat{A}, \hat{B}) = \dfrac{0.00006}{\sqrt{0.00006 \times 0.00006}} = 1.$ Hence $\rho(\hat{A}, \hat{B}) = 1.$

Limitations of Szmidt and Kacprzyk's Correlation Coefficient for IFSs. From Examples I and II, we observe the following drawbacks viz; (i) In Example I, $\rho(\hat{A}, \hat{B}) \notin [0, 1]$ in opposition to Definition 4. (ii) In Example II, $\rho(\hat{A}, \hat{B}) = 1$ even when $\hat{A} \neq \hat{B}$.

Modification of Szmidt and Kacprzyk's Correlation Coefficient for IFSs. To remedy some of the limitations of Szmidt and Kacprzyk's correlation coefficient for IFSs, we give the following definition.

Definition 6 The correlation coefficient $\rho(\hat{A}, \hat{B})$ between IFSs $\hat{A}$ and $\hat{B}$ of X is

$$\rho(\hat{A}, \hat{B}) = \frac{|\Delta_1(\hat{A}, \hat{B})| + |\Delta_2(\hat{A}, \hat{B})| + |\Delta_3(\hat{A}, \hat{B})|}{3}, \tag{11}$$

where $\Delta_1(\hat{A}, \hat{B})$, $\Delta_2(\hat{A}, \hat{B})$ and $\Delta_3(\hat{A}, \hat{B})$ are as in Eq. (9). Applying Eq. (11) to Examples I and II, we get

$$\rho(\hat{A}, \hat{B}) = 0.6646, \ \rho(\hat{A}, \hat{B}) = 1.$$

While $\rho(\hat{A}, \hat{B}) \in [0, 1]$ in Example I, we see that $\rho(\hat{A}, \hat{B}) = 1$ even when $\hat{A} \neq \hat{B}$ (Example II). This is also a limitation, because $\rho(\hat{A}, \hat{B}) = 1$ iff $\hat{A} \neq \hat{B}$. Now, we extend the modified Szmidt and Kacprzyk's correlation coefficient to PFSs to see whether the remaining drawback will be remedied.

4 Modification of Szmidt and Kacprzyk's Correlation Coefficient in PFSs

The definition of correlation coefficient of PFSs is the same as that of IFSs in Definition 4, the difference is the setting. Now, we extend the modified Szmidt and Kacprzyk's correlation coefficient to PFSs as follows.

Definition 7 The correlation coefficient $\rho(\hat{A}, \hat{B})$ between two PFSs $\hat{A}$ and $\hat{B}$ of X is

$$\rho(\hat{A}, \hat{B}) = \frac{|\Delta_1(\hat{A}, \hat{B})| + |\Delta_2(\hat{A}, \hat{B})| + |\Delta_3(\hat{A}, \hat{B})|}{3}, \tag{12}$$

where $\Delta_1(\hat{A}, \hat{B})$, $\Delta_2(\hat{A}, \hat{B})$ and $\Delta_3(\hat{A}, \hat{B})$ are as in (9) with the peculiarity of π in PFS.

Theorem 1 *The function $\rho(\hat{A}, \hat{B})$ is a correlation coefficient of PFSs $\hat{A}$ and $\hat{B}$.*

Proof We show that condition (i) of Definition 4 holds, i.e., $\rho(\hat{A}, \hat{B}) \in [0, 1] \implies 0 \leq \rho(\hat{A}, \hat{B}) \leq 1$. But $\rho(\hat{A}, \hat{B}) \geq 0$ since

$$|\Delta_1(\hat{A}, \hat{B})| \geq 0, \ |\Delta_2(\hat{A}, \hat{B})| \geq 0 \text{ and } |\Delta_3(\hat{A}, \hat{B})| \geq 0.$$

Now, we show that $\rho(\hat{A}, \hat{B}) \leq 1$. Recall that

$$\rho(\hat{A}, \hat{B}) = \frac{|\Delta_1(\hat{A}, \hat{B})| + |\Delta_2(\hat{A}, \hat{B})| + |\Delta_3(\hat{A}, \hat{B})|}{3}.$$

Assume that $|\Delta_1(\hat{A}, \hat{B})| = \eta$, $|\Delta_2(\hat{A}, \hat{B})| = \kappa$ and $|\Delta_3(\hat{A}, \hat{B})| = \lambda$. Then

$$\rho(\hat{A}, \hat{B}) = \frac{\eta + \kappa + \lambda}{3}.$$

Thus,

$$\rho(\hat{A}, \hat{B}) - 1 = \frac{\eta + \kappa + \lambda}{3} - 1$$
$$= \frac{(\eta + \kappa + \lambda) - 3}{3}$$
$$= -\frac{[-(\eta + \kappa + \lambda) + 3]}{3}.$$

Thus $\rho(\hat{A}, \hat{B}) - 1 \leq 0 \Longrightarrow \rho(\hat{A}, \hat{B}) \leq 1$. Hence $0 \leq \rho(\hat{A}, \hat{B}) \leq 1$.

Condition (ii) is trivial, so we omit its proof. Suppose $\hat{A} = \hat{B}$, then

$$\Delta_1(\hat{A}, \hat{B}) = \frac{\sum_{i=1}^{n}(\mu_{\hat{A}}(x_i) - \overline{\mu_{\hat{A}}})(\mu_{\hat{A}}(x_i) - \overline{\mu_{\hat{A}}})}{\sqrt{\sum_{i=1}^{n}(\mu_{\hat{A}}(x_i) - \overline{\mu_{\hat{A}}})^2}\sqrt{\sum_{i=1}^{n}(\mu_{\hat{A}}(x_i) - \overline{\mu_{\hat{A}}})^2}} = 1.$$

Similarly, $\Delta_2(\hat{A}, \hat{B}) = 1$ and $\Delta_3(\hat{A}, \hat{B}) = 1$. Therefore $\rho(\hat{A}, \hat{B}) = 1$, so condition (iii) holds. These complete the proof.   □

Remark 1 It follows that $\Delta_1(\hat{A}, \hat{B})$, $\Delta_2(\hat{A}, \hat{B})$ and $\Delta_3(\hat{A}, \hat{B})$ satisfy the conditions in Definition 4.

4.1 Numerical Verification of the Modified Correlation Coefficient in PFS Context

Now, we verify the authenticity of the modified Szmidt and Kacprzyk's correlation coefficient in PFS context and compare the results with the method of Szmidt and Kacprzyk [40] also in PFS context taking into account the peculiarity of the hesitation margin.

Example III. Suppose $\hat{C}$ and $\hat{D}$ are PFSs of $X = \{x_1, x_2, x_3\}$ for

$$\hat{C} = \{\langle\frac{0.1, 0.2, 0.9747}{x_1}\rangle, \langle\frac{0.2, 0.1, 0.9747}{x_2}\rangle, \langle\frac{0.3, 0.0, 0.9539}{x_3}\rangle\},$$

and

$$\hat{D} = \{\langle\frac{0.3, 0.0, 0.9539}{x_1}\rangle, \langle\frac{0.2, 0.2, 0.9592}{x_2}\rangle, \langle\frac{0.1, 0.6, 0.7937}{x_3}\rangle\}.$$

We have

$$\overline{\mu_{\hat{C}}} = \overline{\mu_{\hat{D}}} = 0.2, \ \overline{\nu_{\hat{C}}} = 0.1, \ \overline{\nu_{\hat{D}}} = 0.2667, \ \overline{\pi_{\hat{C}}} = 0.9678, \ \overline{\pi_{\hat{D}}} = 0.9023.$$

Since the MD and NMD in Example I are same with that in Example III, we get $\Delta_1(\hat{C}, \hat{D}) = -1$ and $\Delta_2(\hat{C}, \hat{D}) = -0.9939$.

For the sake of simplicity in computations, let (Table 7)

$$(\pi_{\hat{C}}(x_i) - \overline{\pi_{\hat{C}}}) = \alpha_3, \ (\pi_{\hat{D}}(x_i) - \overline{\pi_{\hat{D}}}) = \beta_3,$$

Table 7. Computation for hesitation grades

X	α_3	β_3	α_3^2	β_3^2	γ_3
x_1	0.0069	0.0516	0.00005	0.00266	0.00036
x_2	0.0069	0.0569	0.00005	0.00324	0.00039
x_3	−0.0139	−0.1086	0.00019	0.01179	0.00151

$$(\pi_{\hat{A}}(x_i) - \overline{\pi_{\hat{A}}})(\pi_{\hat{B}}(x_i) - \overline{\pi_{\hat{B}}}) = \gamma_3.$$

Thus $\Delta_3(\hat{C}, \hat{D}) = \dfrac{0.00226}{\sqrt{0.00029 \times 0.01769}} = 0.9978.$ Hence $\rho(\hat{C}, \hat{D}) = 0.9972.$
While using Szmidt and Kacprzyk's approach, we have

$$\Delta_1(\hat{C}, \hat{D}) = -1, \ \Delta_2(\hat{C}, \hat{D}) = -0.9939 \text{ and } \Delta_3(\hat{C}, \hat{D}) = 0.9978.$$

Hence $\rho(\hat{C}, \hat{D}) = -0.3320.$

Example IV. Suppose $\hat{C}$ and $\hat{D}$ are PFSs of $X = \{x_1, x_2, x_3\}$ for

$$\hat{C} = \{\langle \frac{0.1, 0.2, 0.9747}{x_1} \rangle, \langle \frac{0.2, 0.1, 0.9747}{x_2} \rangle, \langle \frac{0.29, 0.0, 0.9570}{x_3} \rangle\},$$

and

$$\hat{D} = \{\langle \frac{0.1, 0.3, 0.9487}{x_1} \rangle, \langle \frac{0.2, 0.2, 0.9592}{x_2} \rangle, \langle \frac{0.29, 0.1, 0.9518}{x_3} \rangle\}.$$

It follows that

$$\overline{\mu_{\hat{C}}} = \overline{\mu_{\hat{D}}} = 0.1967, \ \overline{\nu_{\hat{C}}} = 0.1, \ \overline{\nu_{\hat{D}}} = 0.2, \ \overline{\pi_{\hat{C}}} = 0.9688, \ \overline{\pi_{\hat{D}}} = 0.9532.$$

Because the MD and NMD in Example II equal that in Example IV, we get $\Delta_1(\hat{C}, \hat{D}) = \Delta_2(\hat{C}, \hat{D}) = 1$. For the sake of simplicity in computations, let (Table 8)

$$(\pi_{\hat{C}}(x_i) - \overline{\pi_{\hat{C}}}) = \alpha_3, \ (\pi_{\hat{D}}(x_i) - \overline{\pi_{\hat{D}}}) = \beta_3,$$

$$(\pi_{\hat{A}}(x_i) - \overline{\pi_{\hat{A}}})(\pi_{\hat{B}}(x_i) - \overline{\pi_{\hat{B}}}) = \gamma_3.$$

Table 8. Computation for hesitation grades

X	α_3	β_3	α_3^2	β_3^2	γ_3
x_1	0.0059	−0.0045	0.000030	0.000020	−0.000027
x_2	0.0059	0.0060	0.000030	0.000036	0.000035
x_3	−0.0118	−0.0014	0.000140	0.000002	0.000017

Thus $\Delta_3(\hat{C}, \hat{D}) = \dfrac{0.000025}{\sqrt{0.000058 \times 0.00020}} = 0.2321$. Hence $\rho(\hat{C}, \hat{D}) = 0.744$.

The Szmidt and Kacprzyk's approach gives the same value.

Advantages of the Modified Szmidt and Kacprzyk's Approach. The examples considered using Szmidt and Kacprzyk's approach and its modified approach in intuitionistic fuzzy context and Pythagorean fuzzy context, respectively show that PFS curbs fuzziness more accurately than IFS. This can been seen in Examples II and IV; while the correlation coefficient between two different IFSs gives a perfect relation contradicting the axiomatic description of correlation coefficient, the correlation coefficient between the same IFSs generalized as PFSs does not give a perfect correlation coefficient in corroboration to the given axiomatic description of correlation coefficient. In fact, the modified Szmidt and Kacprzyk's approach of measuring correlation coefficient between PFSs is very consistent in satisfying the axiomatic description of correlation coefficient unlike the Szmidt and Kacprzyk's approach.

5   Applicative Examples of Pattern Recognition Problems

Supposing we have m alternatives exemplified in PFSs I_i for $i = 1, \ldots, m$, defined in a universal set X. If there is an alternative sample exemplified in PFS J to be associated with any of I_i. The value

$$\rho(\mathsf{I}_i, \mathsf{J}) = \bigvee \Big[\rho(\mathsf{I}_1, \mathsf{J}), \ldots, \rho(\mathsf{I}_m, \mathsf{J}) \Big],$$

where $\rho(\mathsf{I}_i, \mathsf{J})$ states the grade of correlation index between $(\mathsf{I}_i, \mathsf{J})$. The maximum of $\rho(\mathsf{I}_1, \mathsf{J}), \ldots, \rho(\mathsf{I}_m, \mathsf{J})$ indicates that the alternative J is associated with any of such I_i for $i = 1, \ldots, m$.

Now, we consider two cases of pattern recognition in kinds of mineral fields and classifications of building materials as in [15], to establish the application of the modified Szmidt and Kacprzyk's approach of correlation coefficient in contrast to the Szmidt and Kacprzyk's approach.

5.1   Applicative Example I

Assume we have five mineral fields characterized by six minerals represented by five Pythgorean fuzzy sets,

$$\hat{K} = \{\hat{K}_1, \hat{K}_2, \hat{K}_3, \hat{K}_4, \hat{K}_5\}$$

in the feature, $X = \{x_1, ..., x_6\}$ as seen in Table 9. Suppose there is another kind of hybrid mineral, $\hat{S}$. Our aim is to categorize $\hat{S}$ into any of $\hat{K}_1$, $\hat{K}_2$, $\hat{K}_3$, $\hat{K}_4$ and $\hat{K}_5$.

Table 9. Kinds of mineral fields

Feature space						
PFSs	x_1	x_2	x_3	x_4	x_5	x_6
$\mu_{\hat{K}_1}(x_i)$	0.739	0.033	0.188	0.492	0.020	0.739
$\nu_{\hat{K}_1}(x_i)$	0.125	0.818	0.626	0.358	0.628	0.125
$\mu_{\hat{K}_2}(x_i)$	0.124	0.030	0.048	0.136	0.019	0.393
$\nu_{\hat{K}_2}(x_i)$	0.665	0.825	0.800	0.648	0.823	0.653
$\mu_{\hat{K}_3}(x_i)$	0.449	0.662	1.000	1.000	1.000	1.000
$\nu_{\hat{K}_3}(x_i)$	0.387	0.298	0.000	0.000	0.000	0.000
$\mu_{\hat{K}_4}(x_i)$	0.280	0.521	0.470	0.295	0.188	0.735
$\nu_{\hat{K}_4}(x_i)$	0.715	0.368	0.423	0.658	0.806	0.118
$\mu_{\hat{K}_5}(x_i)$	0.326	1.000	0.182	0.156	0.049	0.675
$\nu_{\hat{K}_5}(x_i)$	0.452	0.000	0.725	0.765	0.896	0.263
$\mu_{\hat{S}}(x_i)$	0.629	0.524	0.210	0.218	0.069	0.658
$\nu_{\hat{S}}(x_i)$	0.303	0.356	0.689	0.753	0.876	0.256

After applying Eq. (6), to determine the hesitation margin values we thus find the mean values of the three parameters in Table 10.

Table 10. Mean values of μ, ν, π

PFSs	$\bar{\mu}$	$\bar{\nu}$	$\bar{\pi}$
$\hat{K}_1$	0.368	0.447	0.705
$\hat{K}_2$	0.125	0.736	0.644
$\hat{K}_3$	0.852	0.114	0.249
$\hat{K}_4$	0.415	0.515	0.685
$\hat{K}_5$	0.398	0.517	0.542
$\hat{S}$	0.385	0.539	0.665

Employing Szmidt and Kacprzyk's correlation coefficient and its modification to find the correlation between the mineral fields $\hat{K}_i$ ($i = 1, 2, 3, 4, 5$) and the sample $\hat{S}$ in Table 9 with regards to Table 10, we obtain the values in Table 11.

Table 11. Correlation coefficient values

ρ	$\rho(\hat{K}_1, \hat{S})$	$\rho(\hat{K}_2, \hat{S})$	$\rho(\hat{K}_3, \hat{S})$	$\rho(\hat{K}_4, \hat{S})$	$\rho(\hat{K}_5, \hat{S})$
Szmidt and Kacprzyk's method	0.0202	0.2236	0.1902	0.6528	0.4882
Modified method	**0.5893**	**0.2236**	**0.5994**	**0.6528**	**0.5724**

From Table 11, it follows that the correlation coefficient value between $\hat{K}_4$ and $\hat{S}$ is the greatest. Hence, we infer that the hybrid mineral $\hat{S}$ can be associated with the mineral field $\hat{K}_4$. Also, the modified method of Szmidt and Kacprzyk's approach yields a consistent results when compare to the Szmidt and Kacprzyk's approach.

5.2   Applicative Example II

Here, we consider a problem of pattern recognition in classification of building materials. Suppose there are four classes of building materials symbolize by Pythagorean fuzzy sets, $\hat{C}_1, \hat{C}_2, \hat{C}_3, \hat{C}_4$ in $X = \{x_1, ..., x_{10}\}$, where X is the feature space, as shown in Table 12. Assume there is another class of building material yet to be classified, say $\hat{B}$. The task is to determine which of $\hat{C}_1, \hat{C}_2, \hat{C}_3, \hat{C}_4$ can be associated with $\hat{B}$.

Table 12. Classes of building materials

PFSs	Feature space									
	x_1	x_2	x_3	x_4	x_5	x_6	x_7	x_8	x_9	x_{10}
$\mu_{\hat{C}_1}(x_i)$	0.173	0.102	0.530	0.965	0.420	0.008	0.331	1.000	0.215	0.432
$\nu_{\hat{C}_1}(x_i)$	0.524	0.818	0.326	0.008	0.351	0.956	0.512	0.000	0.625	0.534
$\mu_{\hat{C}_2}(x_i)$	0.510	0.627	1.000	0.125	0.026	0.732	0.556	0.650	1.000	0.145
$\nu_{\hat{C}_2}(x_i)$	0.365	0.125	0.000	0.648	0.823	0.153	0.303	0.267	0.000	0.762
$\mu_{\hat{C}_3}(x_i)$	0.495	0.603	0.987	0.073	0.037	0.690	0.147	0.213	0.501	1.000
$\nu_{\hat{C}_3}(x_i)$	0.387	0.298	0.006	0.849	0.923	0.268	0.812	0.653	0.284	0.000
$\mu_{\hat{C}_4}(x_i)$	1.000	1.000	0.857	0.734	0.021	0.076	0.152	0.113	0.489	1.000
$\nu_{\hat{C}_4}(x_i)$	0.000	0.000	0.123	0.158	0.896	0.912	0.712	0.756	0.389	0.000
$\mu_{\hat{B}}(x_i)$	0.978	0.980	0.798	0.693	0.051	0.123	0.152	0.113	0.494	0.987
$\nu_{\hat{B}}(x_i)$	0.003	0.012	0.132	0.213	0.876	0.756	0.721	0.732	0.368	0.000

After computing the values of HM using Eq. (6), the mean values of the PFSs are given in Table 13.

Table 13. Mean values of μ, ν, π

PFSs	$\overline{\mu}$	$\overline{\nu}$	$\overline{\pi}$
$\hat{C}_1$	0.418	0.465	0.585
$\hat{C}_2$	0.537	0.345	0.565
$\hat{C}_3$	0.475	0.448	0.537
$\hat{C}_4$	0.544	0.395	0.390
$\hat{B}$	0.537	0.381	0.511

By applying Szmidt and Kacprzyk's correlation coefficient and its modification to find the correlation between $\hat{C}_j$ $(j = 1, 2, 3, 4)$ and the unknown building material $\hat{B}$ using the information in Tables 12 and 13, we obtain the results in Table 14.

Table 14. Correlation coefficient values

ρ	$\rho(\hat{C}_1, \hat{B})$	$\rho(\hat{C}_2, \hat{B})$	$\rho(\hat{C}_3, \hat{B})$	$\rho(\hat{C}_4, \hat{B})$
Szmidt and Kacprzyk's method	−0.1932	−0.0568	0.4827	0.9802
Modified method	**0.1932**	**0.1098**	**0.4827**	**0.9802**

From Table 14, it is observed that the correlation coefficient value between the building material $\hat{C}_4$ and the unknown building material $\hat{B}$ is the greatest. So, it is meet to say that the unclassified building material $\hat{B}$ belongs to $\hat{C}_4$. Although the results of Szmidt and Kacprzyk's method and its modification give the same recognition, the first two correlation coefficient values of Szmidt and Kacprzyk's method do not satisfy the axiomatic definition of correlation coefficient. But, the modified method of Szmidt and Kacprzyk's approach yields a valid/reliable results when compare to the Szmidt and Kacprzyk's approach.

6   Conclusions

In this paper we have modified Szmidt and Kacprzyk's correlation coefficient for IFSs, extended it to Pythagorean fuzzy context and used it to determine problems of pattern recognition. The method of computing correlation coefficient between IFSs proposed by Szmidt and Kacprzyk was reviewed, and it was discovered that; (i) the correlation coefficient violated the axiomatic description of correlation coefficient for IFSs, (ii) the correlation coefficient does not give a reliable result because it asserted that two IFSs have a perfect relation although the two IFSs are not equal in crass opposition to reality (see Example II). After modifying the approach in [40] and tested using IFSs, we found that while the first limitation was resolved, the second remains. Then, the modified Szmidt and Kacprzyk's approach was extended to Pythagorean fuzzy context since PFS is more reliable than IFS. After authenticating the validity of the modified Szmidt and Kacprzyk's approach in Pythagorean fuzzy environment, it was found to completely overcome the hitherto limitations. Mathematically, the modified Szmidt and Kacprzyk's approach in Pythagorean fuzzy setting was also proven to be a reasonable correlation coefficient. Finally, some pattern recognition problems were discussed via the proposed correlation coefficient measure. Incorporating this approach in Pythagorean fuzzy domain with clustering algorithm could be a resourceful research area for future consideration.

Acknowledgments. This work is supported by the Foundations of Chongqing Municipal Key Laboratory of Institutions of Higher Education ([2017]3), Chongqing Development and Reform Commission (2017[1007]), and Chongqing Three Gorges University.

References

1. Atanassov, K.T.: Intuitionistic fuzzy sets. Fuzzy Set Syst. **20**, 87–96 (1986)
2. Atanassov, K.T.: Geometrical interpretation of the elements of the intuitionistic fuzzy objects. Preprint IM-MFAIS-1-89, Sofia (1989)
3. Atanassov, K.T.: Intuitionistic Fuzzy Sets: Theory and Applications. Physica-Verlag, Heidelberg (1999)
4. Bai, S.M., Chen, S.M.: Automatically constructing concept maps based on fuzzy rules for adapting learning systems. Expert Syst. Appl. **35**(1–2), 41–49 (2008)
5. Boran, F.E., Akay, D.: A biparametric similarity measure on intuitionistic fuzzy sets with applications to pattern recognition. Inform. Sci. **255**(10), 45–57 (2014)
6. Chen, S.M., Chang, C.H.: Fuzzy multiattribute decision making based on transformation techniques of intuitionistic fuzzy values and intuitionistic fuzzy geometric averaging operators. Inform. Sci. **352**, 133–149 (2016)
7. Chiang, D.A., Lin, N.P.: Correlation of fuzzy sets. Fuzzy Set Syst. **102**(2), 221–226 (1999)
8. De, S.K., Biswas, R., Roy, A.R.: An application of intuitionistic fuzzy sets in medical diagnosis. Fuzzy Set Syst. **117**(2), 209–213 (2001)
9. Dumitrescu, D.: A definition of an informational energy in fuzzy set theory. Studia Univ. Babes-Bolyai Math. **22**, 57–59 (1977)
10. Dumitrescu, D.: Fuzzy correlation. Studia Univ. Babes-Bolyai Math. **23**, 41–44 (1978)
11. Du, Y.Q., Hou, F., Zafar, W., Yu, Q., Zhai, Y.: A novel method for multiattribute decision making with interval-valued Pythagorean fuzzy linguistic information. Int. J. Intell. Syst. **32**(10), 1085–1112 (2017)
12. Ejegwa, P.A.: Distance and similarity measures for Pythagorean fuzzy sets. Granul. Comput. **5**(2), 225–238 (2018). https://doi.org/10.1007/s41066-018-00149-z
13. Ejegwa, P.A.: Improved composite relation for Pythagorean fuzzy sets and its application to medical diagnosis. Granul. Comput. **5**(2), 277–286 (2019)
14. Ejegwa, P.A.: Pythagorean fuzzy set and its application in career placements based on academic performance using max-min-max composition. Complex Intell. Syst. **5**, 165–175 (2019). https://doi.org/10.1007/s40747-019-0091-6
15. Ejegwa, P.A.: Modified Zhang and Xu's distance measure of Pythagorean fuzzy sets and its application to pattern recognition problems. Neural Comput. Appl. (2019). https://doi.org/10.1007/s00521-019-04554-6
16. Ejegwa, P.A.: Personnel appointments: a Pythagorean fuzzy sets approach using similarity measure. J. Inform. Comput. Sci. **14**(2), 94–102 (2019)
17. Ejegwa, P.A.: Modal operators on Pythagorean fuzzy sets and some of their properties. J. Fuzzy Math. **27**(4), 939–956 (2019)
18. Ejegwa, P.A.: New similarity measures for Pythagorean fuzzy sets with applications. Int. J. Fuzzy Comput. Model. **3**(1), 75–94 (2020)
19. Ejegwa, P.A.: Generalized triparametric correlation coefficient for Pythagorean fuzzy sets with application to MCDM problems. Granul. Comput. (2020). https://doi.org/10.1007/s41066-020-00215-5
20. Ejegwa, P.A., Adamu, I.M.: Distances between intuitionistic fuzzy sets of second type with application to diagnostic medicine. Note IFS **25**(3), 53–70 (2019)
21. Ejegwa, P.A., Awolola, J.A.: Novel distance measures for Pythagorean fuzzy sets with applications to pattern recognition problems. Granul. Comput. (2019). https://doi.org/10.1007/s41066-019-00176-4

22. Ejegwa, P.A., Onasanya, B.O.: Improved intuitionistic fuzzy composite relation and its application to medical diagnostic process. Note IFS **25**(1), 43–58 (2019)
23. Ejegwa, P.A., Tyoakaa, G.U., Ayenge, A.M.: Application of intuitionistic fuzzy sets in electoral system. Int. J. Fuzzy Math. Arch. **10**(1), 35–41 (2016)
24. Feng, Y., Li, C.: Comparison system of impulsive control system with impulse time windows. J. Intell. Fuzzy Syst. **32**(6), 4197–4204 (2017)
25. Garg, H.: A novel correlation coefficients between Pythagorean fuzzy sets and its applications to decision making processes. Int. J. Intell. Syst. **31**(12), 1234–1252 (2016)
26. Garg, H.: A new improved score function of an interval-valued Pythagorean fuzzy set based TOPSIS method. Int. J. Uncertain. Quantif. **7**(5), 463–474 (2017)
27. Garg, H.: A linear programming method based on an improved score function for interval-valued Pythagorean fuzzy numbers and its application to decision-making. Int. J. Uncertain. Fuzz. Knowl. Based Syst **29**(1), 67–80 (2018)
28. Garg, H.: Hesitant Pythagorean fuzzy Maclaurin symmetricmean operators and its applications to multiattribute decision making process. Int. J. Intell. Syst. **34**(4), 601–626 (2019)
29. Gerstenkorn, T., Manko, J.: Correlation of intuitionistic fuzzy sets. Fuzzy Set Syst. **44**(1), 39–43 (1991)
30. Hatzimichailidis, A.G., Papakostas, A.G., Kaburlasos, V.G.: A novel distance measure of intuitionistic fuzzy sets and its application to pattern recognition problems. Int. J. Intell. Syst. **27**, 396–409 (2012)
31. Hung, W.L.: Using statistical viewpoint in developing correlation of intuitionistic fuzzy sets. Int. J. Uncertainty Fuzziness Knowl. Based Syst. Int. J. Uncertainty Fuzziness Knowl. Based Syst. **9**(4), 509–516 (2001)
32. Hung, W.L., Wu, J.W.: Correlation of intuitionistic fuzzy sets by centroid method. Inform. Sci. **144**(1), 219–225 (2002)
33. Liu, P., Chen, S.M.: Group decision making based on Heronian aggregation operators of intuitionistic fuzzy numbers. IEEE Trans. Cybern. **47**(9), 2514–2530 (2017)
34. Liu, B., Shen, Y., Mu, L., Chen, X., Chen, L.: A new correlation measure of the intuitionistic fuzzy sets. J. Intell. Fuzzy Syst. **30**(2), 1019–1028 (2016)
35. Mitchell, H.B.: A correlation coefficient for intuitionistic fuzzy sets. Int. J. Intell. Syst. **19**(5), 483–490 (2014)
36. Murthy, C.A., Pal, S.K., Majumder, D.D.: Correlation between two fuzzy membership functions. Fuzzy Set Syst. **17**, 23–38 (1985)
37. Singh, S., Ganie, A.H.: On some correlation coefficients in Pythagorean fuzzy environment with applications. Int. J. Intell. Syst. (2020). https://doi.org/10.1002/int.22222
38. Szmidt, E., Kacprzyk, J.: Intuitionistic fuzzy sets in some medical applications. Note IFS **7**(4), 58–64 (2001)
39. Szmidt, E., Kacprzyk, J.: Medical diagnostic reasoning using a similarity measure for intuitionistic fuzzy sets. Note IFS **10**(4), 61–69 (2004)
40. Szmidt, Eulalia, Kacprzyk, Janusz: Correlation of intuitionistic fuzzy sets. In: Hüllermeier, Eyke, Kruse, Rudolf, Hoffmann, Frank (eds.) IPMU 2010. LNCS (LNAI), vol. 6178, pp. 169–177. Springer, Heidelberg (2010). https://doi.org/10.1007/978-3-642-14049-5_18
41. Thao, N.X.: A new correlation coefficient of the intuitionistic fuzzy sets and its application. J. Intell. Fuzzy Syst. **35**(2), 1959–1968 (2018)
42. Thao, N.X.: A new correlation coefficient of the Pythagorean fuzzy sets and its applications. Soft Comput. (2019). https://doi.org/10.1007/s00500-019-04457-7

43. Thao, N.X., Ali, M., Smarandache, F.: An intuitionistic fuzzy clustering algorithm based on a new correlation coefficient with application in medical diagnosis. J. Intell. Fuzzy Syst. **36**(1), 189–198 (2019)
44. Wang, W., Xin, X.: Distance measure between intuitionistic fuzzy sets. Pattern Recogn. Lett. **26**, 2063–2069 (2005)
45. Yager, R.R.: Pythagorean membership grades in multicriteria decision making. Technical report MII-3301 Machine Intelligence Institute, Iona College, New Rochelle, NY (2013)
46. Yager, R.R.: Pythagorean membership grades in multicriteria decision making. IEEE Trans. Fuzzy Syst. **22**(4), 958–965 (2014)
47. Yager, Ronald R.: Properties and applications of Pythagorean fuzzy sets. In: Angelov, Plamen, Sotirov, Sotir (eds.) Imprecision and Uncertainty in Information Representation and Processing. SFSC, vol. 332, pp. 119–136. Springer, Cham (2016). https://doi.org/10.1007/978-3-319-26302-1_9
48. Yager, R.R., Abbasov, A.M.: Pythagorean membership grades, complex numbers and decision making. Int. J. Intell. Syst. **28**(5), 436–452 (2016)
49. Xu, Zeshui: On correlation measures of intuitionistic fuzzy sets. In: Corchado, Emilio, Yin, Hujun, Botti, Vicente, Fyfe, Colin (eds.) IDEAL 2006. LNCS, vol. 4224, pp. 16–24. Springer, Heidelberg (2006). https://doi.org/10.1007/11875581_2
50. Xu, S., Chen, J., Wu, J.J.: Cluster algorithm for intuitionistic fuzzy sets. Inform. Sci. **178**, 3775–3790 (2008)
51. Yu, C.: Correlation of fuzzy numbers. Fuzzy Set Syst. **55**, 303–307 (1993)
52. Zadeh, L.A.: Fuzzy sets. Inform. Control **8**, 338–353 (1965)
53. Zhang, X.: A novel approach based on similarity measure for Pythagorean fuzzy multiple criteria group decision making. Int. J. Intell. Syst. **31**, 593–611 (2016)
54. Zhang, X.L., Xu, Z.S.: Extension of TOPSIS to multiple criteria decision making with Pythagorean fuzzy sets. Int. J. Intell. Syst. **29**(12), 1061–1078 (2014)
55. Zeng, W. Li, D., Yin, Q.: Distance and similarity measures of Pythagorean fuzzy sets and their applications to multiple criteria group decision making. Int. J. Intell. Syst. (2018). https://doi.org/10.1002/int.22027

On Position and Attitude Control of Flapping Wing Micro-aerial Vehicle

Dexiu Ma[1,2], Long Jin[2(✉)], Dongyang Fu[3], Xiuchun Xiao[3], and Mei Liu[2]

[1] School of Information Science and Engineering, Lanzhou University,
Lanzhou 730000, China
madx20@lzu.edu.cn

[2] Chongqing Key Laboratory of Big Data and Intelligent Computing, Chongqing Institute of Green and Intelligent Technology, Chinese Academy of Sciences,
Chongqing 400714, China
jinlongsysu@foxmail.com, mliu@lzu.edu.cn

[3] School of Electronics and Information Engineering, Guangdong Ocean University,
Zhanjiang 524025, China
springxxc@163.com, fdy163@163.com

Abstract. The flapping wing micro-air vehicle (FWMAV) is a new type of aerial vehicle which uses mechanical structure to simulate bird flight. The FWMAV possesses the characteristics of small size, light weight, high flight efficiency and no occupation of runway. Firstly, we calculate the aerodynamic lift and drag on the wing surface. Then, through the analysis and summary of the existing FWMAV control technology, the attitude control model of the aerial vehicle is obtained. Besides, the control system based on STM32F103 is designed, which is equipped with a communication module and thus controls the position and flight attitude through the control signals received wirelessly. Finally, the feasibility of the system is verified by MATLAB and Simulink simulation.

Keywords: Flapping wing micro-aerial vehicle (FWMAV) · STM32F103 · Controller · Position and attitude

1 Introduction

Recently, according to the unique flight mode of insects and humming-birds, researchers have successfully developed flapping wing micro-aerial vehicle (FWMAV), in order to serve human beings better [1]. Bionic research shows that flapping wing flight is better than fixed wing and rotor flight for micro-aerial

L. Jin—This work was supported by the National Natural Science Foundation of China (No. 61703189), by CAS Light of West China Program, by the Team Project of Natural Science Foundation of Qinghai Province, China (No. 2020-ZJ-903), by the National Key Research and Development Program of China (No. 2017YFE0118900), in part by the Sichuan Science and Technology Program (No. 19YYJC1656), in part by the Fundamental Research Funds for the Central Universities (No. lzujbky-2019-89).

© Springer Nature Switzerland AG 2020
M. Han et al. (Eds.): ISNN 2020, LNCS 12557, pp. 207–216, 2020.
https://doi.org/10.1007/978-3-030-64221-1_18

vehicles with characteristic size equivalent to birds or insects [2]. Compared with fixed wing and rotor aerial vehicle, FWMAV has the main characteristics of integrating lifting, hovering and propulsion functions into a flapping wing system, which has strong maneuverability and flexibility [3]. At low Reynolds number, it is better than fixed wing and rotor aircraft, and more suitable for miniaturization [4]. The strengths of such aerial vehicles are small size, low energy consumption, and great flexibility during the flight [5]. Furthermore, they possess excellent performances about hovering as well as the low altitude flight and are widely used in defense and civil fields. Thus, advanced FWMAV technologies draw wide attention from researchers around the world [6].

At present, some remarkable achievements have been made in the research of the position and attitude of the FWMAV [7]. More and more advanced technologies have been introduced into the position and attitude research of FWMAV, such as widely used neural networks, adaptive controllers, fuzzy control and so on [8–12]. At the same time, distributed control and effective mathematical model prevailing in other fields provide a reference for FWMAV research [13–15]. In the actual production process, it is difficult to fully achieve the flapping action of insects or birds because of the weight of the aerial vehicle. Therefore, in the design of the micro flapping mechanism, the flapping action on the basis of simulating the flight of birds is simplified. Furthermore, this paper designs a controller with STM32F103 as the core. The controller has a wireless transceiver module, which connects the mobile application (APP) with the esp8266WiFi module to obtain the real-time flight status of the aircraft at any time. The corresponding control instructions are sent through the mobile APP, so as to realize the attitude and position control of the FWMAV.

2   Aerodynamic Lift and Drag

At present, the unsteady aerodynamic theory under low Reynolds coefficient is not fully mature, so we can only use the experimental method for reference to study the aerodynamic force of FWMAV [16]. In order to facilitate the analysis, the coordinate system as shown in Fig. 1 is set.

Figure 1(a) is the frame coordinate system. Specifically, the origin of coordinates is fixed at the center of gravity of the fuselage; Fig. 1(b) is a rotating coordinate system set by the right wing of the body as the research object, in which plane XOY and $\acute{X}\acute{O}\acute{Y}$ coincide. The delayed stall mechanism is that birds will form leading edge vortex attached to the wing surface in the process of flutter, which can increase the lift coefficient and lift thrust [17]. The rotating circulation effect refers to the torsion of the wing around the axis in the span direction while flapping [18]. In [19], the forces produced by the instantaneous aerodynamic delay stall and rotation cycle are obtained as

$$F_t = C_t(\alpha(t))\rho\dot{\phi}(t)^2 \int_0^R c(r)r^2 \, dr/2,$$

$$F_n = C_n(\alpha(t))\rho\dot{\phi}(t)^2 \int_0^R c(r)r^2 \, dr/2 + C_{rot}\rho\dot{\phi}(t) \int_0^R c(r)r^2 \, dr/2,$$

$$(1)$$

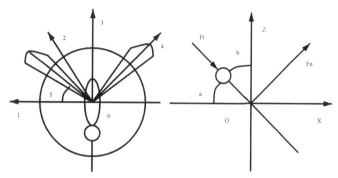

(a) Body coordinate system. (b) Rotating coordinate system.

Fig. 1. Aerodynamic diagram of the FWMAV.

where F_t is the tangential force parallel to the plane of the wing; F_n is the normal force perpendicular to the plane of the wing; ρ is the air density; $c(r)$ is the wing chord length; $\phi(t)$ is the flutter angle; angle of attack $\alpha(t) = \pi/2 - \varphi$ with φ being the rotation angle; R is the wing length; the dimensionless force coefficients $C_t(\alpha(t))$ and $C_n(\alpha(t))$ are functions of attack angle α, of which $C_t(\alpha(t)) = 0.4\cos^2(2\alpha)$ for $0 \le \alpha \le \pi/4$ and otherwise $C_t(\alpha(t)) = 0$, and $C_n(\alpha(t)) = 3.4\sin\alpha$ in view of most insects; $C_{rot} = \pi$. Moreover, by observing the movement of insect wings, the flapping angle $\phi(t)$ of FWMAV and the rotation angle φ can be expressed as follows:

$$\phi = \phi_{max}\cos(2\pi ft),$$
$$\varphi = \varphi_{max}\cos(2\pi ft - \theta),$$

where $\varphi_{max} \in [0, \pi/2]$ is the flapping amplitude; ϕ_{max} denotes the rotation angle amplitude; θ is the rotation phase angle; f is the flapping frequency. Subsequently, the aerodynamic lift and drag caused by the right wing motion are studied. The aerodynamics are divided into lift F_L along the Z axis and resistance F_D along the X axis, which are calculated as follows:

$$F_L(t) = F_n\cos\alpha - F_t\sin\alpha,$$
$$F_D(t) = F_n\sin\alpha + F_t\cos\alpha.$$

In the body coordinate system $XOYZ$, the lift along Z axis is represented by $F_{LZ}(t) = 2F_L(t)$. Decompose the drag and then the drag components of the two wings along the Y axis are formulated as $F_{DY}(t) = 2F_D(t)\cos\phi$ with that along the X axis are offset. The above analysis shows that when the flapping surface is horizontal, the wings of the bionic micro flapping aerial vehicle are flapping on the inclined plane when flying. Assuming that the angle between the flapping surface and the horizontal plane is shown in Fig. 2, lift $\acute{F}_L$ and resistance $\acute{F}_D$ of the flapping wing can be obtained:

$$\acute{F}_L(t) = F_{LZ} \cos \gamma + F_{DY} \sin \gamma,$$
$$\acute{F}_D(t) = -F_{LZ} \sin \gamma + F_{DY} \cos \gamma. \tag{2}$$

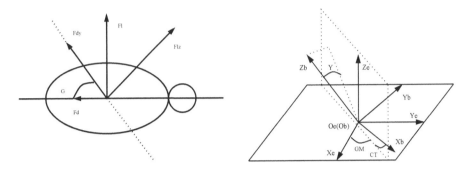

Fig. 2. Tilting and flapping of FWMAV.   **Fig. 3.** Schematic diagram of attitude angle.

3   Attitude Control of FWMAV

To build the basis for FWMAV attitude control, attitude angle is introduced. The attitude control model is also given.

3.1   Control Strategy

The dynamic system of FWMAV not only has the characteristics of nonlinearity, strong coupling and time-varying, but also has many uncertain factors, which brings great difficulties to the research of control technology. In the design of fuzzy controller, developing a control algorithm is essential [20]. In [21], an adaptive fuzzy logic controller is proposed to solve the problem of spacecraft attitude control. The main feature is that there is no need for a clear model to explain the control relationship between output and input. Li proposed an adaptive fuzzy controller for a class of nonlinear multiple-input-multiple-output (MIMO) systems. Such controller has stability and good tracking performance [22]. Yu further designed an adaptive fuzzy PID controller [23]. Attitude control is always the most important and widely used control method in aerial vehicle control. Pitch angle, yaw angle and roll angle are collectively referred to as the attitude angle of the body. Therefore, in order to describe the motion attitude of the aircraft, the attitude angle is introduced, as shown in Fig. 3.

3.2   Attitude Control Model

The angular velocity of the aerial vehicle around the center of mass should be equal to the vector sum of the angular velocity of the aerial vehicle around each axis:

$$\boldsymbol{\omega} = \dot{\boldsymbol{\psi}} + \dot{\boldsymbol{\vartheta}} + \dot{\boldsymbol{\gamma}}, \tag{3}$$

where $\boldsymbol{\omega}$ is the angular velocity of the aircraft around the center of mass and $\boldsymbol{\omega} = [\omega_x \ \omega_y \ \omega_z]^{\mathrm{T}}$ with the symbol ^{T} being the transpose operation of a vector or a matrix; $\dot{\boldsymbol{\vartheta}}$ is the pitch angular velocity and $\dot{\boldsymbol{\vartheta}} = [0 \ 0 \ \dot{\vartheta}]^{\mathrm{T}}$; $\dot{\boldsymbol{\psi}}$ is the yaw rate and $\dot{\boldsymbol{\psi}} = [0 \ \dot{\psi} \ 0]^{\mathrm{T}}$; $\dot{\boldsymbol{\gamma}}$ is the roll angle speed and $\dot{\boldsymbol{\gamma}} = [\dot{\gamma} \ 0 \ 0]^{\mathrm{T}}$. Further, the components of the rotational angular velocity $\boldsymbol{\omega}$ along each axis of the body coordinate system are ω_x, ω_y and ω_z respectively:

$$
\begin{bmatrix} \omega_x \\ \omega_y \\ \omega_z \end{bmatrix} = L(\vartheta)L(\gamma) \begin{bmatrix} 0 \\ 0 \\ \dot{\psi} \end{bmatrix} + L(\gamma) \begin{bmatrix} 0 \\ \dot{\vartheta} \\ 0 \end{bmatrix} + \begin{bmatrix} \dot{\gamma} \\ 0 \\ 0 \end{bmatrix}
$$

$$
= \begin{bmatrix} 1 & 0 & 0 \\ 0 & \cos\gamma & \sin\gamma \\ 0 & -\sin\gamma & \cos\gamma \end{bmatrix} \begin{bmatrix} \cos\vartheta & 0 & \sin\vartheta \\ 0 & 1 & 0 \\ -\sin\vartheta & 0 & \cos\vartheta \end{bmatrix} \begin{bmatrix} 0 \\ 0 \\ \dot{\psi} \end{bmatrix} + \begin{bmatrix} 1 & 0 & 0 \\ 0 & \cos\gamma & \sin\gamma \\ 0 & -\sin\gamma & \cos\gamma \end{bmatrix} \begin{bmatrix} 0 \\ \dot{\vartheta} \\ 0 \end{bmatrix} + \begin{bmatrix} \dot{\gamma} \\ 0 \\ 0 \end{bmatrix}
$$

$$
= \begin{bmatrix} \dot{\psi}\sin\vartheta + \dot{\gamma} \\ \dot{\psi}\cos\vartheta\sin\gamma + \dot{\vartheta}\cos\gamma \\ \dot{\psi}\cos\vartheta\cos\gamma - \dot{\vartheta}\sin\gamma \end{bmatrix} = \begin{bmatrix} 1 & 0 & \sin\vartheta \\ 0 & \cos\gamma & \cos\vartheta\sin\gamma \\ 0 & -\sin\gamma & \cos\vartheta\cos\gamma \end{bmatrix} \begin{bmatrix} \dot{\gamma} \\ \dot{\vartheta} \\ \dot{\psi} \end{bmatrix},
$$

$$\tag{4}$$

Transform Eq. (4) to obtain

$$
\begin{bmatrix} \dot{\vartheta} \\ \dot{\psi} \\ \dot{\gamma} \end{bmatrix} = \begin{bmatrix} 1 & -\sin\gamma\tan\vartheta & -\cos\gamma\tan\vartheta \\ 0 & \cos\gamma & -\sin\gamma \\ 0 & \sin\gamma/\cos\vartheta & \cos\gamma/\cos\vartheta \end{bmatrix} \begin{bmatrix} \omega_x \\ \omega_y \\ \omega_z \end{bmatrix}. \tag{5}
$$

The above formula is the attitude control model of the aerial vehicle, which establishes the corresponding relationship between the rotation angular velocity and the attitude angle of the aerial vehicle in flight.

3.3   Space Description of FWMAV Attitude Control System

Since the aerial vehicle is symmetrical, its product of inertia is zero. Newton-Euler principle is used to establish the dynamic equation. According to the moment of momentum theorem, we can get:

$$
\begin{bmatrix} M_x \\ M_y \\ M_z \end{bmatrix} = \begin{bmatrix} I_x d\omega_x/dt \\ I_y d\omega_y/dt \\ I_z d\omega_z/dt \end{bmatrix} + \begin{bmatrix} \omega_y\omega_z(I_z - I_y) \\ \omega_x\omega_z(I_x - I_z) \\ \omega_x\omega_y(I_y - I_x) \end{bmatrix}, \tag{6}
$$

where M_x, M_y and M_z are the closing external torque of the aerial vehicle in the rolling, yawing and pitching directions; I_x, I_y and I_z are the moment of inertia of the aerial vehicle on the axes of the body coordinate system; $\mathrm{d}\omega_x/\mathrm{d}t$, $\mathrm{d}\omega_y/\mathrm{d}t$ and $\mathrm{d}\omega_z/\mathrm{d}t$ is the component of the rotational angular acceleration vector on each axis of the body coordinate system. In addition, according to formula (6), we can get:

$$\begin{bmatrix} \mathrm{d}\omega_x/\mathrm{d}t \\ \mathrm{d}\omega_y/\mathrm{d}t \\ \mathrm{d}\omega_z/\mathrm{d}t \end{bmatrix} = \begin{bmatrix} (M_x - \omega_y\omega_z(I_z - I_y))/I_x \\ (M_y - \omega_x\omega_z(I_x - I_z))/I_y \\ (M_z - \omega_x\omega_y(I_y - I_x))/I_z \end{bmatrix} \tag{7}$$

Further, the space description of FWMAV attitude control system is introduced:

$$\begin{cases} \dot{x} = f(x) + g(x)u, \\ y = [\vartheta\ \psi\ \gamma]^{\mathrm{T}}, \end{cases} \tag{8}$$

where $f(x) = [\vartheta\ \psi\ \gamma\ \mathrm{d}\omega_x/\mathrm{d}t\ \mathrm{d}\omega_y/\mathrm{d}t\ \mathrm{d}\omega_z/\mathrm{d}t]^{\mathrm{T}}$; $x = [\vartheta\ \psi\ \dot{\gamma}\ \omega_x\ \omega_y\ \omega_z]^{\mathrm{T}}$ is the state vector; M_{wx}, M_{wy}, M_{wz} are the average aerodynamic moment of the wing in a beat period, and $u = [M_{wx}\ M_{wy}\ M_{wz}]^{\mathrm{T}}$ is the control input vector, the details of parameters refer to [24],

$$g(x) = \begin{bmatrix} 0 & 0 & 0 \\ 0 & 0 & 0 \\ 0 & 0 & 0 \\ 1/I_x & 0 & 0 \\ 0 & 1/I_y & 0 \\ 0 & 0 & 1/I_z \end{bmatrix}.$$

4   FWMAV Flight Control System Design and Attitude Control Method

In this section, the main components of the FWMAV control system and the attitude control method used in this paper are introduced.

4.1   Overall Design Scheme

The microprocessor STM32F103 controls the direct current (DC) motor and steering engine according to the real-time flight status of the aerial vehicle and the control command from the mobile APP, so as to realize flapping flight. The control system block diagram is shown in Fig. 4.

4.2   Actuator

The actuator is mainly composed of DC motor and steering engine. The flapping of aerial vehicle wings is mainly realized by DC motor. The speed of DC motor is adjusted by changing the duty cycle of input signal, so as to change the flapping frequency of aerial vehicle. The driving circuit is shown in Fig. 5. The steering gear is a position servo actuator suitable for use in a control system that requires constant change of angle. This paper uses the steering gear to change the force on the body, so as to realize the attitude control of the FWMAV.

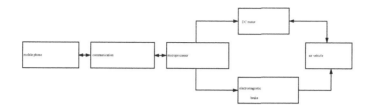

Fig. 4. Flight control system block diagram.

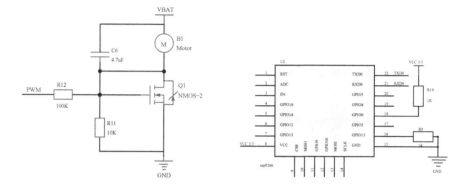

Fig. 5. Drive circuit of motor.   **Fig. 6.** WiFi module.

4.3   Communication Module

In the control system, esp8266WiFi module is used for communication. WiFi technology is selected in wireless communication because of its long signal transmission distance. The main function of the communication module is that the control command sent by the mobile APP is transmitted to the microprocessor through the WiFi module, so as to control the flight status of the FWMAV. The circuit diagram of the esp8266WiFi communication module is shown in Fig. 6.

4.4   Attitude Control Method

In order to achieve different flight attitudes during FWMAV flight, PID control with simple structure and stable performance is adopted in this paper. We propose the control rule of PID controller as follows:

$$u(t) = K_p(e(t) + \frac{1}{T_i} \int_0^t e(t)\, \mathrm{d}t + T_d \frac{\mathrm{d}e(t)}{\mathrm{d}t}), \tag{9}$$

Where $u(t)$ is the output of the controller; K_p is the proportional coefficient; $e(t)$ is the deviation between the system output and the system output; T_i is the integral time constant; T_d is the differential time constant.

The control law of attitude angle can be written as

$$e(k) = \varepsilon_d(k) - \varepsilon(k), \tag{10}$$

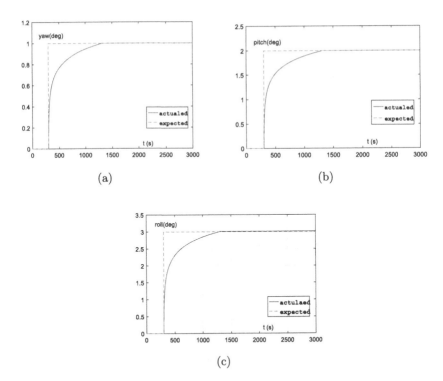

Fig. 7. The simulation control curves of yaw angle, pitch angle and roll angle are shown in Figs. 7(a), 7(b) and 7(c) respectively. (Color figure online)

$$\delta(k) = K_p e(k) + K_i \sum_{n=0}^{k} e(n) + K_d(e(k) - e(k-1)) + K_q q(k), \qquad (11)$$

where K_i is the integral coefficient; K_d is the differential coefficient; K_q is the angular velocity gain of the attitude; ε_d is the expected value of the attitude angle; ε is the actual value of the attitude angle; q is the angular velocity of the actual attitude angle.

5   Simulation Study

To obtain better simulation results, after many experiments, the final control parameters are selected as $K_p = 43$, $K_d = 0.0001$. The initial values of the attitude angles are $0°$. The expected values of yaw angle, pitch angle and roll angle are set to $1°$, $2°$ and $3°$ respectively. The simulation time is set to 3000 s. It is assumed that the FWMAV in this paper needs to keep the flapping wing aerial vehicle level flying at a pitch angle of. The deviation between the actual pitch angle and the expected pitch angle is taken as the input of PID controller, and the output is the deflection of the tail. The simulation results are shown in Fig. 7(a), 7(b) and 7(c).

There are two curves in each figure. The red dashed line represents the expected value of the attitude angle, and the blue implementation represents the actual value of the attitude angle. When the flight attitude of the aerial vehicle changes, the system can respond in time and make the actual value of attitude angle gradually equal to the expected value. Therefore, this system can effectively control the FWAMV to fly smoothly.

6   Conclusion

In this paper, we have established the ground coordinate system and the body coordinate system, and have described the space attitude of the aerial vehicle by combining the ground coordinate system and body coordinate system, in order to control and study the position and attitude of FWMAV. Then, we have presented the attitude control model and state space description of FWMAV. Furthermore, the control system based on STM32 has been designed and the flight control board of the aerial vehicle has been made to realize real-time control of the aerial vehicle. We introduced the attitude control method of FWMAV. Finally, the simulation of attitude angle by MATLAB and Simulink has verified the feasibility of the system.

References

1. Mackenzie, D.: A flapping of wings. Science **335**(6075), 1430–1433 (2012)
2. Deng, X., Schenato, L., Wu, W., Sastry, S.S.: Flapping flight for biomimetic robotic insects: part I-system modeling. IEEE Trans. Robot. **22**(4), 776–788 (2006)
3. Lee, J., Choi, H., Kim, Y.: A scaling law for the lift of hovering insects. J. Fluid Mech. **782**(10), 479–490 (2015)
4. Gerdes, J., Gupta, S., Wilkerson, S.: A review of bird-inspired flapping wing miniature air vehicle designs. J. Mech. Robot. **4**(2), 103–114 (2012)
5. Ellington, C.P.: The novel aerodynamics of insect flight: applications to micro-air vehicles. J. Exp. Biol. **202**(23), 3439–3448 (1999)
6. He, W., Yan, Z., Sun, C., Chen, Y.: Adaptive neural network control of a flapping wing micro aerial vehicle with disturbance observer. IEEE Trans. Cybern. **47**(10), 3452–3465 (2017)
7. He, W., Meng, T., He, X.: Iterative learning control for a flapping wing micro aerial vehicle under distributed disturbances. IEEE Trans. Cybern. **49**(4), 1524–1535 (2019)
8. Li, S., Zhang, Y., Jin, L.: Kinematic control of redundant manipulators using neural networks. IEEE Trans. Neural Netw. Learn. Syst. **28**(10), 2243–2254 (2017)
9. Jin, L., Li, S., Yu, J., He, J.: Robot manipulator control using neural networks: a survey. Neurocomputing **285**(12), 23–34 (2018)
10. Li, D., Chen, X., Xu, Z.: Gain adaptive sliding mode controller for flight attitude control of MAV. Opt. Precis. Eng. **21**(5), 1183–1191 (2013)
11. Jin, L., Zhang, Y., Qiao, T., Tan, M., Zhang, Y.: Tracking control of modified Lorenz nonlinear system using ZG neural dynamics with additive input or mixed inputs. Neurocomputing **196**, 82–94 (2016)

12. Jin, L., Yan, J., Du, X., Xiao, X., Fu, D.: RNN for solving time-variant generalized sylvester equation with applications to robots and acoustic source localization. IEEE Trans. Industr. Inform. **16**(99), 6359–6369 (2020)

13. Zhang, J., Jin, L., Cheng, L.: RNN for perturbed manipulability optimization of manipulators based on a distributed scheme: a game-theoretic perspective. IEEE Trans. Neural Netw. 1–11 (2020)

14. Luo, X., Sun, J., Wang, Z., Li, S., Shang, M.: Symmetric and non-negative latent factor models for undirected, high dimensional and sparse networks in industrial applications. IEEE Trans. Industr. Inform. **13**, 3098–3107 (2017)

15. Luo, X., et al.: Incorporation of efficient second-order solvers into latent factor models for accurate prediction of missing QoS data. IEEE Trans. Cybern. **48**(4), 1216–1228 (2018)

16. Wu, J., Sun, M.: Unsteady aerodynamic forces of a flapping wing. J. Exp. Biol. **207**(7), 1137–1150 (2004)

17. Zou, S., Gao, A., Shi, Y., Wu, J.: Causal mechanism behind the stall delay by airfoils pitching-up motion. Theor. Appl. Mech. Lett. **7**(5), 311–315 (2017)

18. Carr, Z., DeVoria, A., Ringuette, M.: Aspect-ratio effects on rotating wings: circulation and forces. J. Fluid Mech. **767**(10), 497–525 (2015)

19. Schenato, L.: Analysis and control of flapping flight: from biological to robotic insects. Ph.D. dissertation, UC Berkeley (2003)

20. He, W., et al.: Development of an autonomous flapping-wing aerial vehicle. Sci. China Inf. Sci. **60**(6), 1–8 (2017). https://doi.org/10.1007/s11432-017-9077-1

21. Kwan, C., Xu, H., Lewi, F.: Robust spacecraft attitude control using adaptive fuzzy logic. Int. J. Syst. Sci. **31**(10), 1217–1225 (2000)

22. Li, H., Tong, S.: A hybrid adaptive fuzzy control for a class of nonlinear MIMO systems. IEEE Trans. Fuzzy Syst. **11**(1), 24–34 (2003)

23. Yu, W.: Adaptive fuzzy PID control for nonlinear systems with H∞ tracking performance. IEEE International Conference on Fuzzy System, BC, Canada, pp. 1010–1015 (2006)

24. Duan, H.: Flight attitude control of MAV. Ph.D. dissertation, pp. 1–134 (2007)

Supply Chain Financing Model with Data Analysis Under the Third-Party Partial Guarantee

Shengying Zhao and Xiangyuan Lu[✉]

School of Management Science and Engineering, Dongbei University of Finance and
Economics, Dalian 116025, China
xiangyuan_lu@hotmail.com

Abstract. With respect to a two-level supply chain which is comprised of a core
retailer and a capital-constrained manufacturer under the background of digital
economy, this paper use the Stackelberg game model and data simulation analysis
to investigated the influence of the third-party partial guarantee on the decision-
making of accounts receivable financing in the supply chain. With the consideration of retailer's financing risk, this paper points out that the guarantee cost
undertaken by the manufacturer can reduce the financing interest rate provided
by the financial institution and improves the loan-to-value ratio of the financial
institution. Meanwhile, retailer under-taken the guarantee cost will reduce the
double marginalization effects to the supply chain financing operation, and then
promoting the coordination of the supply chain. When the third-party guarantee
company is introduced to the supply chain accounts receivable financing, the optimal decision is that the guarantee cost is undertaken by the retailer. Finally, the
simulation analysis is conducted to verify the results of this research.

Keywords: Financial engineering · Big data · Supply chain finance · Simulation

1 Introduction

Capital constraint is a key factor that commonly affects firms' operations, especially for
the development of small- and medium-sized enterprises (SMEs). Due to the appearance
and development of supply chain finance, financing of SMEs has gained a new driving
force to address their financing difficulty. Specifically, due to easily diffusible and structured, accounts receivable financing has obtained a rapidly development in recent years.
The total net amount of accounts receivable of industrial enterprises in China was 14.34
trillion yuan in August 2018. However, there also appeared a variety of defaults and loan
frauds. For example, the 61 transactions between Chengxing and Nuoya in 2019 involved
around 3.4 billion yuan of accounts receivable in total. Facts showed that Chengxing
fabricated the contract with companies, including JD.com, thus resulting in defaults of
Nuoya. To discover the financing risk control strategies that should be taken to guarantee
supply chain accounts receivable financing and minimize the losses of the supply chain

M. Han et al. (Eds.): ISNN 2020, LNCS 12557, pp. 217–229, 2020.
https://doi.org/10.1007/978-3-030-64221-1_19

finance market, this paper studies the impact of the third-party guarantee on the accounts receivable financing.

As a typical supply chain financing method, many literatures focused on the accounts receivable financing [1, 2]. Koch gives an economic model based on inventory and accounts receivable financing method [3]. Burman and Dunham put forward the operation process and monitoring of financing based on inventory and accounts receivable, respectively [4, 5]. Poe introduced that asset-based financing is an important realization model of the logistics finance [6]. Czternasty and Mikołajczak shows that non-recourse factoring plays a positive role in improving the financial structure of SMEs [7]. Lu et al. investigated two supply chain finance models with partial credit guarantee provided by a third-party or a supplier when a retailer borrows from a bank [8]. However, all these literatures do not research the effect of loan-to-value ratio to the financing model.

As to scholars studying risk management of supply chain finance, scholars include Kouvelis, Cai Olson, etc., have studied risk control of supply chain finance from the perspective of the bankruptcy risks, credit rating and tax [9–13]. Nevertheless, above research focuses on bank credit and trade credit. So far, little attention has been paid to the impact of guarantee on the supply chain financial risk management. Lai and Yu analyzed the guarantee pricing, considering the bankruptcy risk and preferred debt [14]. Cossin and Hricko assumed endogeneity of breach of contract, and discussed the loan to value and the discount rate of the pledge [15]. Gan et al. established the credit line decision-making model of banks resorting to downside risk avoidance [16]. Yan et al. found that some credit guarantee contracts can realize maximization of profits and coordination of channels in the supply chain finance system [17].

As to the accounts receivable model of the supply chain finance, the above scholars mainly focused on the optimal decision-making and then to optimize the benefits of the supply chain through the adjustment of the guarantee coefficient, loan rate, financing limit, etc. However, only a few literatures have investigated the influence of external guarantee enterprises on the supply chain financing behaviors. In response to this research gap, this paper studies a two-level supply chain comprised of a core retailer and a capital-constrained manufacturer through model and simulation analysis. According to the retailer's default probability, this paper established a Stackelberg game model to analyze the influence of the third-party partial guarantee on the decision-making of accounts receivable financing in the supply chain.

The remainder of this paper is organized as follows. Section 2 discusses the notations, assumptions, and sequence of events. Section 3 provides analysis of the financing model with third-party guarantee. Section 4 enumerates the impact of third-party guarantee on the financing model and conducts some simulations. Finally, Sect. 5 presents the conclusions.

2   Model

This paper researched a two-level supply chain which is comprised of a core retailer and a capital-constrained manufacturer. The retailer purchases single products from the manufacturer at certain wholesale price to a random market and delays the payment. After providing goods for the retailer, the capital-constrained manufacturer does not have adequate funds to support its own purchase for further manufacturing. As a result, it needs to borrow funds from financial institutions, such as banks and factoring companies, using receipts of accounts receivable. Considering the guarantee provided by the third-party guarantee company for the financing of the accounts receivable, the guarantee fees should be undertaken by the manufacturer or the retailer. Once the manufacturer (retailer) fails to pay the loan interest on the due date, the financial institutions have the right to ask for compensations from the company providing the guarantee.

2.1 Notations and Assumptions

Notations are summarized in Table 1 below.

Table 1. Parameters and variables.

Parameters	Decision variables
p: Market selling price per unit product	w: Wholesale price
c: Manufacturing cost per unit product	q: Order quantity
θ: loan-to value ration of accounts receivable to be paid by the financial institution, where $\theta \in (0, 1)$	r: Financing interest rate of the financial institution
α: Probability of retailer's breach of contract (referring to not paying the loans)	Function
	Abbreviations
r_f: Average rate of return of the financing market (under the condition of risk-free rate) $r_f \in (0, 1)$	R: Retailer
D: Random market demand	M: Manufacturer
$f(D)$, $F(D)$: Probability density function and cumulative distribution function of the random market demand	FI: Financial institution
	SC: Supply chain
$h(D)$, $H(D)$: Increase failure rate (IFR) and increasing generalized failure rate (IGFR) $h(D) = \frac{f(D)}{1-F(D)}$, $H(D) = \frac{Df(D)}{1-F(D)}$	0, 1, 2: Refer to no guarantee, guarantee cost undertaken by the manufacturer, and guarantee cost undertaken by the retailer

This study's assumptions are as follows.

(1) The retailer faced the random market demand, whose probability density function is $f(D)$ and cumulative distribution function is $F(D)$, and the random market demand is consistent with the nature of the increasing generalized failure rate (IGFR);

(2) The supplier has limited liability, and the internal capital level after providing goods for the retailer is zero;

(3) The information is symmetric in financing, and all financing participants are risk-neutral;

2.2 Sequence of Events

The sequence of events is summarized in Fig. 1, Fig. 2, and Fig. 3 below, respectively. Figure 1 summarized the supply chain accounts receivable financing flow chart when there is no external guarantee. Figure 2 shows the decision sequence when the manufacturer covering the guarantee cost. Figure 3 shows the sequence of events when the retailer covering the guarantee cost.

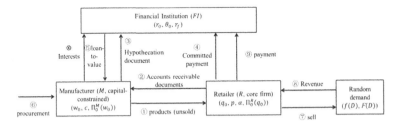

Fig. 1. Accounts receivable financing flow chart (without the guarantee)

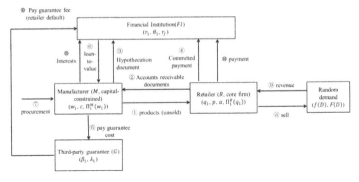

Fig. 2. Accounts receivable financing flow chart (manufacturer undertaken the guarantee cost)

3 Financing Model with Third-Party Guarantee

3.1 Benchmark-Without Guarantee (Issue 0)

The financing decision-making sequence when there is no third-party guarantee company provides partial credit guarantee (PCG) is shown in Fig. 1. First of all, the manufacturer announces the wholesale price w_0. Then the retailer decides the order quantity q_0. Finally, the financial institution announces its financing interest rate r_0. We proceed backwards to derive the optimal decisions.

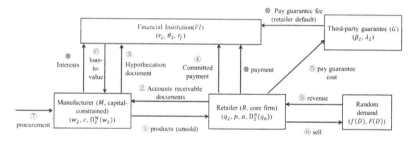

Fig. 3. Accoutns receivable financing flow chart (retailer undertaken the guarantee cost)

First, the problem of financial institution can be used to solve following problem:

$$\theta_0 w_0 q_0 \left(1 + r_f\right) = (1 - \alpha)(w_0 q_0 + w_0 q_0 r_0 - (1 - \theta_0) w_0 q_0) + \alpha w_0 q_0 r_0 \qquad (1)$$

Second, the retailer decides its optimal order quantity q_0 based on Eq. (2):

$$\max_{q_0 \geq 0} \Pi_0^R(q_0) = pE_D \min\{D, q_0\} - w_0 q_0 (1 - \alpha) \qquad (2)$$

Finally, the manufacturer decides the wholesale price w_0 based on Eq. (3):

$$\max_{w_0 \geq c} \Pi_0^M(w_0) = \theta_0 w_0 q_0 + (1 - \alpha)(1 - \theta_0) w_0 q_0 - c q_0 - w_0 q_0 r_0 \qquad (3)$$

We derive Lemma 1 below by integrating Eq. (1), (2), and (3).

Lemma 1: The optimal decisions are as follows:

(1) The optimal order quantity is: $\left(\frac{p(1 - \alpha - \theta_0 r_f)}{1 - \alpha}\right)\left(\bar{F}(q_0^*) - q_0^* f(q_0^*)\right) = c$;

(2) The optimal wholesale price is: $w_0^* = \frac{p}{1 - \alpha}\bar{F}(q_0^*)$;

(3) The optimal financing interest rate is: $r_0^* = \theta_0(r_f + \alpha)$.

(Hereinafter, for all proofs please refer to the Appendix. In addition, we use superscript "*" for the final optimal solutions).

Proposition 1: Under the condition of no guarantee, if θ_0 remains unchanged, then $\alpha \uparrow \Rightarrow w_0^* \uparrow, q_0^* \downarrow, r_0^* \uparrow$. If α remains unchanged, then $\theta_0 \uparrow \Rightarrow w_0^* \uparrow, q_0^* \downarrow, r_0^* \uparrow$.

Proposition 1 reveals that the financing interest rate of the financial institutions, such as banks and factoring companies, will increase correspondingly when the retailer's probability of default increases. When financial institutions such as banks show a growing willingness to the loan-to-value ratio, they will improve their financing interest rate correspondingly. In the process of accounts receivable financing, the manufacturer usually expects the financial institution to provide a higher loan-to value, thus reduce their financing risks that brought by the retailer's default. But Proposition 1 points out that a high loan-to-value ratio will cause a high wholesale price, thus resulting in the decrease of the order quantity and the supply chain's operation efficiency. The rationale behind

Proposition 1 is the financial institution will set a higher financing interest rate to avoid financing risks when providing a higher loan-to-value ratio. Then will improve the manufacturer's financing. Hence, the manufacturer will set a high wholesale price and the retailer will decide a lower order quantity. Therefore, the financing efficiency of supply chain will be reduced. Besides, Proposition 1 also indicates that the financing efficiency and the operation efficiency of the supply chain will both fall when the retailer's default possibility increases.

3.2 Manufacturer Undertaken the Guarantee Cost (Issue 1)

When the guarantee cost is undertaken by the manufacturer, the financing decision-making flow chart is shown in Fig. 2. First of all, the manufacturer announces the wholesale price w_1. Second, the retailer decides the order quantity q_1. Third, the financial institution announces the financing interest rate r_1. Finally, the guarantee announces its guarantee coefficient λ_1 and the guarantee cost rate β_1. We also proceed backwards to derive the optimal decisions.

First, the third-party guarantee's problem can be formulated as Eq. (4):

$$\beta_1 w_1 q_1 = \lambda_1 w_1 q_1 \alpha \tag{4}$$

Then, the problem of financial institution can be used to solve following problem:

$$\theta_1 w_1 q_1 (1 + r_f) = (1 - \alpha)(w_1 q_1 + w_1 q_1 r_1 - (1 - \theta_1)w_1 q_1) + \alpha(w_1 q_1 r_1 + \lambda_1 w_1 q_1) \tag{5}$$

Next, the retailer decides its optimal order quantity q_1 based on Eq. (6):

$$\max_{q_1 \geq 0} \Pi_1^R(q_1) = pE_D \min\{D, q_1\} - w_1 q_1 (1 - \alpha) \tag{6}$$

At last, the manufacturer decides the wholesale price w_1 based on Eq. (7):

$$\max_{w_1 \geq c} \Pi_1^M(w_1) = \theta_1 w_1 q_1 + (1 - \alpha)(1 - \theta_1)w_1 q_1 - cq_1 - w_1 q_1 r_1 - \beta_1 w_1 q_1 \tag{7}$$

We derive Lemma 2 below by integrating Eq. (4), (5), (6) and (7).

Lemma 2: When the third-party guarantee cost is undertaken by the manufacturer, the optimal decisions of each part are as follows:

(1) The optimal order quantity is: $\left(\frac{p(1 - \alpha - \theta_1 r_f)}{1 - \alpha}\right)\left(\bar{F}(q_1^*) - q_1^* f(q_1^*)\right) = c$;
(2) The optimal wholesale price is: $w_1^* = \frac{p}{1 - \alpha}\bar{F}(q_1^*)$;
(3) The optimal financing interest rate is: $r_1^* = \theta_1(r_f + \alpha) - \alpha\lambda_1$;
(4) The guarantee cost rate and guarantee coefficient satisfies: $\beta_1 = \lambda_1 \alpha$.

Lemma 2 points out that the guarantee coefficient and guarantee cost rate of the third-party guarantee will not affect the optimal decisions of the manufacturer and the retailer when the manufacturer undertaken the guarantee cost. As shown in Proposition 2 below, this is due to that the increase of the guarantee coefficient will lead to the decrease of the financial institution's financing interest rate or increase of the financial institution's loan-to-value ratio when there is third-party providing the guarantee. It will cover the manufacturer's guarantee cost. Lemma 2 also points out that both the guarantee cost and guarantee coefficient are consistent with the retailer's default probability. It means that the higher retailer's default probability, the higher guarantee coefficient of the third-party guarantee and the guarantee cost.

Proposition 2: When the third-party guarantee cost is undertaken by the manufacturer, if α and θ_1 remains unchanged, then $\lambda_1 \uparrow \Rightarrow r_1^* \downarrow$, w_1^* and q_1^* remains unchanged. If α and the optimal financing interest rate r_1^* remain unchanged, then $\lambda_1 \uparrow \Rightarrow \theta_1 \uparrow$, $w_1^* \uparrow$ and $q_1^* \downarrow$.

Proposition 2 points out that, when the third-party guarantee company provides guarantee for accounts receivable financing and the guarantee cost is undertaken by the manufacturer, the third-party guarantee will reduce the financing interest rate of the financial institution when the loan-to-value ratio remains unchanged. But it has no effect to the manufacturer or the retailer. The rationale behind this is as follows. The decrease of the financing interest rate caused by guarantee will decrease the manufacturer's financing cost, then the manufacturer will maintain its optimal wholesale price and the retailer maintain its optimal order quantity. If the financing interest rate of the financial institution remains unchanged, then the manufacturer undertaken the guarantee cost can improve the financial institution's loan-to-value ratio, and the manufacturer can maintain a high wholesale price. It is interesting that the manufacturer's profits will increase with the guarantee coefficient even the manufacturer undertaken the guarantee cost. This is because that a higher guarantee coefficient can reduce the financing interest rate or increase the loan-to-value ratio, then it can offset the manufacturer's guarantee cost and improve the manufacturer's profits.

3.3 Retailer Undertaken the Guarantee Cost (Issue 2)

The sequence of events under this issue is shown in Fig. 3. First of all, the manufacturer announces the wholesale price w_2. Second, the retailer decides the order quantity q_2. Third, the financial institution announces the financing interest rate r_2. Finally, the third-party guarantee announces its guarantee coefficient λ_2 and the guarantee cost rate β_2. We also proceed backwards to derive the optimal decisions.

First, the third-party guarantee's problem can be formulated as Eq. (8):

$$\beta_2 w_2 q_2 = \lambda_2 w_2 q_2 \alpha \tag{8}$$

Then, the problem of financial institution can be used to solve following problem:

$$\theta_2 w_2 q_2 (1 + r_f) = (1 - \alpha)(w_2 q_2 + w_2 q_2 r_2 - (1 - \theta_2) w_2 q_2)$$
$$+ \alpha (w_2 q_2 r_2 + \lambda_2 w_2 q_2) \tag{9}$$

Next, the retailer decides its optimal order quantity q_2 based on Eq. (10):

$$\max_{q_2 \geq 0} \Pi_2^R(q_2) = p E_D \min\{D, q_2\} - w_2 q_2 (1 - \alpha) - \beta_2 w_2 q_2 \tag{10}$$

Finally, the manufacturer decides its wholesale price w_2 based on Eq. (11):

$$\max_{w_2 \geq c} \Pi_2^M(w_2) = \theta_2 w_2 q_2 + (1 - \alpha)(1 - \theta_2)w_2 q_2 - cq_2 - w_2 q_2 r_2 \tag{11}$$

Lemma 3: When the third-party guarantee cost is undertaken by the retailer, the optimal decisions of each part are as follows:

(1) The optimal order quantity is: $\left(\frac{p(1-(1-\lambda_2)\alpha-\theta_2 r_f)}{1-(1-\lambda_2)\alpha}\right)\left(\bar{F}(q_2^*) - q_2^* f(q_2^*)\right) = c$;

(2) The optimal wholesale price is: $w_2^* = \frac{p}{1-(1-\lambda_2)\alpha}\bar{F}(q_2^*)$;

(3) The optimal financing interest rate is: $r_2^* = \theta_2(r_f + \alpha) - \alpha\lambda_2$;

(4) The guarantee cost rate and guarantee coefficient satisfies: $\beta_2 = \lambda_2\alpha$.

Different from the manufacturer undertaken the guarantee cost, Lemma 3 points out that the retailer's undertaken the guarantee cost will affect the optimal wholesale price and order quantity, which is shown in Proposition 3 below:

Proposition 3: When the third-party guarantee cost is undertaken by the retailer, and if α and θ_2 remains unchanged, then $\lambda_2 \uparrow \Rightarrow w_2^* \downarrow, q_2^* \uparrow, r_2^* \downarrow$. If α and the optimal financing interest rate r_2^* remain unchanged, then $\lambda_2 \uparrow \Rightarrow \theta_2 \uparrow, w_2^* \downarrow, q_2^* \uparrow$.

When the guarantee cost is undertaken by the retailer, Lemma 3 and Proposition 3 reveals that the financing interest rate of the financial institution will decrease and the loan-to-value ratio will increase when the guarantee coefficient of the third-party guarantee increase. Specially, Different from the manufacturer undertaken the guarantee cost, the guarantee cost undertaken by the retailer will decrease the manufacturer's wholesale price, and then increase the retailer's order quantity. Due to the retailer's default risk is the main source of financing risk in the process of accounts receivable financing, retailer undertaking the guarantee cost can better avoid the financing risk and improve the supply chain financing and supply chain operation efficiency.

4 Impact of Third-Party Guarantee and Simulation Analysis

4.1 Impact of Third-Party Guarantee: Model Analysis

Above we analyzed the optimal decisions of supply chain under the issue that without guarantee, manufacturer undertaken the guarantee cost and retailer undertaken the guarantee cost. Then we analyze how the guarantee cost affects the supply chain accounts receivable financing.

Proposition 4: If $\lambda_1 = \lambda_2 = \lambda \in (0, 1), \theta_0 = \theta_1 = \theta_2 = \theta \in (0, 1)$, then $r_0^* > r_1^* = r_2^*, w_0^* = w_1^* > w_2^*$, and $q_0^* = q_1^* < q_2^*$. Besides, $\Pi_0^R = \Pi_1^R < \Pi_2^R, \Pi_0^M = \Pi_1^M > \Pi_2^M, \Pi_0^{SC} = \Pi_1^{SC} < \Pi_2^{SC}$.

Proposition 4 reveals that manufacturer or retailer undertaken the guarantee cost can reduce the financing interest rate when the guarantee coefficient of the third-party guarantee and the loan-to-value ratio of the financial institution keep unchanged, but has different effects to the financing decisions. When the manufacturer undertaken the guarantee cost, the manufacturer's financing costs will decreased due to the decrease of financing interest rate. Nevertheless, since the financing guarantee cost is undertaken by the manufacturer, the optimal wholesale price remains unchanged. When the retailer undertaken the guarantee cost, the financing interest rate keeps unchanged. Proposition 4 points out that the guarantee cost undertaken by the retailer can reduce the manufacturer's wholesale price, then improve its own profits and supply chain operation efficiency. The rationale behind this is as follows. When the guarantee cost is undertaken by the retailer, the expected profits and financing interest rate of the financial institution remains unchanged, but the third-party guarantee reduced the manufacturer's financing risk and financing cost. Consequently, the manufacturer will set a low wholesale price to improve the supply chain operation efficiency.

Proposition 4 indicates that the manufacturer undertaken guarantee cost has low effect to supply chain financing efficiency when the third-party guarantee is introduced to provide guarantee in the process of accounts receivable financing. However, retailer undertaken guarantee cost can reduce manufacturer's financing risk and cost, and then improving the supply chain financing and operation efficiency. Therefore, it is important to let the retailer pay the guarantee cost to the third-party guarantee in the process of the supply chain accounts receivable financing. Then the financing risk and cost will decrease, and then improve financing risks.

4.2 Simulation Analysis

Concerning the issue that without the third-party guarantee (issue 0), we conduct Case 1 below for simulation. All parameters are summarized from real company's operations, such as "Xi'an Wuxiu commercial factoring Co., Ltd".

Case 1: Assume that the random market demand of products sold by the retailer obeys the normal distribution with the mean as 300 and the variance as 120; the market price of products as $p = 1$; the internal fund level of the capital-restrained manufacturer as 0; the unit product cost as $c = 0.2$.

(1) Assume that the loan-to-value ratio paid by the financial institution is $\theta = 0.8$. The financing interest rate is $r_f = 0.1$. Then the optimal decisions of each part changes with the retailer's default probability is shown in Fig. 4(a) and 4(b) below.
(2) Suppose that the probability of retailer's default probability is $\alpha = 0.12$, The financing interest rate is $r_f = 0.2$. Then the optimal decisions of each part changes with the financial institution's loan-to-value ratio is shown in Fig. 5(a) and 5(b) below.

Analysis of Case 1 reveals that the increase of retailer's default probability will increase he manufacturer's financing cost, thus leading to a higher wholesale price (Fig. 4(a)) and a lower supply chain operation efficiency (Fig. 4(b)) when there is no third-party guarantee. In addition, the manufacturer will undertake a high financing cost and risk if wants a higher loan-to-value ratio, which will financially result in the decrease of financing and supply chain operation efficiency (Fig. 5).

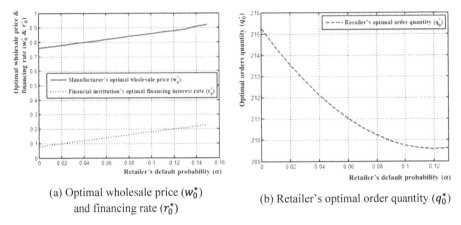

(a) Optimal wholesale price (w_0^*) and financing rate (r_0^*)

(b) Retailer's optimal order quantity (q_0^*)

Fig. 4. Impact of retailer's default probability under without guarantee

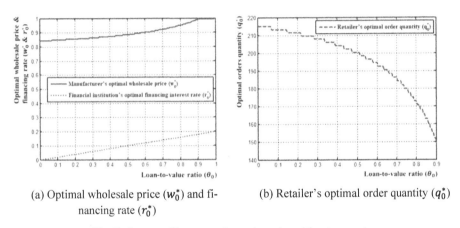

(a) Optimal wholesale price (w_0^*) and financing rate (r_0^*)

(b) Retailer's optimal order quantity (q_0^*)

Fig. 5. Impact of loan-to-value ratio under without guarantee

Regarding the situation where the manufacturer and the retailer pay the guarantee cost (namely issue 1 and issue 2), we conduct Case 2 below for simulation.

Case 2: Assume that the random market demand of products sold by the retailer obeys the normal distribution with the mean as 300 and the variance as 120; the market price of products as $p = 1$; the internal fund level of the manufacturer is 0; the unit product cost is $c = 0.2$. The average yield of the financing market is $r_f = 0.1$. The retailer's default probability is $\alpha = 0.1$. The loan-to-value ratio paid by the financial institution is $\theta = 0.8$. Then the optimal decisions of each part changes with the guarantee coefficient are shown in Fig. 6(a) (financing interest rate), Fig. 6(b) (wholesale price), Fig. 7(a) (order quantity), and Fig. 7(b) (profit) below.

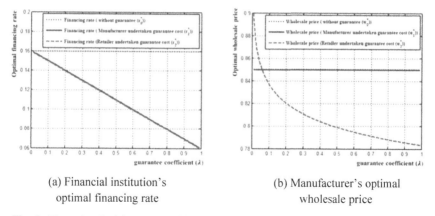

(a) Financial institution's
optimal financing rate

(b) Manufacturer's optimal
wholesale price

Fig. 6. Financing decisions under different situations (interest rate and wholesale price)

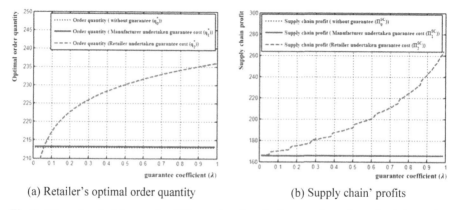

(a) Retailer's optimal order quantity

(b) Supply chain' profits

Fig. 7. Financing decisions under different situations (order quantity and supply chain profits)

Analysis of Case 2 shows that the financing interest rate provided by the financial institution will be decreased when there a third-party guarantee for the accounts receivable financing. And results show that the impact of manufacturer undertaken the guarantee cost on the financing interest rate is consistent with retailer undertaken the guarantee cost (Fig. 6(a)). However, the manufacturer's guarantee cost and financing cost will be decreased if the guarantee cost is undertaken by the retailer. Then the manufacturer will set a lower wholesale price (Fig. 6(b)) and reduce the retailer's order cost. Therefore, the retailer's order quantity will be increased (Fig. 7(a)) and then improve the supply chain operation efficiency (Fig. 7(b)).

5 Conclusions

This paper analyzed the effect of the partial guarantee behavior provided by the third-party guarantee on the decisions of the accounts receivable financing in the supply chain. Under the background of retailer's default, this paper compared three issues: without guarantee, manufacturer undertaken the guarantee cost and retailer undertaken the guarantee cost through model and simulation analysis. The main results are as follows: (1) manufacturer undertaken the guarantee cost can reduce the financing interest rate or improve the loan-to-value ratio of the financial institution, but it cannot improve the supply chain financing efficiency; (2) retailer undertaken the guarantee cost can reduce the financing interest rate of the financial institution and the wholesale price of the manufacturer. Then weak the double marginalization effects of the supply chain operation and improve retailer's and whole supply chain's profits; (3) it is important to let the retailer undertaken the guarantee cost to the third-party guarantee in the process of the supply chain accounts receivable financing.

This paper also has some limitations. For example, this paper assumes that the third-party guarantee is risk-neutral, and not considers the salvage value of unsold products at the end of the sales period.

References

1. Bougheas, S., Mateut, S., Mizena, P.: Corporate trade credit and inventories: new evidence of a trade-off from accounts payable and receivable. J. Bank. Finance **3**(2), 300–307 (2009)
2. Yu, M.: Mode of operation and risk prevention of supply chain financing for SMEs. Adv. Appl. Econ. Finance **3**(2), 532–536 (2012)
3. Koch, A.R.: Economic aspects of inventory and receivables financing. Law Contemp. Probl. **13**(4), 566–578 (1948)
4. Burman, R.W.: Practical aspects of inventory and receivables financing. Law Contemp. Probl. **13**(4), 555–565 (1948)
5. Dunham, A.: Inventory and accounts receivable financing. Harvard Law Rev. **62**(4), 588–615 (1949)
6. Poe, T.R.: Subjective judgments and the asset-based lender. J. Commer. Lend. Rev. **13**(2), 67–70 (1998)
7. Czternasty, W., Mikołajczak, P.: Financing of SME using non-recourse factoring–legal, economic and tax aspects. Management **17**(1), 358–370 (2013)
8. Lu, Q.H., Gu, J., Huang, J.Z.: Supply chain finance with partial credit guarantee provided by a third-party or a supplier. Comput. Ind. Eng. **135**, 440–455 (2019)
9. Kouvelis, P., Zhao, W.H.: Who should finance the supply chain? Impact of credit ratings on supply chain decisions. Manuf. Serv. Oper. Manag. **20**(1), 19–35 (2018)
10. Cai, G.S., Chen, X.F., Xiao, Z.G.: The roles of bank and trade credits: theoretical analysis and empirical evidence. Prod. Oper. Manag. **23**(4), 583–598 (2014)
11. Kouvelis, P., Zhao, W.H.: Financing the newsvendor: supplier vs. bank, and the structure of optimal trade credit contracts. Oper. Res. **60**(3), 566–580 (2012)
12. Jing, B., Chen, X.F., Cai, G.S.: Equilibrium financing in a distribution channel with capital constraint. Prod. Oper. Manag. **21**(6), 1090–1101 (2012)
13. Lu, X.Y., Wu, Z.Q.: How taxes impact bank and trade financing for Multinational Firms. Eur. J. Oper. Res. **286**(1), 218–232 (2020)

14. Lai, V.S., Yu, M.T.: An accurate analysis of vulnerable loan guarantees. Res. Finance **17**(3), 223–248 (1999)
15. Cossin, D., Hricko, T.: A structural analysis of credit risk with risky collateral: a methodology for haircut determination. Econ. Notes **32**(2), 243–282 (2003)
16. Gan, X.H., Sethi, S.P., Yan, H.M.: Channel coordination with a risk-neutral supplier and a downside-risk-averse retailer. Prod. Oper. Manag. **14**(1), 80–89 (2005)
17. Yan, N.N., Sun, B.W., Zhang, H., et al.: A partial credit guarantee contract in a capital-constrained supply chain: financing equilibrium and coordinating strategy. Int. J. Prod. Econ. **173**, 122–133 (2016)

Signal, Image and Video Processing

Detecting Apples in Orchards Using YOLOv3 and YOLOv5 in General and Close-Up Images

Anna Kuznetsova, Tatiana Maleva, and Vladimir Soloviev[✉]

Financial University Under the Government of the Russian Federation,
38 Shcherbakovskaya, Moscow 105187, Russia
{AnAKuznetsova,TVMaleva,VSoloviev}@fa.ru

Abstract. A machine vision system for apple harvesting robot was developed based on the YOLOv3 and the YOLOv5 algorithms with special pre- and post-processing and the YOLOv3 equipped with special pre- and post-processing procedures is able to achieve an a share of undetected apples (FNR) at 9.2% in the whole set of images, 6,7% in general images, and 16,3% in close-up images. A share of objects mistaken for apples (FPR) was at 7.8%. The YOLOv5 can detect apples quite precisely without any additional techniques, showing FNR at 2.8% and FPR at 3.5%.

Keywords: Machine vision · Apple harvesting robot · YOLO

1 Introduction

Mechanization and the use of chemical fertilizers have increased labor productivity in agriculture significantly. However, manual labor is still the main cost component in agriculture [1, 2]. In horticulture, fruits are picked manually, and the share of manual labor exceeds 40% in the total value of grown fruits while the crop shortage reaches 40–50% [3].

The development of intelligent robots for fruit harvesting can increase labor productivity significantly, reducing both crop shortages and the proportion of heavy routine manual harvesting operations.

Prototypes of fruit harvesting robots are being developed since the late 1960s. However, nowadays, not a single prototype is used in agricultural enterprises. The cost of such robots reaches hundreds of thousands of dollars, even though the speed of fruit picking is deficient, and a lot of fruits are left on trees.

The low fruit harvesting speed and the high share of unhandled fruits result from the machine vision systems' insufficient quality [4, 5].

During the last few years, many deep learning models have been trained to detect different fruits on trees, including apples. However, the computer vision systems based on these models in the existing prototypes of fruit-picking robots work too slowly. These systems also take yellow leaves for apples, do not detect apples with a lot of overlapping leaves and branches, darkened apples, green apples on a green background, etc.

© Springer Nature Switzerland AG 2020
M. Han et al. (Eds.): ISNN 2020, LNCS 12557, pp. 233–243, 2020.
https://doi.org/10.1007/978-3-030-64221-1_20

To solve these problems arising in apple detection in orchards, we propose to use the YOLOv3 algorithm with special pre- and post-processing of images taken by the camera placed on the manipulator of the harvesting robot with a special pre- and post-processing procedure.

2   Literature Review

2.1   Color-, Shape-, and Texture-Based Algorithms for Fruits Detection in Images

The first fruit detection algorithms were based on color. Setting the color threshold makes it possible to determine which pixels in the image belong to the fruit. In [6], citrus fruits, in [7–9] apples were detected based on color. The obvious advantage of fruits detection by color is the ease of implementation, but this method detects green and yellow-green apples very poorly. Also, bunches of red apples merge into one large apple.

Spherical fruits (including apples, citrus fruits, etc.) can be detected in images using the analysis of geometric shapes. In [10–12], various modifications of the Hough circular transformation were applied to detect fruits. In [13, 14], fruits were detected in images by identifying convex objects. Systems based on such algorithms work very fast, but complex scenes, especially when fruits are overlapping by leaves or other fruits, as a rule, are not recognized effectively in such systems. Many authors combined color and geometric shapes analysis, which led to an improvement in the quality of fruit detection in uneven lighting, overlapping of fruits by leaves and other fruits, etc. The main advantage of the analysis of geometric shapes is the low dependence of the quality of recognition of objects on the level of lighting [12]. However, this method gives significant errors, since not only apples have a round shape, but also gaps between leaves, leaf silhouettes, spots, and shadows on apples.

Fruits photographed outdoors differ from the leaves and branches in texture, and this can be used to facilitate the separation of fruits from the background. Differences in texture play a particularly important role in fruit detection when the fruits are grouped in clusters or overlapped by other fruits or leaves. In [15], apples were detected using texture analysis combined with color analysis. In [16], apples were detected using texture analysis combined with geometric shapes analysis. In [17, 18], to detect citrus fruits, simultaneous analysis of texture, color, and shape was carried out. Fruit detection by texture only works in close-up images with fair resolution and works very poorly in backlight. The low speed of texture-based fruit detection algorithms and a too-high share of not detected fruits lead to the inefficiency of such techniques in practical use.

2.2   Using Machine Learning for Fruits Detection in Images

The first prototype of a fruit-picking robot that detects red apples against green leaves using machine learning was presented in 1977 [19]. In [20–22], linear and KNN-classifiers were used to detect apples, peaches. bananas, lemons, and strawberries. In [23], apples were detected in images using K-means clustering. All these early-stage machine learning techniques for fruit detection were tested on very limited datasets of several dozens of images, so the results cannot be generalized for practical use. For example, in [24], 92% apple detection accuracy was reported based on a test set of just 59 apples.

Since 2012, with the advent of deep convolutional neural networks, in particular, AlexNet [25], machine vision and its use for detecting various objects in images, including fruits, received an impetus in development. In 2015, VGG16 neural network was developed as an improved version of AlexNet [26]. In 2018, the authors of [27] built a robot for harvesting kiwi fruits with a machine vision system based on VGG16. In the field trials, 76% of kiwi fruits were detected. In 2016, a new algorithm was proposed – YOLO (You Look Only Once) [28]. Before this, in order to detect objects in images, classification models based on neural networks were applied to a single image several times – in several different regions and/or on several scales. The YOLO approach involves a one-time application of one neural network to the whole image. The model divides the image into regions and immediately determines the scope of objects and probabilities of classes for each item. The third version of the YOLO algorithm was published in 2018 as YOLOv3 [29]. The YOLO algorithm is one of the fastest, and it has already been used in robots for picking fruits. In [30, 31] published in 2019, a modification of the YOLO algorithm was used to detect apples. The authors made the network tightly connected: each layer was connected to all subsequent layers, as the DenseNet approach suggests [32]. The IoU (Intersection over Union) indicator turned out to be 89.6%, with an average apple recognition time equal to 0.3 s.

In [33], DaSNet-v2 algorithm was developed to detect apples. This neural network detects objects in images in a single pass, taking into account their superposition, just like YOLO. The IoU in this model was at 86%.

The authors of [34] compared the standard Faster R-CNN algorithm, the proposed by them modification of Faster R-CNN, and YOLOv3 for the detection of oranges, apples, and mangoes. The modification proposed in this paper reveals about 90% of the fruits, which is 3–4% better than the standard Faster R-CNN on the same dataset and at about the same level as YOLOv3. However, the average recognition time for YOLOv3 was 40 ms versus 58 ms for the modified Faster R-CNN network and 240 ms for the standard Faster R-CNN network.

Now the new version of the YOLO algorithm was released, the YOLOv5 [35]. In the next section, we will compare the apple detection efficiency of the YOLOv3 and YOLOv5.

3 Research Methodology

3.1 Basic Pre- and Post-processing of Images for YOLOv3-Based Apple Detection Efficiency Improving

The Department of data analysis and machine learning of the Financial University under the Government of the Russian Federation, together with the Laboratory of machine technologies for cultivating perennial crops of the VIM Federal Scientific Agro-Engineering Center, is developing a robot for harvesting apples.

The VIM Center develops the mechanical component of the robot, while the Financial University is responsible for the intelligent algorithms for detecting fruits and operating the manipulator for their picking.

To detect apples, we compare the use of the YOLOv3 [29] and the YOLOv5 [35], both trained on the COCO dataset [36], which contains 1.5 million objects of 80 categories marked out in images.

Using the standard YOLOv3 algorithm without pre- and post-processing to detect apples in this set of images showed that not all apples in images are detected successfully (Fig. 1).

Fig. 1. Examples of apple detection results without pre- and post-processing.

To improve the detection efficiency of apples, first, we pre-processed images by means of adaptive histogram alignment, thickening of the borders, and slight blur. This led to the mitigation of such negative effects as shadows, glare, minor damages of apples, and overlapping apples by thin branches. During pre-processing we detected images with backlight by the prevailing average number of dark pixels and strongly lightened these images. Images with spots on apples, perianth, as well as thin branches were improved during pre-processing by replacing pixels of brown shades by yellow pixels.

During post-processing, in order to prevent the system from taking yellow leaves for apples, we discarded recognized objects whose ratio of the greater side of the circumscribed rectangle to the smaller one was more than 3. In order not to take the gaps between the leaves for apples, during post-processing, objects were discarded whose area of the circumscribed rectangle was less than the threshold.

The algorithm was tested on a set of apple images of various varieties, including red and green, made by employees of the VIM Federal Scientific Agro-Engineering Center. Among these photographs were both close-up images with one or more apples, and general images with several dozen apples.

In general, the YOLOv3 algorithm, supplemented by the described pre- and post-processing procedures, quite precisely detects both red and green apples [37] (Fig. 2, 3). Green apples are better detected if the shade of the apple is at least slightly different from the shade of the leaves (Fig. 3).

Fig. 2. Examples of red apples detection. (Color figure online)

3.2 Pre-processing for YOLOv3-Based Apple Detection in General Images

In the apple harvesting robot we are developing, the machine vision system is a combination of two stationary Sony Alpha ILCE-7RM2 cameras with Sony FE24-240 mm f/3.5-6.3 OSS lenses, and one Logitech Webcam C930e camera mounted on the second movable shoulder of the manipulator before the grip.

The first two cameras take general shots for detecting apples and drawing up the optimal route for the manipulator to collect them. The camera on the manipulator adjusts the position of the grip relative to the apple during the apple-picking process.

Therefore, it is essential to efficiently detect apples both in general images and in close-up images. It turned out that in general images a lot of apples remain undetected. For example, in the two images shown in Fig. 4 and 5, only 2 and 4 apples were detected respectively among several dozens.

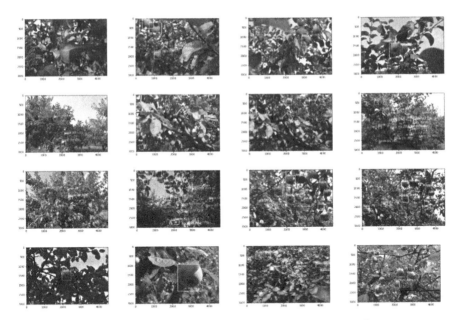

Fig. 3. Examples of green apples detection. (Color figure online)

Fig. 4. 2 apples detected in the general image without special general image pre-processing.

Fig. 5. 4 apples found in the general image without special general image pre-processing.

Dividing general plan images into nine regions with the subsequent application of the algorithm separately for each region made it possible to increase the number of detected apples [38]. So, after applying this procedure to the image presented in Fig. 5, 57 apples were detected (Fig. 6), and after applying this technique to the image in Fig. 5 made it possible to detect 48 apples (Fig. 7).

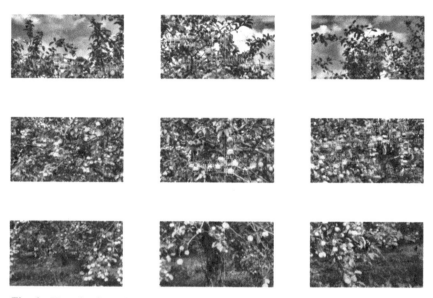

Fig. 6. 57 apples found in the general image after special general image pre-processing.

3.3 Using YOLOv5 for Apple Detection in General and Close-Up Images

Applying the standard YOLOv5 algorithm to detect apples in the test set of images have shown extremely promising results. Without any pre- and post-processing, the YOLOv5 detected 97.2% of apples in images (Fig. 8).

3.4 Using YOLOv5 for Apple Detection in General and Close-Up Images

Applying the standard YOLOv5 algorithm to detect apples in the test set of images have shown extremely promising results. Without any pre- and post-processing, the YOLOv5 detected 97.2% of apples in images (Fig. 8).

4 Results

The apple detection quality was evaluated on a test dataset of 878 images with red and green apples of various varieties (5142 apples in total):

Fig. 7. 48 apples found in the general image after special general image pre-processing.

Fig. 8. YOLOv5 for apple detection without image pre- and post-processing.

- 553 general images (4365 apples);
- 274 close-up images (533 apples).

On average, it takes 19 ms to detect one apple, considering pre- and post-processing.

To assess the fruit detection quality, it is essential to understand what proportion of objects is mistaken by the system for apples (False Positive Rate):

$$FPR = 1 - Precision = \frac{FP}{TP + FP},$$

and what proportion of apples remains undetected (False Negative Rate):

$$FNR = 1 - Recall = \frac{FP}{TP + FN}.$$

Here TP (True Positives), FP (False Positives), and FN (False Negatives) are respectively real apples detected by the algorithm in images, objects mistaken by the algorithm for apples, and undetected apples.

The results of these metrics calculation for YOLOv3 with pre- and post-processing are presented in Table 1.

Table 1. Apple detection quality metrics.

Whole set of images				General images				Close-up images			
No of images	No of apples	FNR	FPR	No of images	No of apples	FNR	FPR	No of images	No of apples	FNR	FPR
878	5142	9,2%	7,8%	552	4358	6,7%	7,8%	274	533	16,3%	6,3%

Both the FPR and the FNR are quite small.

The standard YOLOv5, without any special pre- and post-processing, was able to detect 4998 of 5142 apples in general and close-up images. FNR was at the 2.8% level, FPR was equal to 3.5%. It means that the YOLOv5 is not able to detect only 2.8% of apples, and only 3.5% of detections are false.

5 Conclusion

The proposed technique made it possible to adapt the YOLOv3 for apple picking robot, gividing an average apple detection time of 19 ms with FPR at 7.8% and FNR at 9.2%. Both speed and error fractions were less than in all known similar systems.

But technology is developing, and the YOLOv5 can detect apples quite precisely without any additional techniques: only 2.8% of apples were not detected, and only 3.5% of objects detected as apples actually belong to the background.

References

1. Bechar, A., Vigneault, C.: Agricultural robots for field operations: concepts and components. Biosyst. Eng. **149**, 94–111 (2016)

2. Sistler, F.E.: Robotics and intelligent machines in agriculture. IEEE J. Robot. Autom. **3**(1), 3–6 (1987)

3. Ceres, R., Pons, J., Jiménez, A., Martín, J., Calderón, L.: Design and implementation of an aided fruit-harvesting robot (Agribot). Ind. Robot **25**(5), 337–346 (1998)

4. Edan, Y., Han, S.F., Kondo, N.: Automation in agriculture. In: Nof, S. (ed.) Springer Handbook of Automation. SHB, pp. 1095–1128. Springer, Heidelberg (2009). https://doi.org/10.1007/978-3-540-78831-7_63

5. Grift, T., Zhang, Q., Kondo, N., Ting, K.C.: A review of automation and robotics for the bio-industry. J. Biomech. Eng. **1**(1), 37–54 (2008)

6. Bulanon, D.M., Burks, T.F., Alchanatis, V.: Image fusion of visible and thermal images for fruit detection. Biosyst. Eng. **103**(1), 12–22 (2009)

7. Mao, W.H., Ji, B.P., Zhan, J.C., Zhang, X.C., Hu, X.A.: Apple location method for the apple harvesting robot. In: Proceedings of the 2nd International Congress on Image and Signal Processing – CIPE 2009, Tianjin, China, 7–19 October 2009, pp. 17–19 (2009)

8. Bulanon, D.M., Kataoka, T.: A fruit detection system and an end effector for robotic harvesting of Fuji apples. Agric. Eng. Int. CIGR J. **12**(1), 203–210 (2010)

9. Wei, X., Jia, K., Lan, J., Li, Y., Zeng, Y., Wang, C.: Automatic method of fruit object extraction under complex agricultural background for vision system of fruit picking robot. Optics **125**(12), 5684–5689 (2014)

10. Whittaker, A.D., Miles, G.E., Mitchell, O.R.: Fruit location in a partially occluded image. Trans. Am. Soc. Agric. Eng. **30**(3), 591–596 (1987)

11. Xie, Z.Y., Zhang, T.Z., Zhao, J.Y.: Ripened strawberry recognition based on Hough transform. Trans. Chin. Soc. Agric. Mach. **38**(3), 106–109 (2007)

12. Xie, Z., Ji, C., Guo, X., Zhu, S.: An object detection method for quasi-circular fruits based on improved Hough transform. Trans. Chin. Soc. Agric. Mach. **26**(7), 157–162 (2010)

13. Kelman, E.E., Linker, R.: Vision-based localization of mature apples in tree images using convexity. Biosyst. Eng. **118**(1), 174–185 (2014)

14. Xie, Z., Ji, C., Guo, X., Zhu, S.: Detection and location algorithm for overlapped fruits based on concave spots searching. Trans. Chin. Soc. Agric. Mach. **42**(12), 191–196 (2011)

15. Zhao, J., Tow, J., Katupitiya, J.: On-tree fruit recognition using texture properties and color data. In: IEEE/RSJ International Conference on Intelligent Robots and Systems, Edmonton, Canada, 2–6 August, 2005, pp. 263–268 (2005)

16. Rakun, J., Stajnko, D., Zazula, D.: Detecting fruits in natural scenes by using spatial-frequency based texture analysis and multiview geometry. Comput. Electron. Agric. **76**(1), 80–88 (2011)

17. Kurtulmus, F., Lee, W.S., Vardar, A.: Green citrus detection using 'eigenfruit', color and circular Gabor texture features under natural outdoor conditions. Comput. Electron. Agric. **78**(2), 140–149 (2011)

18. Kurtulmus, F., Lee, W.S., Vardar, A.: An advanced green citrus detection algorithm using color images and neural networks. J. Agric. Mach. Sci. **7**(2), 145–151 (2011)

19. Parrish, E.A., Goksel, J.A.K.: Pictorial pattern recognition applied to fruit harvesting. Trans. Am. Soc. Agric. Eng. **20**(5), 822–827 (1977)

20. Sites, P.W., Delwiche, M.J.: Computer vision to locate fruit on a tree. Trans. Am. Soc. Agric. Eng. **31**(1), 257–263 (1988)

21. Bulanon, D.M., Kataoka, T., Okamoto, H., Hata, S.: Development of a real-time machine vision system for apple harvesting robot. In: Society of Instrument and Control Engineers Annual Conference, Sapporo, Japan, 4–6 August 2004, pp. 595–598 (2004)

22. Seng, W.C., Mirisaee, S.H.: A new method for fruits recognition system. In: Proceedings of the 2009 International Conference on Electrical Engineering and Informatics – ICEEI 2009, Selangor, Malaysia, 5–7 August, 2009, vol. 1, pp. 130–134 (2009)

23. Wachs, J.P., Stern, H.I., Burks, T., Alchanatis, V.: Low and high-level visual feature-based apple detection from multi-modal images. Precis. Agric. **11**, 717–735 (2010). 10.1007/s11119-010-9198-x
24. Tao, Y., Zhou, J.: Automatic apple recognition based on the fusion of color and 3D feature for robotic fruit picking. Comput. Electron. Agric. **142**(A), 388–396 (2017)
25. Krizhevsky, A., Sutskever, I., Hinton, G.E.: ImageNet classification with deep convolutional neural networks. In: Advances in Neural Information Processing Systems 25 – NIPS 2012, Harrahs and Harveys, Lake Tahoe, Canada, 3–8 December 2012, pp. 1–9 (2012)
26. Simonyan, K., Zisserman, A.: Very deep convolutional networks for large-scale image recognition. In: International Conference on Learning Representations – ICLR 2015, San Diego, California, USA, 7–9 May 2015, pp. 1–14 (2015)
27. Williams, H.A.M., Jones, M.H., Nejati, M., Seabright, M.J., MacDonald, B.A.: Robotic kiwifruit harvesting using machine vision, convolutional neural networks, and robotic arms. Biosyst. Eng. **181**, 140–156 (2019)
28. Redmon, J., Divvala, S., Girshick, R., Farhadi, A.: You only look once: unified, real-time object detection. In: 29th IEEE Conference on Computer Vision and Pattern Recognition – CVPR 2016, Las Vegas, Nevada, USA, 26 June–1 July 2016, pp. 779–788 (2016)
29. Redmon, J., Divvala, S., Girshick, R., Farhadi, A.: YOLOv3: an incremental improvement. In: 31th IEEE Conference on Computer Vision and Pattern Recognition – CVPR 2018, Salt Lake City, Utah, USA, 18–22 June 2018, pp. 1–6 (2018)
30. Tian, Y., Yang, G., Wang, Z., Wang, H., Li, E., Liang, Z.: Apple detection during different growth stages in orchards using the improved YOLO-V3 model. Comput. Electron. Agric. **157**, 417–426 (2019)
31. Tian, Y., Yang, G., Wang, Zh., Li, E., Liang, Z.: Detection of apple lesions in orchards based on deep learning methods of CycleGAN and YOLO-V3-Dense. J. Sens. 1–14 (2019). Special issue: Sensors in Precision Agriculture for the Monitoring of Plant Development and Improvement of Food Production
32. Huang, G., Liu, Zh., van der Maaten, L., Weinberger, K.Q.: Densely connected convolutional networks. In: 30th IEEE Conference on Computer Vision and Pattern Recognition – CVPR 2017, Honolulu, Hawaii, USA, 22–25 July 2017, pp. 1–9 (2017)
33. Kang, H., Chen, C.: Fruit detection, segmentation and 3D visualization of environments in apple orchards. Comput. Electron. Agric. **171** (2020). Article no: 105302
34. Wan, S., Goudos, S.: Faster R-CNN for multi-class fruit detection using a robotic vision system. Comput. Netw. **168**, 107036 (2020)
35. Jocher, G., Nishimura, K., Mineeva, T., Vilariño, R.: YOLOv5 (2020). https://github.com/ultralytics/yolov5. Accessed 10 July 2020
36. COCO: Common Objects in Context Dataset. http://cocodataset.org/#overview. Accessed 19 Apr 2020
37. Kuznetsova, A., Maleva, T., Soloviev, V.: Detecting apples in orchards using YOLOv3. In: Gervasi, O., et al. (eds.) ICCSA 2020. LNCS, vol. 12249, pp. 923–934. Springer, Cham (2020). https://doi.org/10.1007/978-3-030-58799-4_66
38. Kuznetsova, A., Maleva, T., Soloviev, V.: Using YOLOv3 algorithm with pre- and post-processing for apple detection in fruit-harvesting robot. Agronomy **10**(7), 1016 (2020)

Robust Graph Regularized Non-negative Matrix Factorization for Image Clustering

Xiangguang Dai[1], Keke Zhang[1], Juntang Li[2], Jiang Xiong[1], and Nian Zhang[3(✉)]

[1] Key Laboratory of Intelligent Information Processing and Control of Chongqing Municipal Institutions of Higher Education, Chongqing Three Gorges University, Chongqing 40044, China
daixiangguang@163.com, xk_zhang0924@126.com, xjcq123@126.com
[2] State Grid Chongqing Yongchuan Electric Power Supply Branch, Chongqing, China
398340239@qq.com
[3] Department of Electrical and Computer Engineering, University of the District of Columbia, Washington, D.C. 20008, USA
nzhang@udc.edu

Abstract. Non-negative matrix factorization and its variants have been utilized for computer vision and machine learning, however, they fail to achieve robust factorization when the dataset is corrupted by outliers and noise. In this paper, we propose a roust graph regularized non-negative matrix factorization method (RGRNMF) for image clustering. To improve the clustering effect on the image dataset contaminated by outliers and noise, we propose a weighted constraint on the noise matrix and impose manifold learning into the low-dimensional representation. Experimental results demonstrate that RGRNMF can achieve better clustering performances on the face dataset corrupted by Salt and Pepper noise and Contiguous Occlusion.

Keywords: Noise · Graph regularization · Dimensionality reduction · Non-negative matrix factorization

1 Introduction

Clustering for computer vision and subspace learning is a challenging work. Many clustering methods were proposed for image retrieval [1], image indexing [2] and image classification [3]. To achieve image clustering effectively, a widely

This work is supported by Foundation of Chongqing Municipal Key Laboratory of Institutions of Higher Education ([2017]3), Foundation of Chongqing Development and Reform Commission (2017[1007]), Scientific and Technological Research Program of Chongqing Municipal Education Commission (Grant Nos. KJQN201901218 and KJQN201901203), Natural Science Foundation of Chongqing (Grant No. cstc2019jcyj-bshX0101), Foundation of Chongqing Three Gorges University and National Science Foundation (NSF) grant #2011927 and DoD grant #W911NF1810475.

© Springer Nature Switzerland AG 2020
M. Han et al. (Eds.): ISNN 2020, LNCS 12557, pp. 244–250, 2020.
https://doi.org/10.1007/978-3-030-64221-1_21

used approach is to discover an effective low-dimensional representation for the original data. Therefore, a lot of researches were presented to dig out the geometrical structure information of the original data, which can lead to a more discriminative representation.

In the past decades, many dimensionality reduction techniques were proposed including principal components analysis (PCA) [4] and non-negative matrix factorization (NMF) [5]. Among these methods, the non-negative property of the learned representation leads to be more meaningful in image representation. NMF decomposes the original data matrix into two low-dimensional matrices (i.e. a basis matrix and an encoding matrix), whose product can be best approximate to the original data matrix. Due to the excellent property of NMF, some variants [6–19] were proposed to improve the clustering accuracy from different views.

Although traditional NMF performs very well in learning a parts-based representation for clustering, it fails to achieve clustering while the original data is heavily corrupted. In the view of recent researches, the loss function of traditional NMF are very sensitive to outliers. In other words, the Frobenius norm enlarges the approximation error between the original data matrix and the product of the decomposed matrices. To address this issue, some studies [6–12] proposed some robust loss functions to minimize the reconstruction error. These proposed methods can reduce outliers of the representation, but they cannot remove outliers. Moreover, the learned representation cannot respect the geometrical structure of the original data contaminated by outliers and noise.

To address above-mentioned problems, we present a robust graph regularized non-negative matrix factorization (RGRNMF) for image clustering. Firstly, we propose a robust framework to measure the approximation error. Secondly, we construct a weighted graph to encode the geometrical information of the original data. Our achievements are as follows:

- We propose a robust non-negative matrix factorization framework to remove outliers, and we incorporate the geometrical information of the original data into the learned representation.
- Extensive experiments demonstrate that our proposed framework can achieve image clustering from the original data corrupted by Salt and Pepper noise or Contiguous Occlusion.

2 Related Works

Supposed that there are n sample images $\{x_i\}_{i=1}^n$ and any image x_i has m features. Thus, we denote the original data matrix by $V \in R^{m \times n}$. Due to the high-dimensional property of V, it is a challenging task to achieve image clustering. Generally, NMF is utilized to find two low-dimensional matrices $W \in R^{m \times r}$ and $H \in R^{r \times n}$ such that the product of W and H can be approximately equal to V. There, we have

$$V \approx WH, \tag{1}$$

where r is a factorization rank and $r << \min\{m, n\}$. Generally, problem (1) can be transformed into a non-convex optimization problem as follows:

$$\min_{W,H} \quad Error(V, WH)$$
$$s.t. \quad W \geq 0, H \geq 0. \tag{2}$$

where the loss function $Error$ can be the Frobenius norm, L_1 norm, $L_{2,1}$ norm or Huber. Recently, Guan et al. [12] proposed a Truncated Cauchy loss (CauchyNMF) to reduce outliers, which can be summarized as follows:

$$\min_{W \geq 0, H \geq 0} F(W, H) = \sum_{i=1}^{m} \sum_{j=1}^{n} g\left(\frac{(V - WH)_{ij}}{\gamma}\right), \tag{3}$$

where $g(x) = \begin{cases} ln(1 + x), & 0 \leq x \leq \sigma \\ ln(1 + \sigma), & x > \sigma \end{cases}$; σ and γ denote the scale parameter and the truncation parameter. σ can be obtained by three-sigma-rule, and γ is given by the Nagy algorithm [12]. Traditional NMF utilizes the different loss functions to reduce outliers, but they cannot remove outliers. Therefore, a robust NMF framework was proposed to eliminate outliers as follows:

$$\min_{W,H,E} \quad loss(M, WH, E) + \lambda \Omega(E, W, H)$$
$$s.t. \quad W \geq 0, H \geq 0, \tag{4}$$

where M is the original data matrix corrupted by noises, E is an error matrix, λ is a hyper-parameter, and the function Ω is the constraint term. Zhang et al. [11] proposed the Frobenius norm as the loss function and the L_1 norm as the constraint on E, which can be described as follows:

$$\min_{W,H,E} \parallel M - WH - E \parallel_F^2 + \lambda \parallel E \parallel_M$$
$$s.t. \quad W \geq 0, H \geq 0, \tag{5}$$

where $\parallel E \parallel_M = \sum_{ij} |e_{ij}|$.

3   Robust Graph Regularized Non-negative Matrix Factorization

3.1   Model Formulation

Previous NMF models have some defects: 1) They cannot remove outliers from the dataset corrupted by Salt and Pepper noise or Contiguous Occlusion. 2) While the dataset is corrupted by noises, the learned representation H cannot preserve the geometrical structure information.

In (5), Zhang et al. [11] supposed that the error matrix E is sparse, but the outliers in E are neglected. If the error matrix contains some outliers, then the constraint $\parallel E \parallel_M$ is not inappropriate for outliers. Supposed that all outliers of the corrupted image matrix $M \in R^{m \times n}$ produced by Salt and Pepper noise or Contiguous Occlusion are detected. A weight graph S can be utilized to label these outliers as follows:

$$S_{ij} = \begin{cases} 0, & \text{if the pixel } M_{ij} \text{ is an outlier,} \\ 1, & \text{otherwise,} \end{cases} \tag{6}$$

Thus, we propose the constraint on E by the following form:

$$\parallel E \otimes S \parallel_M \tag{7}$$

To learn the geometrical structure information of the original data, manifold regularization is proposed to construct the relation between the original data and the low-dimensional representation. A widely used manifold regularization term [20] can be described as follows:

$$tr(H(D - U)H^T), \tag{8}$$

where $U_{jl} = e^{-\frac{\parallel x_j - x_l \parallel^2}{\sigma}}$ and $D_{ii} = \sum_j W_{ij}$. In summary, combining (7), (8) and (5) results in our robust graph regularized non-negative matrix factorization (RGRNMF), which can be summarized into the following optimization problem

$$\min_{W,H,E} F(W, H, E)$$
$$= \parallel M - WH - E \parallel_F^2 + \lambda \parallel E \otimes S \parallel_F^2$$
$$+ \gamma tr(H(D - U)H^T) \tag{9}$$
$$s.t. \quad W \geq 0, H \geq 0,$$

where λ and γ are hyper-parameters.

4   Optimization Scheme

It is obvious that problem (9) is non-convex. Therefore, the global optimal solution cannot be searched. Supposed that the k−th solution of problem (9) is obtained. We can have the $k + 1$−th solution by optimizing the following problems

$$E^{k+1} = \arg\min_E \parallel M - W^k H^k - E \parallel_F^2$$
$$+ \lambda \parallel E \otimes S \parallel_F^2 \tag{10}$$

and

$$W^{k+1} = \arg\min_W \parallel M - WH^k - E^{k+1} \parallel_F^2$$
$$s.t. \quad W \geq 0 \tag{11}$$

and

$$H^{k+1} = \arg\min_H \parallel M - W^{k+1}H - E^{k+1} \parallel_F^2$$
$$+ \gamma tr(H(D - U)H^T) \tag{12}$$
$$s.t. \quad H \geq 0.$$

It is easy to obtain the solution of problems (10), (11) and (12) as follows:

$$e_{ij} \leftarrow \frac{m_{ij} - (WH)_{ij}}{1 + \lambda s_{ij}}. \tag{13}$$

$$w_{il} \leftarrow w_{il} \frac{(MH^T)_{il} - (EH^T)_{il}}{(WHH^T)_{il}}, \tag{14}$$

$$h_{lj} \leftarrow h_{lj} \frac{(W^T M)_{lj} - (W^T E)_{lj} + \gamma H U_{lj}}{(W^T W H)_{lj} + \gamma H D_{lj}}. \tag{15}$$

5   Experimental Results

We compare our proposed method (RGRNMF) with NMF [5], RNMF [9], Mah-NMF [8] and CauchyNMF [12] on the clustering performances of the ORL dataset. To verify the clustering ability on the corrupted data, we propose two corruptions including Salt and Pepper noise and Contiguous Occlusion. For Salt and Pepper noise, there are several percentages of corrupted pixels from 1% to 25%. Similarly, we vary the corrupted block size for Contiguous Occlusion from 1 to 16.

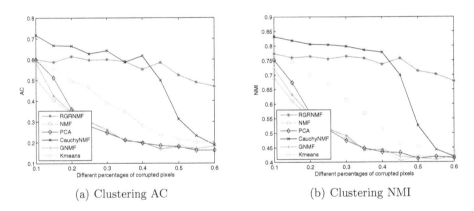

(a) Clustering AC   (b) Clustering NMI

Fig. 1. The clustering performances on the ORL dataset corrupted by Salt and Pepper noise.

To evaluate the clustering effect of all methods, we propose Accuracy (AC) and Normalized Mutual Information (NMI) [21]. Let $\lambda = 100$ and $\gamma = 100$. Figure 1 and 2 show the clustering performances on the ORL dataset contaminated by Salt and Pepper noise and Contiguous Occlusion. From these figures, we observe that:

- CauthyNMF achieves satisfactory clustering ACs and NMIs from the ORL dataset corrupted by Salt and Pepper noise and Contiguous Occlusion in the beginning, however, it obtains the poor clustering effect finally. This phenomenon indicates that CauthyNMF cannot handle heavy outliers.

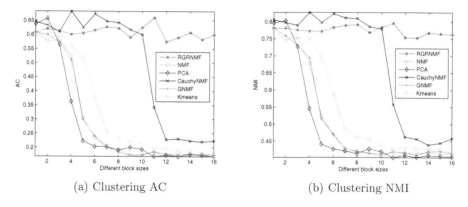

(a) Clustering AC   (b) Clustering NMI

Fig. 2. The clustering performances on the ORL dataset corrupted by Contiguous Occlusion.

- NMF, PCA Kmeans and GNMF fail to achieve clustering. This means that They cannot handle outliers.
- RGRNMF has relatively stable clustering performances on the Salt and Pepper noise and Contiguous Occlusion, that is to say, RGRNMF is more robust to outliers.

6   Conclusion

This paper proposed robust graph regularized non-negative matrix factorization (RGRNMF) to handle Salt and Pepper noise and Contiguous Occlusion. Clustering results demonstrate that our proposed NMF framework has the following properties. Firstly, RGRNMF can learn a more effective and discriminative parts-based representation from the ORL dataset corrupted by handle Salt and Pepper noise or Contiguous Occlusion. Secondly, RGRNMF is more robust to outliers than existing NMF methods.

References

1. Chen, L., Xu, D., Tsang, I.W., Li, X.: Spectral embedded hashing for scalable image retrieval. IEEE Trans. Cybern. **44**(7), 1180–1190 (2014)
2. Datta, R., Joshi, D., Jia, L.I., Wang, J.Z.: Image retrieval: ideas, influences, and trends of the new age. ACM Comput. Surv. **40**(2), 35–94 (2008)
3. Banerjee, B., Bovolo, F., Bhattacharya, A., Bruzzone, L., Chaudhuri, S., Mohan, B.K.: A new self-training-based unsupervised satellite image classification technique using cluster ensemble strategy. IEEE Geoence Remote. Sens. Lett. **12**(4), 741–745 (2015)
4. Turk, M., Pentland, A.: Eigenfaces for recognition. J. Cogn. Neurosci. **3**(1), 71–86 (1991)
5. Lee, D.D., Seung, H.S.: Learning the parts of objects by non-negative matrix factorization. Nature **401**(6755), 788–791 (1999)

6. Hamza, A.B., Brady, D.J.: Reconstruction of reflectance spectra using robust non-negative matrix factorization. IEEE Trans. Signal Process. **54**(9), 3637–3642 (2006)
7. Kong, D., Ding, C., Huang, H.: Robust nonnegative matrix factorization using L21-norm. In: Proceedings of the 20th ACM International Conference on Information And Knowledge Management, pp. 673–682 (2011)
8. Guan, N., Tao, D., Luo, Z., Shawetaylor, J.: MahNMF: Manhattan non-negative matrix factorization. J. Mach. Learn. Res. arXiv:1207.3438v1 (2012)
9. Gao, H., Nie, F., Cai, W., Huang, H.: Robust capped norm nonnegative matrix factorization. In: ACM International on Conference on Information and Knowledge Management, pp. 871–880 (2015)
10. Du, L., Li, X., Shen, Y.: Robust nonnegative matrix factorization via half-quadratic minimization. In: IEEE International Conference on Data Mining, pp. 201–210 (2012)
11. Zhang, L., Chen, Z., Zheng, M., He, X.: Robust non-negative matrix factorization. Front. Electr. Electron. Eng. China **6**(2), 192–200 (2015)
12. Guan, N., Liu, T., Zhang, Y., Tao, D., Davis, L.: Truncated cauchy non-negative matrix factorization. IEEE Trans. Pattern Anal. Mach. Intell. **41**(1), 246–259 (2018)
13. Casalino, G., Del Buon, N., Mencar, C.: Subtractive clustering for seeding non-negative matrix factorizations. Inf. Sci. **257**, 369–387 (2014)
14. Wu, W., Jia, Y., Kwong, S., Hou, J.: Pairwise constraint propagation-induced symmetric nonnegative matrix factorization. IEEE Trans. Neural Netw. Learn. Syst. **29**(12), 6348–6361 (2018)
15. Li, H., Li, K., An, J., Zhang, W., Li, K.: An efficient manifold regularized sparse non-negative matrix factorization model for large-scale recommender systems on GPUs. Inf. Sci. (2018). https://doi.org/10.1016/j.ins.2018.07.060
16. Liu, X., Wang, W., He, D., Jiao, P., Jin, D., Cannistraci, C.V.: Semi-supervised community detection based on non-negative matrix factorization with node popularity. Inf. Sci. **381**, 304–321 (2017)
17. Peng, X., Chen, D., Xu, D.: Hyperplane-based nonnegative matrix factorization with label information. Inf. Sci. **493**, 1–9 (2019)
18. Kang, Z., Pan, H., Hoi, S., Xu, Z.: Robust graph learning from noisy data. IEEE Trans. Cybern. (2019)
19. Li, Z., Tang, J., He, X.: Robust structured nonnegative matrix factorization for image representation. IEEE Trans. Neural Netw. Learn. Syst. **29**(5), 1947–1960 (2018)
20. Cai, D., He, X., Han, J., Huang, T.S.: Graph regularized nonnegative matrix factorization for data representation. IEEE Trans. Pattern Anal. Mach. Intell. **33**(8), 1548–1560 (2011)
21. Cai, D., He, X., Han, J.: Document clustering using locality preserving indexing. IEEE Trans. Knowl. Data Eng. **17**(12), 1624–1637 (2005)

ContourRend: A Segmentation Method for Improving Contours by Rendering

Junwen Chen[1,2,3], Yi Lu[2,3], Yaran Chen[2,3(✉)], Dongbin Zhao[2,3],
and Zhonghua Pang[1,3]

[1] Key Laboratory of Fieldbus Technology and Automation of Beijing, North China University
of Technology, Beijing 100144, China
ohhthxplz@gmail.com, zhpang@ncut.edu.cn
[2] State Key Laboratory of Management and Control for Complex Systems,
Institute of Automation, Chinese Academy of Sciences, Beijing 100190, China
{luyi2017,chenyaran2013,dongbin.zhao}@ia.ac.cn
[3] University of Chinese Academy of Sciences, Beijing 101408, China

Abstract. A good object segmentation should contain clear contours and complete regions. However, mask-based segmentation can not handle contour features well on a coarse prediction grid, thus causing problems of blurry edges. While contour-based segmentation provides contours directly, but misses contours' details. In order to obtain fine contours, we propose a segmentation method named ContourRend which adopts a contour renderer to refine segmentation contours. And we implement our method on a segmentation model based on graph convolutional network (GCN). For the single object segmentation task on cityscapes dataset, the GCN-based segmentation contour is used to generate a contour of a single object, then our contour renderer focuses on the pixels around the contour and predicts the category at high resolution. By rendering the contour result, our method reaches 72.41% mean intersection over union (IoU) and surpasses baseline Polygon-GCN by 1.22%.

Keywords: Image segmentation · Convolution neural networks · Contour renderer · Graph convolutional network

1 Introduction

Convolutional neural network (CNN) methods bring various breakthroughs to the field of computer vision, improve the accuracy in the tasks of image classification [1, 2], image classification and location [3], object detection [4], image segmentation [5], and even surpass the human performance. More and more image processing tasks begin to rely on the rich features provided by CNN.

This work is supported partly by National Key Research and Development Plan under Grant No. 2017YFC1700106, and National Natural Science Foundation of China under Grant 61673023, Beijing University High-Level Talent Cross-Training Project (Practical Training Plan).

© Springer Nature Switzerland AG 2020
M. Han et al. (Eds.): ISNN 2020, LNCS 12557, pp. 251–260, 2020.
https://doi.org/10.1007/978-3-030-64221-1_22

In image segmentation task, semantic segmentation predicts the label of every pixel. And CNN can also easily provide the encoding of segmentation information for kinds of usages. Full convolution network (FCN) [6] uses a fully convolutional structure for segmentation and builds a skip architecture to connect semantic information in different depth of the convolution layers. In FCN, the features $8\times$, $16\times$, $32\times$ smaller than the input are used by the transposed convolution to predict mask result. U-Net [7] is also a fully convolutional network and has a symmetric architecture in encoding and decoding feature maps. U-Net concatenates the feature maps with the same resolution in the encoder and decoder and uses transposed convolution to restore these features to output mask results at a higher resolution. In instance segmentation task, segmentation focuses on distinguishing between pixel regions of different objects. Mask R-CNN [8] as the baseline of this task, adds a CNN segmentation branch on Faster R-CNN [9]. Faster R-CNN provides the feature map of the object in a 14×14 grid for the CNN branch, and the CNN branch predicts a 28×28 mask result.

Although these methods utilize the excellent feature extraction ability of the convolution operator, the feature maps with 8 times or 16 times smaller than the input are too coarse for segmentation. While upsampling or resizing these coarse masks to the results with the same size of the input images, there are blurry edges on the mask results' contours which limits the segmentation performance. To reduce this limitation, some methods focus on modifying the convolutional operator and pooling operator to lessen the down-sampling effect in the mask-based models. Pyramid scene parsing network (PSPNet) [10] uses pyramid pooling module to fuse the global context information and reduces false positive results. DeepLab family [11–13] and DenseASPP [14] use dilated convolution to expand the size of receptive field and improves the resolution of segmentation.

To avoid down-sampling effect in the mask-based models, some contour-based segmentation models that distinguish the object by the contour formed by the contour vertices are proposed. These models can obtain clear contours at the same resolution as the input image by determining the coordinates of the contour vertices. PolarMask [15] learns to predict dense distance regression of contour vertices from the object's center position in a polar coordinate. Polygon-RNN [16] and Polygon-RNN++ [17] utilize recurrent neural network (RNN) to find contour vertices one by one. Curve-GCN [18] implements graph convolutional network (GCN) to obtain the coordinates of the contour vertices by regression. Curve-GCN can simultaneously adjust the coordinates of a fixed number of vertices from the initial contour to the target.

Although the above contour-based segmentation methods avoid the effect by downsampling and directly restore the resolution, they are unable to provide complex edges due to the limitation of the fixed number of contour vertices. To this end, we propose a segmentation method to reconsider the segmentation process. The proposed method focuses on improve the contour-based segmentation models by adding a contour renderer, so it is called ContourRend. Rendering on the segmentation results has been learned in the work PointRend [19], which using mask scores to select the unclear points around the contour. Different from PointRend, the contour renderer of our method directly obtains rendering points by offsetting the contour vertices from the contour-based model, and directly renders on the mask with the same resolution as the input image. ContourRend achieves an effective way to refine an object's contour in instance segmentation task.

Our contributions in this paper are two folds,

1. We propose a segmentation method to improve the accuracy of contour-based segmentation models by adding a contour renderer, named ContourRend.
2. The experimental results of ContourRend on the single object segmentation task with cityscapes dataset show the improvement both in training and testing. ContourRend reaches 72.41% mean IoU, and surpasses baseline Polygon-GCN by 1.22%.

2 Method

Our method ContourRend consists of a contour generator and a contour renderer, completes the segmentation problem in two steps as shown in Fig. 1. First, the contour generator generates an initial contour prediction; then, the contour renderer optimizes the contour prediction in pixel level. The contour generator is a contour-based segmentation models which provides the backbone feature map and the initial contour vertex for the contour renderer. The contour renderer optimizes the initial contour like rendering and outputs the mask results with refined edges. Section 2.1 introduces the architecture and the function of the generator, and Sect. 2.2 details how the contour render module refines the initial contour.

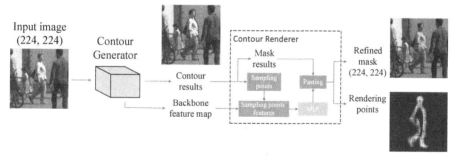

Fig. 1. Inference process of ContourRend. The contour generator provides contour results and the backbone feature map for the contour renderer, and the contour renderer optimizes the contour results by using a MLP to classify the sampled points around the contour.

2.1 Contour Generator

Contour generator aims to generate the backbone feature map and contour vertices for contour renderer. And the contour generator is built according to Tian's GCN-based segmentation model [20] which is similar to Curve-GCN. Graph neural network (GNN) is powerful at dealing with graph structure data and exploring the potential relationship, and using GCN in the mask-based model could improve the features' expression [21, 22] and the result's accuracy [23]. Tian and Curve-GCN utilize GCN to predict contour vertices, and Tian's model implements DeepLab-ResNet to provide the backbone feature map. Tian uses the model on magnetic resonance images and out-performs several state-of-the-art segmentation methods. Figure 2 shows the architecture of our contour generator. The DeepLab-ResNet provides a $512 \times 28 \times 28$ backbone feature map, and two branches of the networks consist of a 3×3 convolution layer and a fully connected layer after the backbone, respectively. The two branches provide a $1 \times 28 \times 28$ edge feature map and a $1 \times 28 \times 28$ vertex feature map, and concatenate the backbone feature map. Before the GCN modules, a 3×3 convolution layers processes the $514 \times 28 \times 28$ feature map provided by the backbone and the two branches, and outputs the $320 \times 28 \times 28$ feature map.

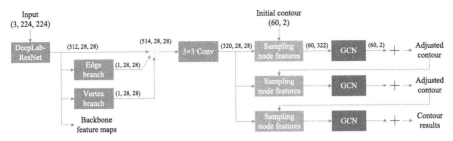

Fig. 2. Contour generator's architecture.

The GCN module use a fixed topology graph to represent the contour as same as the Curve-GCN. The relationship between nodes and edges can be regarded as a ring composed of nodes. Each node is connected to two adjacent nodes on the left side and two on the right side. Figure 3 illustrates the graph by an example with eight nodes.

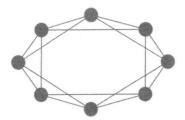

Fig. 3. Example of GCN's fixed topology graph. Every node connects with other four adjacent nodes.

Node features are composed of corresponding coordinate positions on the contour and the contour vertex features extracted from the $320 \times 28 \times 28$ feature map, the contour vertex features are extracted by bilinear interpolation according to their 0–1 positions. After GCN propagates and aggregates nodes' features on the graph, the output of the nodes are offsets of the contour vertices. By scaling result points' coordinates from 0–1 to the input size, the contour generator simply obtains contour segmentation results with the same resolution as the input. And the contour generator is trained by point matching loss (MLoss) according to Curve-GCN. During the training, the L2 loss is calculated as MLoss, and the predicted contour vertices and the target vertices are both sampled to K points in a clockwise order. And the p and p' are the sets of K predicted points and K target points represented by the x and y coordinates from 0 to 1. The loss function is shown as:

$$L_{\text{match}} = (p, p') = \min_{j \in [0, \cdots, K-1]} \sum_{i=0}^{K-1} \left\| p_i - p'_{(j+i)\%K} \right\|_2 \tag{1}$$

where p_i represents the i th predicted points, $p'_{(j+i)\%K}$ represents the matching point of p_i while the index offset is j, the $\%$ indicates modulus operation, and $\left\| p_i - p_{(j+i)\%K} \right\|_2$ indicates the L2 distance between p_i and $p'_{(j+i)\%K}$.

Finally, the contour generator provides 60 contour vertices and the $512 \times 28 \times 28$ backbone feature map for the contour renderer.

2.2 Contour Renderer

The contour renderer samples points based on the contour vertices provided by the contour generator, extracts the points' features by bilinear interpolation according to their positions, then predicts the category scores of these sampled points by a multilayer perceptron (MLP) consisted of a 1×1 convolution layer, and finally gets the refine mask result by pasting the sample point categories to the initial contour.

For the process of sampling points, we develop two methods to select points for the contour renderer during training and testing respectively. During the training, contour vertices are used to represent segmentation results, as opposed to the case of mask scores, the random points around edges can be naturally obtained by offsetting the contour results. The output of the contour generator is a fixed number of points in the range 0–1. By randomly offsetting the x and y coordinates by -0.09–0.09, n offset points for each output points are generated. Then the points' targets are sampled from the mask represented by contour results and cross entropy loss is used as the loss of the renderer. The contour generator and the contour renderer both use points' features, and the renderer loss can be viewed as an auxiliary loss. Figure 4 shows the process of the contour renderer during the training. During the testing, there is no need to calculate the gradient, so more points are used to obtain a dense prediction. For every contour vertex, an $N \times N$ ($N \geq 1$) grid is generated, and the contour vertex is located at the center of the grid. N^2 points evenly cover a $s \times s$ ($s \in [0, 1]$) square area with the gap of $s/(N-1)$ in both x and y coordinates.

Then, the renderer optimizes segmentation results by reclassifying single points around the output contour vertices of the generator. Specifically, the input of the MLP are $(60 \times N \times N, 512)$ point features, and MLP predicts the $(60 \times N \times N, 2)$ category scores (background and foreground scores) of the corresponding points. After changing the contour result of the contour generator into a mask result with the same size as input and pasting the contour renderer's generated points to the mask, the renderer restores the high resolution while retain the complex edge's details. Figure 5 shows the process of the contour renderer during the testing.

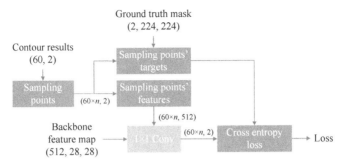

Fig. 4. Contour renderer in training. The ground truth mask has two categories (background and foreground).

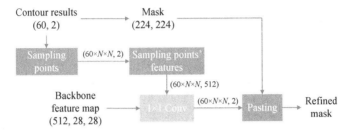

Fig. 5. Contour renderer in testing.

3 Experiments

We conduct a contrast experiment and an ablation experiment to verify the advantages of our method and test the contour renderer's effect in single object segmentation task on cityscapes dataset. In the contrast experiment, we train our ContourRend and compare with other contour-based segmentation methods. Furthermore, we train our contour generator separately as an ablation experiment to explore the contour renderer's effect.

3.1 Dataset

For single object segmentation task, we use the cityscapes dataset and have the same data preprocessing as Curve-GCN [18]. The input is a 224×224 single object image with background, and the object is in the center of the image. The dataset is divided into train set 45984 images, validation set 3936 images and test set 9784 images. And our contour generator's goal is to match the target contour vertices of the single object, our contour renderer's goal is to correctly classify the sampled points' categories (background and foreground).

3.2 Implementation

For the contour renderer, 3 ($n = 3$) rendering points are randomly sampled around every vertex of the contour generator's result in training process, and use a 1×1 convolution layer to classify the 512 dimension features to 2 categories (background and foreground). In testing process, 15×15 ($N = 15$) grid rendering points with the size of 0.09×0.09 ($s = 0.09$) are used to improve the contour, and if the foreground's scores of the renderer's result points are higher than 0.3, the points are considered as the foreground points.

We train the entire network end-to-end on four GTX 1080 Ti GPUs with batch size of 8, set the learning rate begin with $3e-4$, and 0.1 learning rate decay every 10 epochs. And we use $1e-5$ weight decay [24] to prevent overfitting.

3.3 Results

3.3.1 Contrast Experiment

In the contrast experiment, we train our ContourRend and compare the IoU by categories and the mean IoU with other contour-based methods, Polygon-RNN++ and Polygon-GCN. Table 1 shows the results. Lines 1, 2 refer to Polygon-RNN++ [17], Line 3 refers to Curve-GCN's Polygon-GCN [18]. Our contour generator is similar to Curve-GCN's Polygon-GCN, so we choose Polygon-GCN as our experiment's baseline. From the results, our method surpasses the Polygon-GCN by 1.22%.

Table 1. This table shows the mean IoU of some contour-based methods and our method in the single object segmentation task on cityscapes dataset.

Methods	Bicycle	Bus	Person	Train	Truck	Motorcycle	Car	Rider	Mean
Polygon-RNN++ [17]	57.38	75.99	68.45	59.65	76.31	58.26	75.68	65.65	67.17
Polygon-RNN++ (with BS) [17]	63.06	81.38	72.41	64.28	78.90	62.01	79.08	69.95	71.38
Polygon-GCN [18]	63.68	**81.42**	72.25	61.45	**79.88**	60.86	79.84	70.17	71.19
CountourRend (ours)	**65.18**	80.90	**74.16**	**64.40**	78.26	**63.30**	**80.69**	**72.36**	**72.41**

Figure 6 shows some of the results to visualize the renderer's effect. The first column is the 224×224 input image, the second column is the contour result of the contour generator which is trained separately, the third column is the ground truth contour, the fourth column is the contour result of the contour generator in ContourRend which is trained with the contour renderer, the fifth column is the mask result predicted by the contour renderer in ContourRend, the sixth column is the ground truth mask and which is use to calculate IoU, the last column is a visualization of the contour renderer's output. For the man in the first line's pictures, the separately trained contour generator can not fit the man's feet and back well (column 2), and the contour generator trained in ContourRend predicts a better contour result (column 4), then after the contour renderer, the man's shoes can also be segmented (column 5). Besides, ContourRend can also predict the women's bag in line 2 column 4, 5 and the truck's tyre in line 3 column 4,

5. ContourRend improves the original contour generator's performance and also refines the details by the contour renderer.

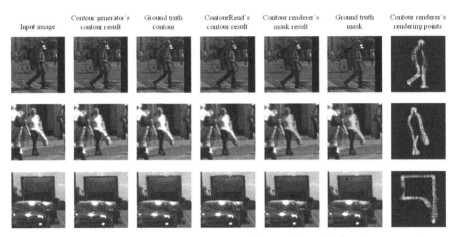

Fig. 6. Contrast experiment results. Column 1: Input image, column 2: Contour generator's contour result (trained separately), column 3: Ground truth contour, column 4: ContourRend's contour result, column 5: Rendered mask, column 6: Ground truth mask, column 7: Contour Renderer's point results. ContourRend improves the segmentation result by refining the details around the contour.

3.3.2   Ablation Experiment

In the ablation experiment, we train the contour generator separately and calculate the IoU by converting the contour result to mask result, then split the contour generator in ContourRend which has been trained in the contrast experiment and also calculate the IoU by it's contour result. Table 2 line 1 shows the result of the contour generator which is not trained with our contour renderer, and line 2 is the result of ContourRend's contour generator which is trained with our contour renderer in the contrast experiment, line 3 is ContourRend's result after the contour renderer improves the contour. From the results of line 1 and line 2, the contour renderer improves the contour generator's mean IoU by 2.67% in the training, and compare with the line 1 and line 3, ContourRend makes 7.16% improvement on the mean IoU by improving the contour in the testing. According to the ablation experiment, our method improves both the contour-based model's accuracy in the training and testing.

The contour renderer can improve the accuracy during training, because the renderer resamples the segmented pixels and makes the model focus on the object's contour. This phenomenon has also led to a conjecture that the evenly participation of all pixels in the original image in the training may cause computational waste and even lead to the decline of segmentation accuracy. The renderer loss reset the weights of the pixels to participate the segmentation which improves the performance of the baseline model. This can deduce that the pixels around contour play more important role in segmentation than other pixels.

Table 2. This Table shows the effect of the contour renderer in ContourRend, the result is the mean IoU in the single object segmentation task on cityscapes dataset.

Methods	Bicycle	Bus	Person	Train	Truck	Motorcycle	Car	Rider	Mean
Contour generator	57.74	73.71	66.76	56.52	71.70	56.14	75.42	64.02	65.25
ContourRend-ablation	59.69	76.67	69.93	59.77	75.14	57.03	77.19	67.91	67.92
ContourRend	**65.18**	**80.90**	**74.16**	**64.40**	**78.26**	**63.30**	**80.69**	**72.36**	**72.41**

4 Conclusion

In order to tackle the problem of blurry edges in mask-based segmentation models' results and improve the accuracy of contour-based segmentation models, we propose a segmentation method by combining a contour-based segmentation model and a contour renderer. In the single object segmentation task on cityscapes dataset, our method reaches 72.41% mean IoU and surpasses Polygon-GCN by 1.22%. And the proposed contour renderer enhanced contour-based segmentation mechanism is also effective to improve the performance of other kinds of contour based segmentation methods, as Polar-Mask and Curve-GCN, in the future work, the combinations between these methods and ContourRend are expect to study.

References

1. Krizhevsky, A., Sutskever, I., Hinton G.E.: ImageNet classification with deep convolutional neural networks. In: Advances in Neural Information Processing Systems, pp. 1097–1105 (2012)
2. Zhao, D., Chen, Y., Lv, L.: Deep reinforcement learning with visual attention for vehicle classification. IEEE Trans. Cogn. Dev. Syst. **9**(4), 356–367 (2017)
3. Simonyan, K., Zisserman, A.: Very deep convolutional networks for large-scale image recognition. In: International Conference on Learning Representations, pp. 1–14 (2015)
4. Chen, Y., Zhao, D., Lv, L., Zhang, Q.: Multi-task learning for dangerous object detection in autonomous driving. Inf. Sci. **432**, 559–571 (2018)
5. He, K., Zhang, X., Ren, S., Sun, J.: Deep residual learning for image recognition. In: IEEE Conference on Computer Vision and Pattern Recognition, pp. 770–778 (2016)
6. Long, J., Shelhamer, E., Darrell, T.: Fully convolutional networks for semantic segmentation. In: IEEE Conference on Computer Vision and Pattern Recognition, pp. 3431–3440 (2015)
7. Ronneberger, O., Fischer, P., Brox, T.: U-Net: convolutional networks for biomedical image segmentation. Int. Conf. Med. Image Comput. Comput. Assist. Intervention. **9351**, 234–241 (2015)
8. He, K., Gkioxari, G., Dollar, P., Girshick, R.: Mask R-CNN. In: IEEE International Conference on Computer Vision, pp. 2961–2969 (2017)
9. Ren, S., He, K., Girshick, R., Sun, J.: Faster R-CNN: towards real-time object detection with region proposal networks. In: Advances in Neural Information Processing Systems, pp. 91–99 (2015)
10. Zhao, H., Shi, J., Qi, X., Wang, X., Jia, J.: Pyramid scene parsing network. In: IEEE Conference on Computer Vision and Pattern Recognition, pp. 2881–2890 (2017)

11. Chen, L.C., Papandreou, G., Kokkinos, I., Murphy, K., Yuille, A.L.: DeepLab: semantic image segmentation with deep convolutional nets, atrous convolution, and fully connected CRFs. IEEE Trans. Pattern Anal. Mach. Intell. **40**(4), 834–848 (2017)

12. Chen, L.C., Papandreou, G., Schroff, F., Adam, H.: Rethinking atrous convolution for semantic image segmentation. arXiv preprint arXiv:1706.05587 (2017)

13. Chen, L.-C., Zhu, Y., Papandreou, G., Schroff, F., Adam, H.: Encoder-decoder with atrous separable convolution for semantic image segmentation. In: Ferrari, V., Hebert, M., Sminchisescu, C., Weiss, Y. (eds.) ECCV 2018. LNCS, vol. 11211, pp. 833–851. Springer, Cham (2018). https://doi.org/10.1007/978-3-030-01234-2_49

14. Yang, M., Yu, K., Zhang, C., Li, Z., Yang, K.: DenseASPP for semantic segmentation in street scenes. In: IEEE Conference on Computer Vision and Pattern Recognition, pp. 3684–3692 (2018)

15. Xie, E., et al.: PolarMask: single shot instance segmentation with polar representation. In: IEEE Conference on Computer Vision and Pattern Recognition, pp. 12193–12202 (2020)

16. Castrejon, L., Kundu, K., Urtasun, R., Fidler, S.: Annotating object instances with a Polygon-RNN. In: IEEE Conference on Computer Vision and Pattern Recognition, pp. 5230–5238 (2017)

17. Acuna, D., Ling, H., Kar, A., Fidler, S.: Efficient interactive annotation of segmentation datasets with Polygon-RNN++. In: IEEE conference on Computer Vision and Pattern Recognition, pp. 859–868 (2018)

18. Ling, H., Gao, J., Kar, A., Chen, W., Fidler, S.: Fast interactive object annotation with Curve-GCN. In: IEEE Conference on Computer Vision and Pattern Recognition, pp. 5257–5266 (2019)

19. Kirillov, A. Wu, Y., He, K., Girshick, R.: PointRend: image segmentation as rendering. In: IEEE Conference on Computer Vision and Pattern Recognition, pp. 9799–9808 (2020)

20. Tian, Z., Li, X., Zheng, Y., Chen, Z., Shi, Z., Liu, L., Fei, B.: Graph-convolutional-network-based interactive prostate segmentation in MR images. Med. Phys. **47**(9), 4164–4176 (2020)

21. Lu, Y., Chen, Y., Zhao, D., Liu, B., Lai, Z., Chen, J.: CNN-G: convolutional neural network combined with graph for image segmentation with theoretical analysis. IEEE Trans. Cogn. Dev. Syst. (2020). https://doi.org/10.1109/TCDS.2020.2998497

22. Lu, Y., Chen, Y., Zhao, D., Li, Dong.: MGRL: graph neural network based inference in a Markov network with reinforcement learning for visual navigation. Neurocomputing (2020). https://doi.org/10.1016/j.neucom.2020.07.091

23. Lu, Y., Chen, Y., Zhao, D., Chen, J.: Graph-FCN for image semantic segmentation. In: Lu, H., Tang, H., Wang, Z. (eds.) ISNN 2019. LNCS, vol. 11554, pp. 97–105. Springer, Cham (2019). https://doi.org/10.1007/978-3-030-22796-8_11

24. Loshchilov, I., Hutter, F.: Decoupled weight decay regularization. In: International Conference on Learning Representations, pp. 1–11 (2019)

Edge Information Extraction of Overlapping Fiber Optical Microscope Imaging Based on Modified Watershed Algorithm

Cheng Xing[1], Jianchao Fan[2(✉)], Xinzhe Wang[3], and Jun Xing[3]

[1] Institute of Science, Beijing Jiaotong University, Beijing 100044, China
19272023@bjtu.edu.cn
[2] Department of Ocean Remote Sensing, National Marine Environmental Monitoring Center,
Dalian 116023, Liaoning, China
jcfan@nmemc.org.cn
[3] Institute of Information Science and Engineering,
Dalian Polytechnic University, Dalian 116024, Liaoning, China
wxzagm@dlpu.edu.cn, xingjun@dlpu.edu.cn

Abstract. In the microscopic fields of materials, biology, *etc.*, the research based on the images under the optical microscope is an important part in experiments. Because of the specific characteristics along with various materials, the automatic identification, calculation and statistical techniques of particles in the microscope imaging are facing barriers. Though a number of methods have been invented for edge segmentation, the watershed algorithm tends to display good performance in image feature analysis. However, its result of processing the fiber edge images under the microscope does not meet the requirements based on the experiments. This paper aims to propose an improved watershed algorithm in order to better solve the problem of over-segmentation and insufficient segmentation in the field of overlapping optical fiber images. The morphological algorithm is essentially processed as a pretreatment in the beginning. Then after the reinforcement, the OTSU algorithm is demanded as a more accurate binarization process in order to better distinguish the images. On this basis, the distance between the feature pixels is supposed to be calculated aiming to reconstruct the pixel value corresponding to each point. Experimental results are compared with the version obtained by the foreground background marking watershed algorithm.

Keywords: Watershed algorithm · Overlapping fibers · Pixel reconstruction · Morphological · Foreground background marking

---

J. Fan—The work described in the paper was supported in part by the National Natural Science Foundation of China under Grant 41706195, Grant 41876109, in part by the National Key Research and Development Program of China under Grant 2017YFC1404902 and Grant 2016YFC1401007, in part by the National High Resolution Special Research under Grant 41-Y30B12-9001-14/16, in part by the Key Laboratory of Sea-Area Management Technology Foundation under Grant 201701, and in part by the Grant of China Scholarship Council.

© Springer Nature Switzerland AG 2020
M. Han et al. (Eds.): ISNN 2020, LNCS 12557, pp. 261–269, 2020.
https://doi.org/10.1007/978-3-030-64221-1_23

1 Introduction

In many microscopic fields such as materials, biology, *etc.*, the target image under the optical microscope is regarded as a crucial link in the research process, which includes the identification of targets in the image, the number of targets, and the measurement of sizes [1]. However, as for the optical fiber edge image in the actual experimental process, the target usually has problems of relatively complicated shape, high degree of overlap, and weak boundary information [2]. In most cases nowadays, such problems can only be effectively screened by manual recognition [3], but this method not only consumes a lot of manpower and time, but also usually has problems such as inaccurate recognition. Based on the problems above, assisting target edge segmentation algorithm through computer image processing technology has important practical significance [4]. Edge segmentation plays an important role in image analysis, and the watershed algorithm is one of the most representative methods for edge segmentation. However, the traditional watershed algorithm exists the disadvantages of segmentation phenomenon which includes error segmentation of noise and small particles [5–7]. Li.G, Shuyuan.Y [8] proposed a new improved marker-based watershed image segmentation algorithm, which can mark the object segmentation more quickly and accurately compared with traditional watershed algorithm, but in the calculation of the gradient, it takes no consideration of the texture gradient which leads to the unsatisfactory segmentation results for image with rich texture details [9]. At present, except from the foreground background marking watershed algorithm, even though there are more mature edge segmentation algorithms including morphological methods Sobel, Prewitt, Roberts first-order differential operator edge detection algorithm, Canny second-order differential operator edge detection algorithm Hough transform, *etc.* [9], which are mostly applied in edge segmentation, these methods have the shortcomings of long time, complicated steps, low measurement accuracy and great influence on the objective factors. Furthermore, they have poor performance when processing the optical fiber images. Considering many factors from the perspective of image processing, with the inspiration from some of the previous image segmentation algorithm and the characteristics of the optical fiber images, an improved watershed algorithm based on the distance pixel value reconstruction has been invented.

The remainder of this paper is organized as follows. The instruction of the materials used for the experiments and the theorem of the methods in the paper are presented in Sect. 2. The experiments and the comparison are presented in Sect. 3. Finally, the conclusion based on this paper is in Sect. 4.

2 Materials and Methods

2.1 Materials

The materials chosen in the experiment are fiber optical microscope imaging in the form of cross-section. Most of them are seriously overlapped with each other, it turns out that the edge boundary is hard to be distinguished. In addition, the cross-section of the fibers almost cover the entire picture and their distribution is uneven, which increases the recognized difficulty. In order to show the rigor and diversity of the experiment, the images are chosen including typical and untypical characteristics, which aims to illustrate the complexity that the improved watershed algorithm is able to process.

2.2 Theory of the Traditional Watershed Algorithm

The traditional watershed segmentation algorithm mainly adopts the idea of "simulated flooding method", and with the continuous improvement of the watershed algorithm and the widespread use of the watershed algorithm, the "raindrop method" and the process of "overflow" are now mainly used [10].

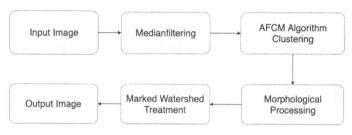

Fig. 1. Processing flow

The calculation idea of the watershed algorithm is to first find the gradient image, and then use the obtained gradient image as the input image of the watershed algorithm, and finally perform the corresponding processing. The gradient image solution formula is as follows [11]:

$$g(x, y) = grad(f(x, y))$$
$$g(x, y) = \left\{ [f(x, y) - f(x - 1, y)]^2 + [f(x, y) - f(x, y - 1)]^2 \right\}^{0.5} \quad (1)$$

where $f(x, y)$ is the input image, $grad(.)$ is expressed as the gradient operator of the gradient image, and $g(x, y)$ is the output image after the gradient operator calculation. Because the image processed in practice is often more complicated, there are more gray-scale minimum points in the image, which leads to over-segmentation of the image. Therefore, the above shortcomings need to be improved to reduce the existence of pseudo-minimum values.

2.3 Watershed Algorithm Based on Background and Background Markers

This method is one of the most effective watershed improvement methods so far. The basic principle is to distinguish the foreground and background of the image from each other [12].

The selection of concrete material has great influence on the characteristics of concrete images in late stage. Therefore, the Watershed Algorithm based on Background and Background Markers adopts the method combining AFCM algorithm with marked watershed algorithm, and the two algorithms could learn from each other to get better image segmentation, the detail steps are shown above in Fig. 1. The method is tried to identify the image edges of the optical fiber cross-section. Through the specific experiment, the algorithm makes effective segmentation for adhesion particles in image. And do a crucial step for statistics of particle size. With this improved marking-based method, the differences among the background and foreground are able to distinguished as well as the boundaries. Consequently, when trying to analyze the edge between the fibers with the watershed algorithm after the marking process, it tends to be better take advantage of the marking information in order to achieve the segmentation operation.

2.4 Improved Watershed Segmentation Algorithm Based on Distances

In order to better improve the over-segmentation problem, especially for the correspond different overlapping cases, this paper uses an algorithm based on the combination of OTSU algorithm [13] and pixel distance process to improve the watershed algorithm. The basic steps can be divided into four specific parts, and the algorithm flow chart is as follows (Fig. 2):

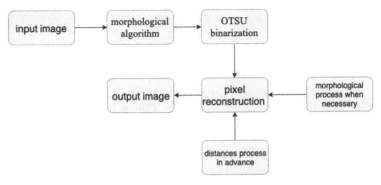

Fig. 2. Improved watershed segmentation algorithm flow figure.

The specific description of each step of the algorithm is as follows.

Step 1. The morphological step is not complicated which includes the basic erosion operation and open operation. After this step, the image will complete a relatively effective but not comprehensive separation, laying the foundation for the subsequent steps.

Step 2. The OTSU algorithm is used to determine the optimal threshold through the grayscale characteristics of the image, so that the variance between the two classes of the target and the background takes the maximum value to ensure the maximum variance between the classes and minimize the probability of misclassification [13]. Set the target and background segmentation threshold of the picture as t; the average gray value of the target pixels is μ_1, which accounts for the entire image pixel ratio ω_1; the average gray value of the background pixels is μ_2, which accounts for the entire image pixel ratio ω_2; The average gray value of all pixels in the image is μ, and the variance between classes is σ^2. The mathematical expression is shown in (2) [14].

$$\sigma^2 = \omega_1(\mu_1 - \mu)^2 = \omega_1\omega_2(\mu_2 - \mu_1)^2$$
$$s.t. \, \omega_1 + \omega_2 = 1, \omega_1\mu_1 + \omega_2\mu_2 = \mu \tag{2}$$

Since the larger the variance between the classes, the larger the difference between the image target and the background pixels, so the threshold t corresponding to when σ^2 takes the maximum value is the optimal threshold.

Step 3. Through mathematical operation, calculate the distance from the pixel point with a pixel value of one to its nearest pixel whose value is zero, and transform to its opposite value. And then set the pixel whose value is zero to positive infinity. The value of each pixel is reset in this way. Definite the horizontal coordinate of the i-th pixel in the white area be X_i and the vertical coordinate of Y_i; its original pixel value is W_i; its updated pixel value is V_i; the horizontal coordinate of the black pixel closest to it is X_0; the ordinate is Y_0, then the expression of V_i is as shown in (3):

$$V_i = -(\sqrt{(X_i - X_0)^2 + (Y_i - Y_0)^2})$$
$$s.t. \, i \in \{i | 1 \leq i \leq n, W_i = 1\} \tag{3}$$

Take a part of the binarized image in the sample, and express its pixel values in a matrix, further explain the algorithm, as shown in (4)

$$\begin{pmatrix} 1 & 0 & 0 & 1 \\ 1 & 1 & 1 & 0 \\ 1 & 1 & 0 & 0 \end{pmatrix} \rightarrow \begin{pmatrix} -1 & \infty & \infty & -1 \\ -\sqrt{2} & -1 & -1 & \infty \\ -2 & -1 & \infty & \infty \end{pmatrix} \tag{4}$$

In (4), the first matrix exemplifies the pixel value corresponding to each part of the binary image, and the pixel value corresponding to the second matrix is obtained by distance transformation. In the second matrix, the pixel located at (3, 1) has the smallest pixel value in its connected area, and it gradually expands outward as the source pixel of the watershed.

Step 4. The watershed algorithm is used to process the new image after each pixel is updated: the image is regarded as the terrain surface, the edge area (high gradient) represents the "watershed", and the low gradient area represents the "water basin". All pixels located in the same catchment basin decrease monotonously along the path to the minimum point of gray value in the catchment basin [10].

This process is a recursive process, defined as follows:

$$X_{h_{min}} = T_{h_{min}}(I)$$
$$\forall h \in [h_{min}, h_{max} - 1] \tag{5}$$

$$X_{h+1} = min_{h+1} \cup C_{xh}(X_h \cap X_{h+1}) \tag{6}$$

Equation (6) is a recursive process, and Eq. (5) is the initial condition of the recursive process. h, h_{min} and h_{max} represent the range of gray value, minimum gray value and maximum gray value, X_{h+1} represents all pixels with gray value $h + 1$, min_{h+1} represents the minimum point, $X_h \cap X_{h+1}$ represents the intersection of X_h and X_{h+1}, C_{xh} represents the basin where X_h is located, $C_{xh}(X_h \cap X_{h+1})$ represents the point where $X_h \cap X_{h+1}$ is in the same basin C_{xh}. Through this iterative process, all pixels in image I are assigned to basins, where points belonging to more than two basins are points in the watershed.

3   Experiment and Simulation

In this paper, the cross-sectional imaging of the fiber under the microscope is selected as the experimental target. A large number of experimental studies have been performed on images with different characteristics which includes the characteristics of different dense arrangement, special shapes, and different overlapping methods.

The following is the original image of the sample (Fig. 3).

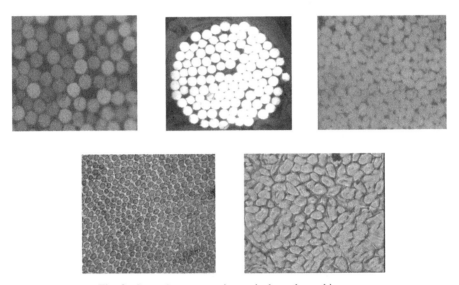

Fig. 3. Several representative optical overlapped images

3.1   Implementation Effect of Traditional Watershed Algorithm

Traditional Watershed Algorithm is based on the gradient of the gray image. After an image is converted into the gray image, it is common that a great number of uncontrollable pixel values exist. Due to the severe over-segmentation of the algorithm, the accuracy of the experimental results obtained is very low, like what is shown in Fig. 4.

3.2   Experimental Simulation of Improved Watershed Segmentation Algorithm

With the help of MATLAB software, the experimental simulation is carried out according to the improved watershed algorithm, and the experimental results were obtained. Here, five representative sample pictures are selected for display. Each of these pictures has its own distinct internal characteristics, making the results more authentic and convincing.

Following the flow figure shown in Fig. 5, the result is as shown in the figure below.

The MATLAB software is used to calculate the number of fibers based on the successful segmentation graph obtained in Fig. 5, and the accuracy of the algorithm is checked by this method. The results are shown in Table 1:

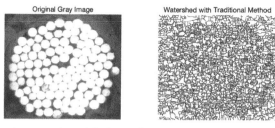

Fig. 4. Comparison between the original gray image and method of the traditional watershed algorithm.

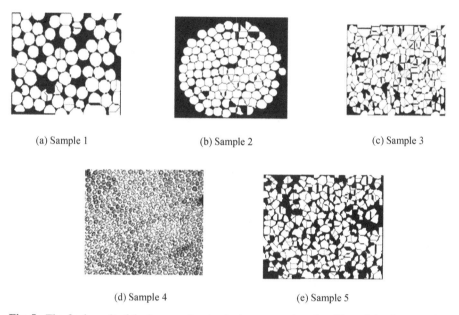

(a) Sample 1   (b) Sample 2   (c) Sample 3

(d) Sample 4   (e) Sample 5

Fig. 5. The final result of the improved watershed segmentation algorithm of the five samples

Table 1. Statistics of the quantities of the cells calculated by MATLAB

	Test quantity	Actual quantity	Error	Absolute error percentage (%)
Sample1	90	87	3	3.4482759
Sample2	122	119	3	2.5210084
Sample3	275	289	14	4.8442906
Sample4	396	396	0	0
Sample5	180 ± 8	202 ± 5	9–35	4.3478261–17.7664975
Overall (average)	211–214	218–220	6–11	2.7272727–5.0458716

In addition, the overall average result based on the whole dataset is calculated in order to figure out the advance more clearly.

Based on the image after watershed segmentation, the above table counts the number of fiber filaments of each sample picture, and calculates the difference between the computer statistics and the actual number of fiber filaments in the third column of the table, and finally calculates the absolute error. Judging from the results, although there are still errors, the errors can be controlled within a certain range, and the expected effect is achieved. The time required for the algorithm to complete the above picture is around 0.2 s, which shows that the algorithm is acceptable at the time level.

3.3 Comparison with Watershed Algorithm Based on Background and Foreground Markers

The results of the modified watershed algorithm based on the background and background markers are shown in Fig. 6 below, here only one sample is shown since the simulation result is not good enough to be displayed for the other samples.

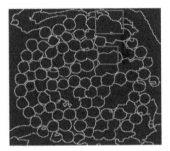

Fig. 6. The result of the watershed algorithm based on marker. The red rectangles show the unexpected segmentation places. (Color figure online)

Through the horizontal comparison of Fig. 5 and Fig. 6, although the improved watershed algorithm based on the foreground and background markers can divide most connected regions, there are still many regions in the result graph obtained by this method that cannot be clearly divided. Therefore, the processing effect of the improved watershed algorithm based on background markers in this experiment is not as good as the improved watershed algorithm in this paper.

4 Conclusion

In this paper, an improved watershed algorithm based on the distance reconstruction is invented in order to overcome the edge recognition difficulties in optical overlapped cross-sectional fibers. Combined with the OTSU algorithm, the original watershed algorithm and the distances construction, the recognition effect is greatly enhanced. More-over, the time complexity of this method is not high, and the desired result can be quickly obtained. However, it was found from the subsequent image test of a large number of fiber microscope images that this method still has a little certain error for the adhesion of a large number of densely connected areas. In the future, further research and improvement are needed.

References

1. Pincus, Z., Theriot, J.: Comparison of quantitative methods for cell shape analysis. J. Microsc. **227**, 140–156 (2007)
2. An, X., Liu, Z., Shi, Y., Li, N., Wang, Y., Joshi, S.H.: Modeling dynamic cellular morphology in images. In: Ayache, N., Delingette, H., Golland, P., Mori, K. (eds.) MICCAI 2012. LNCS, vol. 7510, pp. 340–347. Springer, Heidelberg (2012). https://doi.org/10.1007/978-3-642-33415-3_42
3. Tareef, A., Yang, S., Cai, W.: Automatic segmentation of overlapping cervical smear cells based on local distinctive features and guided shape deformation. Neurocomputing **221**, 94–107 (2017)
4. Liao, M., Zhao, Y.Q., Li, X.H.: Automatic segmentation for cell images based on bottleneck detection and ellipse fitting. Neurocomputing **173**, 615–622 (2016)
5. Deng, C., Wang, Z.H.: Improved gradient and adaptive marker extraction of watershed algorithm. Comput. Eng. Appl. **49**, 136–140 (2013)
6. Gao, L., Yang, S.Y., Li, H.Q.: A new image segmentation algorithm based on marked watershed. Chin. J. Image Graph., 1025–1032 (2007)
7. Cousty, J., Bertrand, G., Najman, L., Couprie, M.: Watershed cuts: minimum spanning forests and the drop of water principle. IEEE Trans. Pattern Anal. Mach. Intell. **31**(8), 1362–1374 (2009)
8. Hamed, M., Keshavarz, A., Dehghani, H., Pourghassem, H.: A clustering technique for remote sensing images using combination of watershed algorithm and gustafson-kessel clustering. In: 2012 Fourth International Conference on Computational Intelligence and Communication Networks (CICN), pp. 222–226 (2012)
9. Chang, J.F, Zhang, F.Z., Zhou, Q.Q.: An algorithm for environmental awareness recommendation algorithm based on fuzzy C-mean clustering. Comput. Res. Dev., 2185–2194 (2013)
10. Jun, J., Chun, X.: Particle size analysis of concrete materials based on AFCM and marked watershed algorithm. In: 2016 International Conference on Industrial Informatics, pp. 294–297 (2016)
11. Liu, Y., Wang, X.P., Wang, J.Q.: Watershed algorithm for brain tumor segmentation based on morphological reconstruction and gradient layered modification. Appl. Res. Comput. **32**, 2487–2491 (2015)
12. Menaka, E., Kumar, S.S., Parameshwari, P.: Assessment of sparse forest and fire detection using threshold watershed algorithm. In: IET Chennai 3rd International on Sustainable Energy and Intelligent Systems (SEISCON 2012), pp. 1–6 (2012)
13. Wang, S., Huang, Y., Li, D.: Multilevel thresholding methods for image segmentation with improved-Otsu based on ant colony algorithm. J. Net. New Media **29**, 25–28 (2008)
14. Wu, C.M., Tian, X.P., Tan, T.: Fast iterative algorithm for two-dimensional Otsu thresholding method. Pattern Recognit. Artif. Intell. **21**, 746–757 (2008)

A Visually Impaired Assistant Using Neural Network and Image Recognition with Physical Navigation

On-Chun Arthur Liu, Shun-Ki Li, Li-Qi Yan, Sin-Chun Ng$^{(\boxtimes)}$,
and Chok-Pang Kwok$^{(\boxtimes)}$

School of Science and Technology, The Open University of Hong Kong,
30 Good Shepherd Street, Ho Man Tin, Kowloon, Hong Kong
scng@ouhk.edu.hk, cpkwok@study.ouhk.edu.hk

Abstract. In Hong Kong, over 2.4% of the total population suffered from visual impairment. They are facing many difficulties in their daily lives, such as shopping and travelling from places to places within the city. For outdoor activities, they usually need to have an assistant to guide their ways to reach the destinations. In this paper, a mobile application assisting visually impaired people for outdoor navigation is proposed. The application consists of navigation, obstacle detection and scene description functions. The navigation function assists the user to travel to the destination with the Global Positioning System (GPS) and sound guidance. The obstacle detection function alerts the visually impaired people for any obstacles ahead that may be avoided for collision. The scene description function describes the scene in front of the users with voice. In general, the mobile application can assist the people with low vision to walk on the streets safely, reliably and efficiently.

Keywords: Visually impaired · Neural network · Image recognition · Navigation

1 Introduction

According to the statistics in the year 2015, Vos stated in [1] that over 940 million people around the world are classified as visually impaired or blind. In Hong Kong, over 174,800 are considered as visually impaired, which is 2.4% of the Hong Kong population [2]. They are facing many difficulties in their daily lives, such as daily consumption and travelling. Due to the disability problem, visually impaired people are seldom to have outdoor activities. The main reason is the unfamiliar routes and the traffic conditions that lead to the destinations. Although there is a cane to assist visually impaired people for outdoor activities, they need to tap everywhere to detect any obstacles around them. Hence, it has chances to hit other pedestrians accidentally and may in turn annoy other peoples. For people who are with low vision, they may not have a cane to assist

© Springer Nature Switzerland AG 2020
M. Han et al. (Eds.): ISNN 2020, LNCS 12557, pp. 270–281, 2020.
https://doi.org/10.1007/978-3-030-64221-1_24

them for outdoor activities. As they are hard to see, they may hit obstacles like temperate traffic signs or fire hydrants easily. For this group of disabilities, they like to stay at home all the time because of insufficient support to them. Moreover, visually impaired people are unable to have "sightseeing" activities because they need a tour guide to "describe" the actual landscape or scenes to them.

To solve the above issues, this paper proposes an intelligent application installed on mobile devices for visually impaired people to assist their outdoor activities. By applying neural network and image recognition technologies, an obstacle detector has been developed to alert the user for any obstacles close to them. Furthermore, the application can describe the conditions around the user so that they can make their right decisions easier. With this function, the application can act as a "tour guide" to describe the actual scenes in front of the users. Also, navigation is an important function for visually impaired people. With the use of navigation guidance, visually impaired people can reach the destinations safely, reliably and more efficiently.

2 The Existing Solutions

To provide a safe outdoor travelling assistance for visually impaired people, the navigation and detection functions are essential to satisfy their specific needs. Due to the development of artificial intelligence (AI) and derived technologies, there are several solutions in the market [7–10]. Although those applications are performing well, those applications contain different scopes of weaknesses, especially in costs, language support, and compatibilities that they are not applicable in Hong Kong [3–6]. There are two main types of programs currently available in the market to assist the visually impaired. One is a simple mobile phone application, and the other one is to combine hardware such as a blind cane to interact with the users.

Eye-D. Eye-D[1] is a mobile application. Its main function is to assist visually impaired people to find where they are, search nearby facilities, and describe their surroundings. The advantage of this application is that it gives users a clear idea of where they are and allows the visually impaired know what is in front of them. But the disadvantage is that there is no voice indication, relying upon an on-screen text, and only describes the object ahead, does not explain the action, making it difficult for the visually impaired to know what is happening ahead.

WeVoice. WeVoice[2] is a mobile application proposed by InnoTech Association[3]. The main function is to read the text aloud. Users need to take a picture with their phone, and the software will then read the text in the photo. The disadvantage is that it is difficult for the visually impaired to target objects accurately.

---

[1] https://eye-d.in/.

[2] https://play.google.com/store/apps/details?id=hk.com.redso.read4u&hl=zh_HK.

[3] https://www.facebook.com/InnoTechAssociation/.

Smart City Walk. Smart City Walk[4] is proposed by Hong Kong Blind Union[5]. The main function is to display the current location, search nearby facilities with voice input, and navigate the user to the destination via voice and text. Although it can be accurately positioned indoors, it can only be used in several buildings with iBeacon [11] installed.

WeWalk. WeWalk[6] is a mobile application with a blind cane. It has a built-in ultrasonic detector that alerts the user to obstacles in front of them. Besides, it has voice navigation to guide visually impaired people to their destinations. But its disadvantages are expensive, and only in English and Turkish, and language navigation is limited to Sweden and parts of Europe.

Although many enterprises developed different types of applications for the visually impaired with the advanced technologies for travelling, the users are still inconvenient to perform outdoor activities with no application describing the environment ahead, prompting with obstacles, and providing navigation at once. The existing applications have many functions, but they are not fulfilling the needs of the visually impaired in Hong Kong. For example, some of the applications require the users to take photographs, and it is difficult for the visually impaired to know the exact location of the objects. These applications also cannot remind the visually impaired before there are obstacles. The applications with these functions do not support Chinese or cater for the Hong Kong market.

According to the insufficient supports to visually impaired people, this paper proposes an application with low-cost hardware that uses voice to communicate with visually impaired people by describing the scene in front of them and providing obstacles prompt with navigation.

3   The System Design

The system has three main parts: image processing, navigation, and voice interaction. The image processing part is done on a Raspberry Pi 3 Model B+ and a machine learning server. The image comes from the Raspberry Pi 3 Model B+, with the camera attached which is put on by the user to allow hand-free video capturing feature. The image processing part includes scene description system and obstacle prompt system based on the latest convolutional neural networks (CNN), which will identify the objects from the camera and output as the description in a sentence, the object names, and the object location on the image. The navigation part is processed on a smartphone, which will provide a way to get to the destination based on the existing global positioning system (GPS) on the smartphone. The voice interaction part is processed on a smartphone and using an existing speaker and microphone, which is using offline voice recognition and text-to-speech technology to have interaction for the visually impaired user.

---

[4] https://play.google.com/store/apps/details?id=com.hkblindunion.smartcitywalk. android&hl=zh_HK.

[5] https://www.hkbu.org.hk/.

[6] https://wewalk.io/en/.

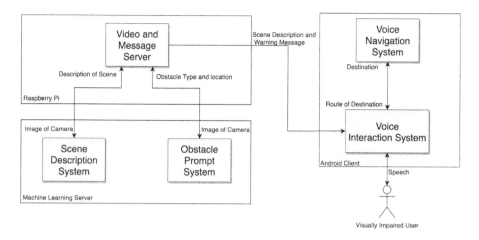

Fig. 1. Components of the system.

The application is supported by client and server devices as shown in Fig. 1 and Fig. 2. The operations of each component are listed below.

1. Mobile device (Android client)
   - Perform voice interaction with the user.
   - Return navigation information upon user request.
   - Use MQTT[7] clients to connect the Raspberry Pi and prompt the user when dangerous information or scene description is received.
   - Open a WIFI hotspot for Raspberry Pi connection.
2. Raspberry Pi (Operation component)
   - Connected to Raspberry Pi WIFI hotspot.
   - Host an MQTT server for data transferring between Android client.
   - Be a hands-free camera device that can easily be installed on the walking stick or worn on the neck since the user may have to pick up the walking stick.
   - Host a video streaming server which will do motion detection and serve images to clients by MJPEG.
3. Machine Learning Server (Server)
   - Serve the request through Flask[8] framework.
   - Process the request of object detection and image caption separately due to the inference time of deep neural networks (DNN) is relatively large.
   - Response the object name and relative direction from the object detection server when the confidence of the object in the image is high enough and the object may be dangerous to users.
   - Return the description in understanding sentences from the image caption server.

---

[7] http://mqtt.org/.
[8] https://flask.palletsprojects.com/en/1.1.x/.

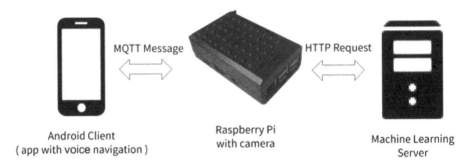

MQTT Message   HTTP Request

Android Client
(app with voice navigation)

Raspberry Pi
with camera

Machine Learning
Server

Fig. 2. Hardware components of the application system.

4   The System Implementation

4.1   Overview of the Mobile Application System

The application uses both hardware and software to achieve the goal. The device can be mounted into the walking stick or mounted into the neck ring. In terms of software, the user interface will be a mobile application that connects to the hardware (Raspberry Pi) via WIFI and searches Raspberry Pi using multicast DNS. This program is mainly responsible for interacting with the user. After the program starts, it will connect with the hardware and then start obstacle recognition, environment description, and map functions at the same time. The user can activate the speech recognition function through buttons to give instructions to the system, such as navigation. The system will convert the results of the three functions into speech, as shown in Fig. 3 based on the priority order as shown in Fig. 4.

Figure 4 shows the judgment process when the text-to-speech function receives instruction. This process ensures the system reads one type of information only to the user at the same time without overlapping. The system will handle the obstacle detection at first because it will directly affect the safety of the user if the user is close to the obstacle. After that, the user will receive the navigation instruction when the user arrives at the intersection, and the navigation system will suitably prompt the user. In general, the system will describe the surrounding environment for the user, but it will only describe the environment when the message is received within five seconds, avoiding to describe the environment that has passed.

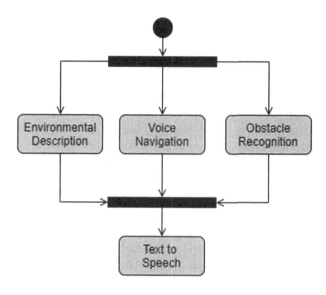

Fig. 3. Active diagram of the application.

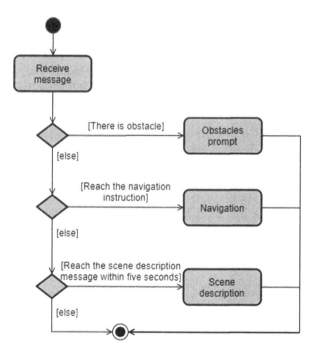

Fig. 4. Priority order for the text to speech function.

4.2 Voice Navigation

Voice navigation aims to make map navigation easier for the visually impaired people through the voice recognition function that allows users to confirm the destination by voice instead of typing through the keyboard. To achieve the goal, the cloud speech recognition service provided by Google[9] is applied, which is built into all Android devices by default. Through the cloud technology, it can convert the speech from the user into text within two seconds, and it has an automatic debugging function to revise the possible errors in speech recognition.

This feature uses dialog flow to analyze and filter the voice input by the user, such as the user saying "Take me to the Open University of Hong Kong" or "Go to the Open University of Hong Kong" and dialog flow responds the action "navigation" and the destination "The Open University of Hong Kong".

Three-Dimensional (3D) Direction Prompt. To let the visually impaired users know the correct direction, rather than the turn left or turn right of a traditional navigation service like Google Map, the 3D direction prompt was implemented. 3D direction prompt applies the latitude and longitude of the user and the next navigation point to calculate the correct direction. Then, it determines whether the user is oriented in that direction by using a gyroscope on the mobile phone. Figure 5 shows that the Mapbox[10] API will return the route and the next navigation point on the route (redpoint). The yellow-green-blue circle represents the user, where blue and the red line represents the user, where blue and the red line represents the direction the user is oriented in heading to the place.

Figure 6 shows that the user is facing the wrong direction since the right (or left) direction is the yellow line, but the face direction of the user is the red line. When the yellow line falls into the green (or yellow) area, the right (or left) channel of the earpiece will alert the user that turns back to the correct direction.

The volume of the left or right channels of the 3D direction prompt is linear as shown in Fig. 7, a reminder and different levels of "beep" sound in left-right channels to navigate the user back to the correct path. To avoid excessive volume, the maximum volume is set to 50, as shown in Fig. 8. Since it is impossible for the user exactly facing the correct direction, there is a buffer of 50° in the correct direction to prevent the user from still hearing the beep when facing the correct direction. To keep the user facing the correct direction, the system uses beep sound on the left-right channels and the small-large volume to inform the user.

---

[9] https://cloud.google.com/speech-to-text.
[10] https://www.mapbox.com/.

Fig. 5. A sample output of the Mapbox route API service. (Color figure online)

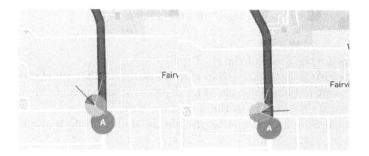

Fig. 6. A sample output of the user facing the wrong direction. (Color figure online)

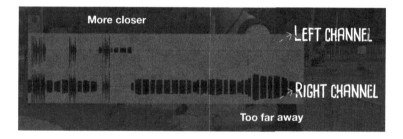

Fig. 7. Volume track for the direction indicator.

4.3 Obstacles Prompt and Scene Description

The design of the obstacles prompt and scene description system as shown in Fig. 9. Fast detection of obstacles is a definite criterion for the users to avoid any dangerous situation. To provide the accurate information to the users, the captioned function includes the following procedures:

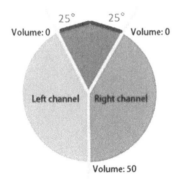

Fig. 8. The 3D direction prompt design.

Step 1. Capture the scene of the road.
Step 2. Filter irrelevant information and adjust the confidence of objects.
Step 3. Sort out the obstacles based on depth estimation algorithm.
Step 4. Illustrate the results to the user with voice.
Step 5. Repeat Step 1 to Step 4 to provide the updated information.

Figure 10 shows a sample image after depth estimation. For example, if an obstacle is detected as shown in Fig. 11, the obstacle prompt will give an immediate warning to the user with the scene description in voice. By post-processing the result of object detection, users will be prompted with the upfront obstacles. Scene description using image caption technology can produce a humanized text description, which can help the users know the scenes ahead. For example, a possible scene description can be "a group of people standing around a bus stop." The processes are similar to obstacle detection, while the application is kept responding to the scene captured from time to time.

Scene description using image caption technology can produce a humanized text description, which can help the visually impaired to know what the world is. For example, a possible scene description can be "a group of people standing around a bus stop." The processes are similar to obstacle detection, while the application is kept responding to the scene captured from time to time.

When the user faces the wrong direction at first, the user is prompted by the voice "You are in the wrong direction, please adjust the facing direction according to the beep sound". And when the user is facing the right direction, the system will also use the voice prompt "You are now in the right direction".

To describe the obstacles ahead, the system will detect the objects on the captured images from the camera as fast as possible.

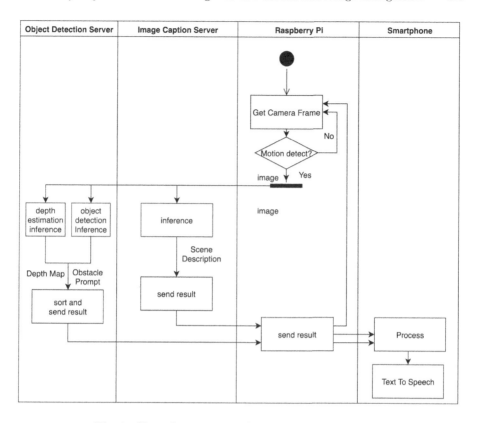

Fig. 9. Obstacles prompt and scene description design.

Fig. 10. Image after depth estimation.

Fig. 11. Image after obstacle detection.

5   Evaluation Results

Observation tests will be used to assess whether the solution helps the users go out and reach the destinations efficiently and safely. A number of the visually impaired people are invited to the observation test by using the usual way, and the application proposed in this paper to walk from Fat Kwong Street (at Homantin) to The Open University of Hong Kong. It will be evaluated by counting the number of objects and people that collided with the users, and the time taken for the users reach the destination. Table 1 shows the average results of user evaluation.

Table 1. Average result of user evaluation.

	Using the usual way	Using our solution
Average number of times colliding with objects	23.5	15.5
Average number of times colliding with people	5	0.5
Time to find the correct direction (in seconds)	8.5	4
Time taken to reach the destination (in minutes)	3.4	2.5

According to the results, the number of object collisions has been reduced by about one-third (from 23.5 to 15.5), and the number of human collisions has dropped significantly to one-tenth (from 5 to 0.5). The data indicated that the use of the new application can reduce the chance of accidents and thus improve the user safety during outdoor activities. Besides, the users can reach the destination more efficiently (from 3.4 minutes to 2.5 minutes) when the proposed application with 3D direction prompts is used.

6   Conclusions

This paper introduces a mobile application as an outdoor assistant to help people with low vision go outside efficiently and safely. On this basis, a real-time environment description system has been developed to let the users know the surrounding in front of them through the voice description. An obstacle prompt system has been implemented for avoiding the users to collide with any object in front of them during the journeys. The 3D direction prompts of the navigation system have been designed to allow the users to recognize the correct direction by using sound from the left-right channels. From the evaluation results, it is shown that the use of the proposed mobile application can assist the people in need with visual disability to travel outside in a safe and efficient way.

References

1. Vos, T., et al.: Global, regional, and national incidence, prevalence, and years lived with disability for 310 diseases and injuries, 1990–2015: a systematic analysis for the Global Burden of Disease Study 2015. Lancet **388**(10053), 1545–1602 (2016)
2. Statistics on People with Visual Impairment. https://www.hkbu.org.hk/en/knowledge/statistics/index
3. Image Description-Computer Vision - Azure Cognitive Services. https://docs.microsoft.com/zh-tw/azure/cognitive-services/computer-vision/concept-describing-images
4. Pricing-Computer Vision API. https://azure.microsoft.com/en-gb/pricing/details/cognitive-services/computer-vision/
5. Xu, K., et al.: Show, attend and tell: neural image caption generation with visual attention. In: International Conference on Machine Learning, pp. 2048–2057 (2015)
6. Vinyals, O., Toshev, A., Bengio, S., Erhan, D.: Show and tell: a neural imagecaption generator. In: 2015 IEEE Conference on Computer Vision and Pattern Recognition (CVPR) (2015). https://doi.org/10.1109/cvpr.2015.7298935
7. Graves, A., Navdeep, J.: Towards end-to-end speech recognition with recurrent neural networks. In: International Conference on Machine Learning, pp. 1764–1772 (2014)
8. He, K., Gkioxari, G., Dollár, P., Girshick, R.: Mask R-CNN. In: Proceedings of the IEEE International Conference on Computer Vision, pp. 2961–2969 (2017)
9. Sandler, M., Howard, A., Zhu, M., Zhmoginov, A., Chen, L.C.: MobileNetV2: inverted residuals and linear bottlenecks. In: Proceedings of the IEEE Conference on Computer Vision and Pattern Recognition, pp. 4510–4520 (2018)
10. Alhashim, I., Wonka, P.: High quality monocular depth estimation via transfer learning. arXiv preprint arXiv:1812.11941 (2018)
11. Newman, N.: Apple iBeacon technology briefing. J. Direct Data Digit. Mark. Pract. **15**(3), 222–225 (2014). https://doi.org/10.1057/dddmp.2014.7

Author Index

Printed in the United States
By Bookmasters